CITROËN
Traction Avant Gold Portfolio 1934~1957

Compiled by
R.M. Clarke

ISBN 1 85520 0317

Distributed by
Brooklands Book Distribution Ltd.
'Holmerise', Seven Hills Road,
Cobham, Surrey, England

Printed in Hong Kong

BROOKLANDS BOOKS

BROOKLANDS ROAD TEST SERIES
AC Ace & Aceca 1953-1983
Alfa Romeo Alfasud 1972-1984
Alfa Romeo Alfetta Coupes GT, GTV, GTV6 1974-1987
Alfa Romeo Giulia Berlinas 1962-1976
Alfa Romeo Giulia Coupes 1963-1976
Alfa Romeo Spider 1966-1987
Allard Gold Portfolio 1937-1958
Alvis Gold Portfolio 1919-1969
American Motors Muscle Cars 1966-1970
Aston Martin Gold Portfolio 1972-1985
Austin Seven 1922-1982
Austin A30 & A35 1951-1962
Austin Healey 3000 1959-1967
Austin Healey 100 & 3000 Col No.1
Austin Healey 'Frogeye' Sprite Col No.1 1958-1961
Austin Healey Sprite 1958-1971
Avanti 1962-1983
BMW Six Cylinder Coupes 1969-1975
BMW 1600 Col. 1 1966-1981
BMW 2002 1968-1976
Bristol Cars Gold Portfolio 1946-1985
Buick Automobiles 1947-1960
Buick Muscle Cars 1965-1970
Buick Riviera 1963-1978
Cadillac Automobiles 1949-1959
Cadillac Automobiles 1960-1969
Cadillac Eldorado 1967-1978
High Performance Capris Gold Portfolio 1969-1987
Chevrolet Camaro & Z-28 1973-1981
High Performance Camaros 1982-1988
Chevrolet Camaro Col No.1 1967-1973
Camaro Muscle Cars 1966-1972
Chevrolet 1955-1957
Chevrolet Impala & SS 1958-1971
Chevrolet Muscle Cars 1966-1971
Chevelle and SS 1964-1972
Chevy EL Camino & SS 1959-1987
Chevy II Nova & SS 1962-1973
Chrysler 300 1955-1970
Citroen Traction Avant Gold Portfolio 1934-1957
Citroen DS & ID 1955-1975
Citroen 2CV 1949-1988
Shelby Cobra Gold Portfolio 1962-1969
Cobras & Replicas 1962-1983
Corvair 1959-1968
Chevrolet Corvette Gold Portfolio 1953 1962
Corvette Stingray Gold Portfolio 1963-1967
High Performance Corvettes 1983-1989
Datsun 240Z 1970-1973
Datsun 280Z & ZX 1975-1983
De Tomaso Collection No.1 1962-1981
Dodge Charger 1966-1974
Dodge Muscle Cars 1967-1970
Excalibur Collection No.1 1952-1981
Ferrari Cars 1946-1956
Ferrari Cars 1973-1977
Ferrari Dino 1965-1974
Ferrari Dino 308 1974-1979
Ferrari 308 & Mondial 1980-1984
Ferrari Collection No.1 1960-1970
Fiat-Bertone X1/9 1973-1988
Fiat Pininfarina 124 + 2000 Spider 1968-1985
Ford Automobiles 1949-1959
Ford GT40 Gold Portfolio 1964-1987
Ford Fairlane 1955-1970
Ford Falcon 1960-1970
High Perfomance Mustangs 1982-1988
Ford Cortina 1600E & GT 1967-1970
Ford RS Escorts 1968-1980
High Performance Escorts Mk1 1968-1974
High Performance Escorts Mk II 1975-1980
Honda CRX 1983-1987
Hudson & Railton 1936-1940
Jaguar Cars 1957-1961
Jaguar Cars 1961-1964
Jaguar Mk2 1959-1969
Jaguar E-Type Gold Portfolio 1961-1971
Jaguar E-Type 1966-1971
Jaguar E-Type V-12 1971-1975
Jaguar XKE Collection No.1 1961-1974
Jaguar XJ6 1968-1972
Jaguar XJ6 Series II 1973-1979
Jaguar XJ6 & XJ12 Series III 1979-1985
Jaguar XJ12 1972-1980
Jaguar XJS Gold Portfolio 1975-1988
Jaguar XK120,XK140,XK150 Gold Portfolio 1948-1960
Jeep CJ5 & CJ6 1960-1976
Jeep CJ5 & CJ7 1976-1986
Jensen Cars 1946-1967
Jensen Cars 1967-1979
Jensen Interceptor Gold Portfolio 1966-1986
Jensen Healey 1972-1976
Lamborghini Cars 1964-1970
Lamborghini Cars 1970-1975
Lamborghini Countach Col No.1 1971-1982
Lamborghini Countach & Urraco 1974-1980
Lamborghini Countach & Jalpa 1980-1985
Lancia Stratos 1972-1985
Land Rover 1948-1973 - A Collection
Land Rover Series II & IIa 1958-1971
Land Rover Series III 1971-1985
Land Rover 90 & 110 1983-1989
Lincoln Gold Portfolio 1949-1960
Lincoln Continental 1961-1969
Lotus and Caterham Seven Gold Portfolio 1957-1989
Lotus Elan Gold Portfolio 1962-1974
Lotus Elan Collection No.2 1963-1972
Lotus Elite 1957-1964
Lotus Elite & Eclat 1974-1982
Lotus Turbo Esprit 1980-1986
Lotus Europa 1966-1975
Lotus Europa Collection No.1 1966-1974
Lotus Seven Collection No.1 1957-1982
Marcos Cars 1960-1988
Maserati 1965-1970
Maserati 1970-1975
Mazda RX-7 Collection No.1 1978-1981
Mercedes 190 & 300SL 1954-1963
Mercedes 230/250/280SL 1963-1971
Mercedes Benz SLs & SLCs Gold Portfolio 1971-1989
Mercedes Benz Cars 1949-1954
Mercedes Benz Cars 1954-1957
Mercedes Benz Cars 1957-1961
Mercedes Benz Competition Cars 1950-1957
Mercury Muscle Cars 1966-1971
Metropolitan 1954-1962
MG TC 1945-1949
MG TD 1949-1953
MG TF 1953-1955
MG Cars 1959-1962
MGA Roadsters 1955-1962
MGA Collection No.1 1955-1982
MGB Roadsters 1962-1980
MGB GT 1965-1980
MG Midget 1961-1980
Mini Moke 1964-1989
Mini Muscle Cars 1961-1979
Mopar Muscle Cars 1964-1967
Mopar Muscle Cars 1968-1971
Morgan Three-Wheeler Gold Portfolio 1910-1952
Morgan Cars 1960-1970
Morgan Cars Gold Portfolio 1968-1989
Morris Minor Collection No.1
Mustang Muscle Cars 1967-1971
Oldsmobile Automobiles 1955-1963
Oldsmobile Muscle Cars 1964-1971
Oldsmobile Toronado 1966-1978
Opel GT 1968-1973
Packard Gold Portfolio 1946-1958
Pantera Gold Portfolio 1970-1989
Plymouth Barracuda 1964-1974
Plymouth Muscle Cars 1966-1971
Pontiac Tempest & GTO 1961-1965
Pontiac GTO 1964-1970
Pontiac Firebird 1967-1973
Pontiac Firebird and Trans-Am 1973-1981
High Performance Firebirds 1982-1988
Pontiac Fiero 1984-1988
Pontiac Muscle Cars 1966-1972
Porsche 356 1952-1965
Porsche Cars in the 60's
Porsche Cars 1960-1964
Porsche Cars 1964-1968
Porsche Cars 1968-1972
Porsche Cars 1972-1975
Porsche Turbo Collection No.1 1975-1980
Porsche 911 1965-1969
Porsche 911 1970-1972
Porsche 911 1973-1977
Porsche 911 Carrera 1973-1977
Porsche 911 Turbo 1975-1984
Porsche 911 SC 1978-1983
Porsche 914 Gold Portfolio 1969-1976
Porsche 914 Collection No.1 1969-1983
Porsche 924 Gold Portfolio 1975-1988
Porsche 928 1977-1989
Porsche 944 1981-1985
Range Rover Gold Portfolio 1970-1988
Reliant Scimitar 1964-1986
Riley 11/2 & 21/2 Litre Gold Portfolio 1945-1955
Rolls Royce Silver Cloud 1955-1965
Rolls Royce Silver Shadow 1965-1981
Rover P4 1949-1959
Rover P4 1955-1964
Rover 3 & 3.5 Litre 1958-1973
Rover 2000 + 2200 1963-1977
Rover 3500 1968-1977
Rover 3500 & Vitesse 1976-1986
Saab Sonett Collection No.1 1966-1974
Saab Turbo 1976-1983
Shelby Mustang Muscle Cars 1965-1970
Stubebaker Gold Portfolio 1947-1966
Stubebaker Hawks & Larks 1956-1963
Sunbeam Tiger & Alpine Gold Portfolio 1959-1967
Thunderbird 1955-1957
Thunderbird 1958-1963
Thunderbird 1964-1976
Toyota MR2 1984-1988
Triumph 2000, 2.5, 2500 1963-1977
Triumph GT6 1966-1974
Triumph Spitfire 1962-1980
Triumph Spitfire Col No.1 1962-1982
Triumph Stag 1970-1980
Triumph Stag Collection No.1 1970-1984
Triumph TR2 & TR3 1952-60
Triumph TR4-TR5-TR250 1961-1968
Triumph TR6 1969-1976
Triumph TR6 Collection No.1 1969-1983
Triumph TR7 & TR8 1975-1982
Triumph Vitesse & Herald 1959-1971
TVR Gold Portfolio 1959-1988
Volkswagen Cars 1936-1956
VW Beetle Collection Col No.1 1970-1982
VW Golf GTi 1976-1986
VW Karmann Ghia 1955-1982
VW Kubelwagen 1940-1975
VW Scirocco 1974-1981
VW Bus, Camper, Van 1954-1967
VW Bus, Camper, Van 1968-1979
VW Bus, Camper, Van 1979-1989
Volvo 120 1956-1970
Volvo 1800 1960-1973

BROOKLANDS ROAD & TRACK SERIES
Road & Track on Alfa Romeo 1949-1963
Road & Track on Alfa Romeo 1964-1970
Road & Track on Alfa Romeo 1971-1976
Road & Track on Alfa Romeo 1977-1989
Road & Track on Aston Martin 1962-1984
Road & Track on Auburn Cord and Duesenburg 1952-1984
Road & Track on Audi & Auto Union 1952-1980
Road & Track on Audi 1980-1986
Road & Track on Austin Healey 1953-1970
Road & Track on BMW Cars 1966-1974
Road & Track on BMW Cars 1975-1978
Road & Track on BMW Cars 1979-1983
Road & Track on Cobra, Shelby & GT40 1962-1983
Road & Track on Corvette 1953-1967
Road & Track on Corvette 1968-1982
Road & Track on Corvette 1982-1986
Road & Track on Datsun Z 1970-1983
Road & Track on Ferrari 1950-1968
Road & Track on Ferrari 1968-1974
Road & Track on Ferrari 1975-1981
Road & Track on Ferrari 1981-1984
Road & Track on Fiat Sports Cars 1968-1987
Road & Track on Jaguar 1950-1960
Road & Track on Jaguar 1961-1968
Road & Track on Jaguar 1968-1974
Road & Track on Jaguar 1974-1982
Road & Track on Jaguar 1983-1989
Road & Track on Lamborghini 1964-1985
Road & Track on Lotus 1972-1981
Road & Track on Maserati 1952-1974
Road & Track on Maserati 1975-1983
Road & Track on Mazda RX7 1978-1986
Road & Track on Mercedes 1952-1962
Road & Track on Mercedes 1963-1970
Road & Track on Mercedes 1971-1979
Road & Track on Mercedes 1980-1987
Road & Track on MG Sports Cars 1949-1961
Road & Track on MG Sprots Cars 1962-1980
Road & Track on Mustang 1964-1977
Road & Track on Peugeot 1955-1986
Road & Track on Pontiac 1960-1983
Road & Track on Porsche 1961-1967
Road & Track on Porsche 1968-1971
Road & Track on Porsche 1972-1975
Road & Track on Porsche 1975-1978
Road & Track on Porsche 1979-1982
Road & Track on Porsche 1982-1985
Road & Track on Porsche 1985-1988
Road & Track on Rolls Royce & B'ley 1950-1965
Road & Track on Rolls Royce & B'ley 1966-1984
Road & Track on Saab 1955-1985
Road & Track on Toyota Sports & GT Cars 1966-1984
Road & Track on Triumph Sports Cars 1953-1967
Road & Track on Triumph Sports Cars 1967-1974
Road & Track on Triumph Sports Cars 1974-1982
Road & Track on Volkswagen 1951-1968
Road & Track on Volkswagen 1968-1978
Road & Track on Volkswagen 1978-1985
Road & Track on Volvo 1957-1974
Road & Track on Volvo 1975-1985
Road & Track - Henry Manney at Large and Abroad

BROOKLANDS CAR AND DRIVER SERIES
Car and Driver on BMW 1955-1977
Car and Driver on BMW 1977-1985
Car and Driver on Cobra, Shelby & Ford GT 40 1963-1984
Car and Driver on Corvette 1956-1967
Car and Driver on Corvette 1968-1977
Car and Driver on Corvette 1978-1982
Car and Driver on Corvette 1983-1988
Car and Driver on Datsun Z 1600 & 2000 1966-1984
Car and Driver on Ferrari 1955-1962
Car and Driver on Ferrari 1963-1975
Car and Driver on Ferrari 1976-1983
Car and Driver on Mopar 1956-1967
Car and Driver on Mopar 1968-1975
Car and Driver on Mustang 1964-1972
Car and Driver on Pontiac 1961-1975
Car and Driver on Porsche 1955-1962
Car and Driver on Porsche 1963-1970
Car and Driver on Porsche 1970-1976
Car and Driver on Porsche 1977-1981
Car and Driver on Porsche 1982-1986
Car and Driver on Saab 1956-1985
Car and Driver on Volvo 1955-1986

BROOKLANDS PRACTICAL CLASSICS SERIES
PC on Austin A40 Restoration
PC on Land Rover Restoration
PC on Metalworking in Restoration
PC on Midget/Sprite Restoration
PC on Mini Cooper Restoration
PC on MGB Restoration
PC on Morris Minor Restoration
PC on Sunbeam Rapier Restoration
PC on Triumph Herald/Vitesse
PC on Triumph Spitfire Restoration
PC on VW Beetle Restoration
PC on 1930s Car Restoration

BROOKLANDS MOTOR & THOROGHBRED & CLASSIC CAR SERIES
Motor & T & CC on Ferrari 1966-1976
Motor & T & CC on Ferrari 1976-1984
Motor & T & CC on Lotus 1979-1983

BROOKLANDS MILITARY VEHICLES SERIES
Allied Mil. Vehicles No.1 1942-1945
Allied Mil. Vehicles No.2 1941-1946
Dodge Mil. Vehicles Col. 1 1940-1945
Military Jeeps 1941-1945
Off Road Jeeps 1944-1971
Hail to the Jeep
US Military Vehicles 1941-1945
US Army Military Vehicles WW2-TM9-2800

BROOKLANDS HOT ROD RESTORATION SERIES
Auto Restoration Tips & Techniques
Basic Bodywork Tips & Techniques
Basic Painting Tips & Techniques
Camaro Restoration Tips & Techniques
Custom Painting Tips & Techniques
Engine Swapping Tips & Techniques
How to Build a Street Rod
Mustang Restoration Tips & Techniques
Performance Tuning - Chevrolets of the '60s
Performance Tuning - Ford of the '60s
Performance Tuning - Mopars of the '60s
Performance Tuning - Pontiacs of the '60s

CONTENTS

Page	Article	Publication	Date
5	New Front-Drive Citroens	Motor	Feb. 26 1935
7	The "Super Modern" Citroen	Light Car	Sept. 28 1934
8	The New Front-Drive Citroen	Motor	Sept. 25 1934
12	The Front-Drive Citroen Road Test	Motor	March 19 1935
14	The Citroen Range	Autocar	Oct. 4 1935
16	Citroen Super Modern Twelve Saloon Road Test	Autocar	Feb. 14 1936
19	The Front-Drive Citroen Twelve Saloon Road Test	Motor	Feb. 25 1936
21	Citroen Super Modern Fifteen Road Test	Autocar	June 19 1936
24	Improved Citroens	Autocar	Sept. 25 1936
26	A New "Family Fifteen" Citroen	Motor	Sept. 29 1936
29	The Citroen "Twelve" Road Test	Practical Motorist	March 13 1937
31	1938 Cars — The Citroens	Motor	Aug. 10 1937
35	Front Wheel Drive and Torsional Suspension	Motor Sport	July 1938
38	The Front-Drive Citroens	Motor	Aug. 16 1938
40	Latest Popular Citroens	Autocar	Aug. 19 1938
42	Citroen Light Fifteen Saloon Road Test	Autocar	Sept. 9 1938
44	A Six-Cylinder F.W.D. for French Market	Motor	Sept. 20 1938
46	Citroen Light Fifteen Road Test	Motor	May 16 1939
48	Citroen De Luxe Saloon	Autocar	June 2 1939
50	1940 Cars — Citroen	Motor	Aug. 15 1939
54	Citroen Six Now Here	Autocar	Aug. 18 1939
57	The 1940 Citroen Cars	Practical Motorist	Aug. 19 1939
58	Anglo-French Cars for Export	Motor	Feb. 21 1940
60	250,000 Miles in 369 Days	Motor	July 22 1942
61	The F.W.D. Citroen	Motor	May 24 1944
62	Across America Part One	Autocar	Aug. 17 1945
66	Across America Part Two	Autocar	Aug. 24 1945
69	Comment on the F.W.D. Citroen	Motor Sport	Oct. 1945
71	The Citroen Light Fifteen	Motor	March 20 1946
75	The Citroen Light Fifteen Road Test	Motor	May 5 1948
78	The Citroen Light Fifteen	Autocar	Feb. 22 1946
80	Citroen Six for Britain	Autocar	Sept. 24 1948
81	Citroen Light Fifteen Saloon Road Test	Autocar	May 21 1948
84	Front-Wheel-Drive Citroens — in Two Sizes	Motor	Sept. 29 1948
86	The Citroen Fifteen Road Test	Motor World	Oct. 22 1948
87	British Citroens Road Test	Autocar	Aug. 19 1949
89	The Citroen Six	Motor	Nov. 30 1949
91	Citroen Six Saloon Road Test	Autocar	Dec. 30 1949
95	The Citroen Light Fifteen Road Test	Motor	March 7 1951
98	Citroen 11CV "Normale" Owners Impression	Motor Sport	Dec. 1951
99	Citroen Continued	Autocar	Aug. 22 1952
100	A Truly Excellent Motor Car Road Test	Motor Sport	May 1952
102	British Citroen's Big Fifteen	Autocar	Oct. 17 1952
106	Three Citroen Models	Motor	Oct. 15 1952
108	The Citroen Big Fifteen	Motor	Nov. 26 1952
112	The Front-Drive Citroen Road Test	Motor World	Feb. 27 1953
114	Citroen Big Fifteen Saloon Road Test	Autocar	March 20 1953
117	Citroen Six Saloon	Autocar	June 26 1953
120	The Citroen Six Road Test	Motor	March 24 1954
124	Citroen Light Fifteen Road Test	Wheels	March 1954
128	The Citroen Six Road Test	Motor	March 23 1955
132	Suspension Under Scrutiny	Autocar	Aug. 19 1955
136	Progressive Maintenance of the Citroen 'Light Fifteen'	Practical Motorist	June 1957
139	1955 Citroen Light Fifteen Buying Used	Autocar	Aug. 14 1959
140	French without Tears	Wheels	Sept. 1971
144	Citroen — A Traction Avant	Car Classics	Aug. 1973
152	25 Years in Front	Thoroughbred & Classic Cars	March 1977
162	Heavy Metal	Collectors Car	Sept. 1981
164	Avant Garde	Autocar	March 3 1984
168	Traction Attraction	Classic and Sportscar	August 1983

ACKNOWLEDGEMENTS

Our first book devoted to the Traction Avant appeared many years ago, and has been out of print for some time now. In the meantime, the car's fiftieth anniversary has come and gone, and enthusiasm for it remains unabated. We thought, therefore, that it deserved new and expanded coverage in a Brooklands Books *Gold Portfolio*.

This book is dedicated to owners and enthusiasts of the Traction alike, and I am sure they will join me in thanking the managements of the following motoring magazines for their permission to reprint original materials: *Autocar, Car Classics, Classic and Sportscar, Collector's Car, Motor, Motor Sport, Motor World, Practical Motorist, Thoroughbred and Classic Cars,* and *Wheels*.

Once again, motoring writer and Citroen enthusiast James Taylor has kindly agreed to pen a few words of introduction for us.

R.M. Clarke

In the English-speaking countries, most people did not know what to make of Citroen's Traction Avant at first. Contemporary reports suggest either polite tolerance of a peculiarly Gallic piece of quirkiness, or a total incomprehension of the significance of its design advances. Make no mistake, though: the Citroen was very much ahead of its time when the first examples were shown at the 1934 Paris Salon. In those days, the automotive norms were a separate chassis, rear-wheel-drive and beam axles; and most cars had side-valve engines. The Citroen broke with all these traditions, for it had unitary construction, front-wheel-drive, independent front suspension and an advanced OHV wet-liner engine. Twenty-three years later, when the last examples left the production line in Paris, most of its pioneering advances had been widely adopted by other manufacturers.

Not only was the Citroen technically advanced, however. It also *looked* quite different from most of its contemporaries. In place of the upright and unimaginative lines of the typical 1930s family saloon, it had long, sleek and low styling, drawn up by an Italian sculptor who had turned his hand to car design. Yet the Traction Avant was still a family saloon, and still affordable by the man in the street. It was perhaps hardly surprising that the enormous cost of introducing such a revolutionary new car brought about Citroen's bankruptcy in the mid-1930s, but the Traction Avant carried on and some 750,000 had been built by the time production ended in 1957. By then, the car had come to symbolise the France of the 1930s, 1940s and 1950s.

I have always loved the Traction Avant, and I am delighted to see it commemorated in this Gold Portfolio from Brooklands Books, as handy and comprehensive a collection of writing about the car as any enthusiast or motoring journalist could wish for.

James Taylor
Woodcote, Oxon.,
February 1990

New Front-drive Citroëns

Two 15 h.p. Four-cylinder Cars of Different Wheelbase Lengths Available as Five and Seven-seater Saloons

Photographs showing the roomy and attractive cars—the six-light and the four-light saloons.

FOLLOWING upon the success of the front-drive "Super Modern Twelve" introduced last year, Citroën Cars, Ltd., have decided to produce a larger four-cylinder model rated at 15 h.p., of which we are able to give the first full description. The new car is being put into production at the Citroën works at Slough, Buckinghamshire, and was displayed to a gathering of the trade and Press at the factory on Friday last.

Low Centre of Gravity

In all the essentials the new Fifteen is designed as a counterpart of the smaller 12 h.p. model. It displays the same "Monoshell" body, which is structurally so strong as to make a separate chassis unnecessary, the drive is taken through the front wheels, the rear axle simply trails and the suspension is by four torsion bars. Despite the low centre of gravity and modest overall height, the body provides plenty of headroom because the absence of the usual propeller shaft enables the flat floor to be set close to the road. Furthermore, a considerable degree of streamlining has been achieved.

The Fifteen is available in two lengths of wheelbase, measuring 10 ft. 1½ ins. and 10 ft. 9 ins. respectively. The track is 4 ft. 8 ins. in each case. The shorter car is a four-light saloon priced at £315, with de luxe equipment and leather upholstery. The body is exceptionally wide so that three people can be carried comfortably on the rear seat. The car is fitted with four doors and bears a strong family resemblance to the smaller 12 h.p. saloon.

The longer Fifteen is termed a seven-seater saloon, although in actual fact the bench-type front seat is sufficiently wide to carry three people, so that the car can be made to hold eight persons in all. Two occasional seats of the folding type are provided, a good point being that the clearance between the seats and the doors, together with the spacing of the seats themselves, enables three rear passengers to stretch their legs in comfort when the occasional seats are in use. This body is provided with quarter windows and is moderately priced at £345 complete.

The equipment of both cars includes Lucas 12-volt lighting and starting sets, Trafficators, sliding roofs, leather upholstery, etc.

As in the 12 h.p. front-drive model the general construction of both the new models is unorthodox but highly practical. The advantages of a low centre of gravity and roomy coachwork have already been mentioned, another most important point being the material reduction in weight as compared with an ordinary chassis. The steel body shell is a double structure of great strength and rigidity, suitably reinforced by cross tubes at the front and back. The rear axle is connected by radius arms to torsion bars, which replace the conventional springs.

Torsion-bar Suspension

The engine, gearbox and final drive form a compact unit housed in the scuttle structure of the body, and the front wheels (which are individually supported by transverse radius arms) are also sprung by torsion bars. An interesting point is that the drive is taken from the clutch to a gearbox just behind the radiator and is then brought back to

There is little in frontal appearance to distinguish the new models from the existing 12 h.p. model. Clean and simple lines predominate. Note the neat way in which the horns are recessed into the front wings.

NEW CITROËN FRONT-DRIVE MODELS . . . . Contd.

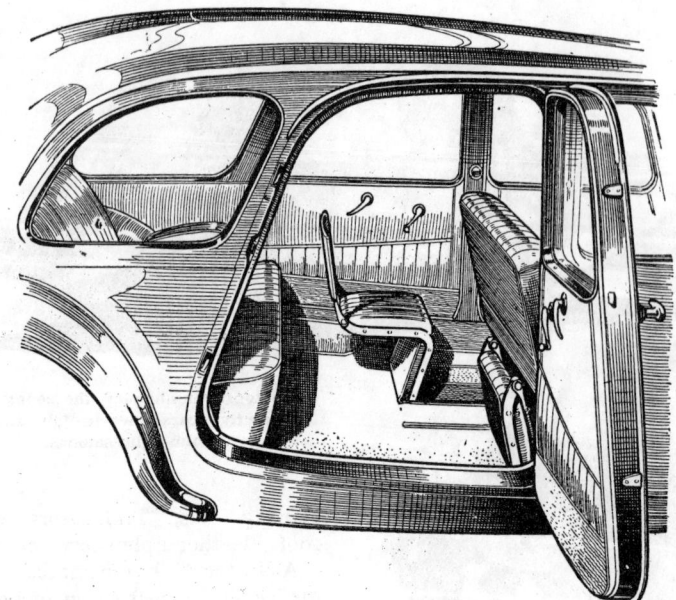

Above is seen the layout of the neat occasional seats which provide uncramped accommodation.

On the right is a view of the front drive and independent wheel suspension assembly.

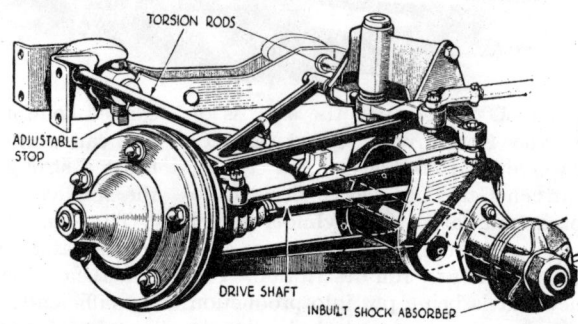

the final-drive bevels. As a result the engine is shifted much farther forward than would otherwise be possible, so providing extra room for the body. The rear seats are actually ahead of the back axle and there is a large space for luggage in the tail, although the overhang is only small.

The engine itself displays many points of interest and is built on the same lines as the 12 h.p. unit. Detachable cylinder barrels, technically known as "wet liners," are fitted. These ensure uniform cooling and are made of a long-wearing iron. Overhead valves are fitted, operated by push rods. The bore and stroke are 78 mm. and 100 mm. respectively, giving a capacity of 1,910 c.c., and a rating of 15 h.p. (tax £11 5s.). The complete power unit is flexibly mounted.

The gearbox provides three forward speeds, through helical-toothed gears, and a synchromesh engagement is employed for the top and middle ratios. The gear lever projects through the facia and, as the hand brake is also located near to the instrument panel, the front compartment is entirely free from obstructions. The steering gear is divided to suit the independent front suspension, and affords the correct geometry necessary for accurate control. The braking system is of the Lockheed hydraulic type with an independent mechanical coupling between the rear shoes and the hand brake.

Coachwork Features

From the point of view of coachwork only, that is, leaving the technical aspect out of it, both models make a very strong appeal by virtue of their comfort and general roominess. The appearance has an air of practical design and smartness.

Above the arm-rests the rear seat measures over 50 ins., indicating that three people can be carried almost regularly. There is real head room when wearing a hat, except for the abnormally tall, and all the leg room that any lounger could wish.

The additional wheelbase length of the six-light model is used to accommodate two occasional seats. Each of these has corresponding toe room beneath the front seat, so that the rear compartment carries five passengers, when required, without undue crowding, whilst for normal use the folded occasional seats form foot-rests.

The driver is well provided for. He has excellent visibility (he can see both wings), correct seating, and is within easy reach of the controls. There is a visor in front of him, in front of the sliding roof, behind which is a parcel net. Companion cubby holes are fitted in the rear quarters. Tools are carried below the driving seat, and there is little fault to find with the disposition of anything.

The "Super Modern Twelve" is now available as a sports saloon, fitted with the 15 h.p. engine (as used in these new 15 h.p. models), at a price of £285. This is additional to the existing range, which remains unchanged.

This off-side view of the engine shows the Solex carburetter and air silencer.

The "Super Modern" Citroën

Torsional Suspension and F.W.D.

"TEN" TO HAVE A BIGGER ENGINE

THE most interesting feature of the Citroën programme for 1935 is the inclusion in the range of the "Super Modern" Twelve. This car has already been described in this journal (in our issue of May 4 this year), but it has not actually been obtainable in this country. Now it takes its normal place in the range and is available as a semi-streamlined saloon at £250, a Roadster at £270, and a fixed-head coupé at £275.

All these cars, although not fully streamlined, have exceptionally modern lines and an interesting point about the roadster is that, although built on two-seater lines, it will hold three people in the front seat and has additional accommodation for two in a roomy dickey.

The three outstanding features of the general design are the use of a system of torsional suspension, the adoption of front-wheel drive, and the use of Monoshell coachwork, which does away with the need for a separate chassis frame.

The torsional suspension, for example, is entirely unlike any normal system of springing, independent or otherwise. Actually, the wheels at the front are carried on hinged arms attached to the front of a pressed-steel cradle, which fits round the engine, and to the king pins. The upper hinged arm on each side is entirely free to move up and down, whilst the lower one is attached to the end of a long torsion bar which runs parallel with the engine; thus, when the car strikes a bump the wheel is deflected upwards, but its movement is opposed by the resistance to twisting of the torsional bar.

At the rear a system of torsional suspension employing exactly the same principle is used, but in this case the wheels are connected by an axle and are not independently sprung.

It is claimed for this torsional suspension system that the flexibility of the torsional bar is constant as opposed to the varying flexibility of the springs, that the whole system represents a considerable saving in weight and that no attention in the matter of lubrication or otherwise is needed.

The Monoshell construction represents a further step forward in the Citroën policy of attaining absolute rigidity and enormous strength in the car as a whole. In place of the conventional chassis frame an extremely rigid metal structure is used and to it are attached steel body panels welded together.

So far as the engine is concerned, a four-cylinder overhead valve unit of 72 mm. by 80 mm. (1,302 c.c.) is employed, the tax at present being £13, and, after January 1 next, £9 15s.

The drive from the engine is taken via a single dry-plate clutch to a three-speed gearbox with synchromesh for second and top, and thence through a spiral bevel drive to the front wheels.

The 10 h.p. model which has proved so successful is being continued for 1935 with only detail changes, most important of which is a slight increase of the stroke of the four-cylinder engine; the dimensions are now 69 mm. by 100 mm., which gives a capacity of 1,496 c.c. as opposed to the old capacity of 1,452 c.c.

Other changes include more attractive trimming for the bodywork, a new layout for the facia board, anti-splash wings, a folding arm rest for the rear seat, safety glass all round, a new radiator shell with a greater slope and the addition of an air silencer to the carburetter. The price of the saloon, which is the only model on the Ten chassis, is £225.

(Above) Despite its semi-streamlined body, the Super Modern saloon has distinctly roomy rear seats.

The new Super Modern Citroen in ghost form (above) to show the Monoshell construction and as a complete car (below). The car has torsional suspension.

The New Front-drive Citroën

NOW IN PRODUCTION AT THE SLOUGH WORKS

An Interesting Car with Independent Torsion-bar Springing, Novel Frameless Construction and Streamlined Coachwork. Other Models Continued with Minor Alterations

AN outstanding feature of the Citroën programme for 1935 is the front-drive model, which was described in *The Motor* a few months ago. This exceptionally interesting car is now in production at the Citroën works at Slough, and is listed at £250 in streamlined saloon form. A fixed-head coupé and an open roadster are also available.

Known as the Super-modern Twelve, this four-cylinder front-drive car is built up from steel pressings in such a way that the conventional chassis frame is eliminated. The advantages gained include a low build and a great saving in weight, while the car has also been shaped to reduce air resistance to a minimum. Additionally, a novel form of independent suspension has been adopted in which torsion bars are used in place of the more conventional types of spring.

Stability Proved in Road Trial

As reported in *The Motor* last May, our road trial of this model showed it to possess great stability. Bends could be taken extremely fast and on a rough road the riding comfort and absence of pitching were exceptional.

In addition to this interesting newcomer, the Citroën Co. is continuing to market the existing range of conventional rear-drive cars. The smallest of these is the Ten, introduced in 1932, which secured a remarkable series of long-distance records at Montlhéry track not long ago. Other four-cylinder models are the Big Twelve and the Light Twelve.

All these models embody well-known Citroën features, such as the floating power method of engine suspension, four-speed gearboxes with synchromesh and free wheel, rigid box-section frames and all-steel coachwork. The 10 h.p. model has been improved by a more artistic radiator, more comfortable interior trimming, special wings, a folding central armrest for the back seat, a new instrument panel and an air silencer. The Big Twelve is of somewhat similar design, but this larger chassis carries wider coachwork. The only six-cylinder model listed is the Light Twenty, available with either saloon or sports coachwork. As the name indicates, this car has a high power to weight ratio, enabling a good performance to be obtained on a fairly high top gear. In saloon form this

Unlike other Citroen engines the power unit of the new model is fitted with overhead valves.

The streamlined tail carries the spare wheel.

Gear Control Rods
Overhead-valve Engine
Divided Steering Mechanism
Inbuilt Shock Absorbers
Front-wheel Drive

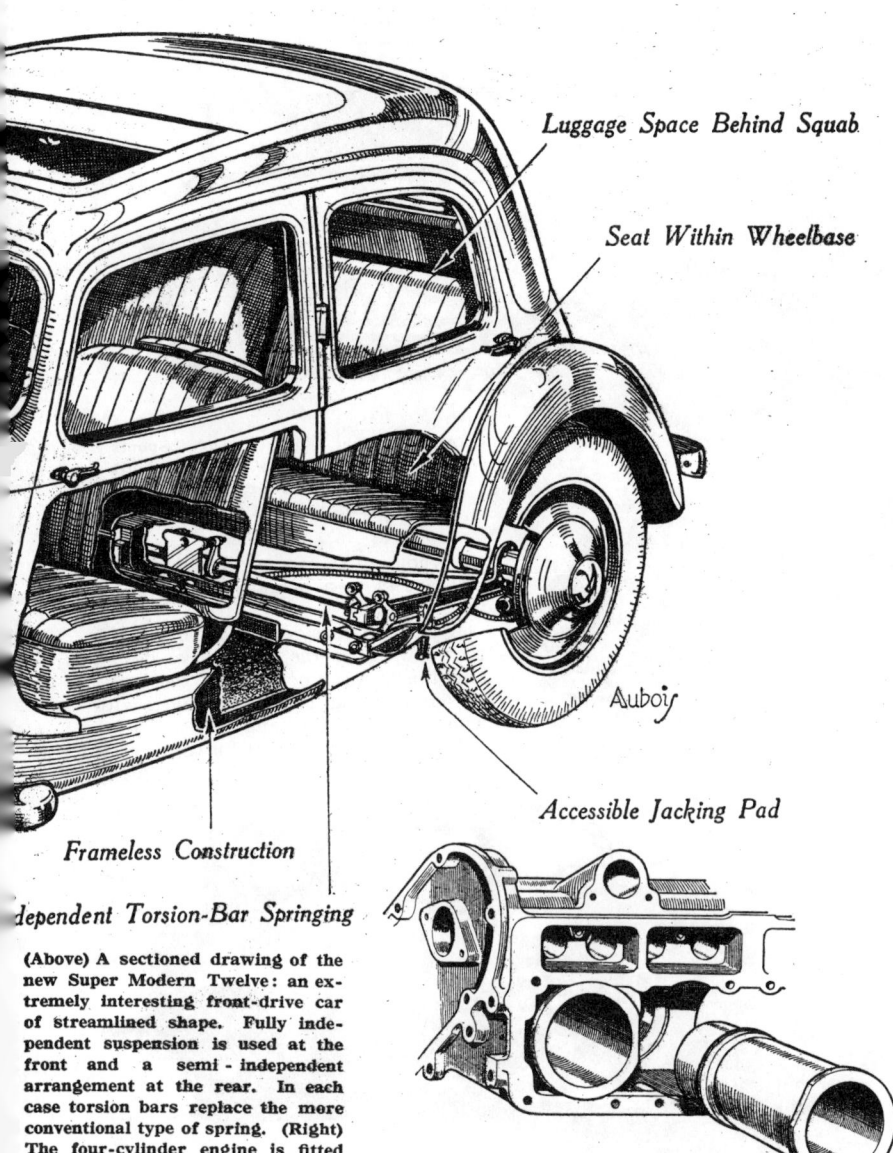

Luggage Space Behind Squab

Seat Within Wheelbase

Accessible Jacking Pad

Frameless Construction

Independent Torsion-Bar Springing

(Above) A sectioned drawing of the new Super Modern Twelve: an extremely interesting front-drive car of streamlined shape. Fully independent suspension is used at the front and a semi-independent arrangement at the rear. In each case torsion bars replace the more conventional type of spring. **(Right)** The four-cylinder engine is fitted with readily detachable liners of durable iron.

car costs only £285. The engine has a capacity of 2,650 c.c. and an R.A.C. rating of 21 h.p.; the cylinder dimensions are 75 mm. by 100 mm.

Reverting to the new front-drive car, this contains so many interesting features of construction as to make a concise description difficult. Most important, perhaps, is the plan of building the car from two steel shells, one within the other, electrically welded to form box-sections of great structural rigidity. Only a trained engineer can perhaps fully appreciate the many ingenuities which have made this scheme a success. Used in conjunction with the front-drive principle, it results in a very low yet roomy car with flat floors, easy access to the seats, ample headroom, and a total weight which is about 25 per cent. less than that of a conventional car of equivalent passenger accommodation.

Next there is the method of springing by torsion bars, which can readily be followed from the sectioned drawing reproduced. The load on each wheel reacts upon a bar of circular section, the elastic twisting action of which replaces the deflection of a more conventional coil or leaf spring.

At the front, each wheel is supported by transverse radius arms, the bearings of which are carried in an extremely strong U-shaped steel cradle. Frictional shock absorbers are built into the pivots of the lower arms, and these arms are also connected by dog couplings to torsion rods which run rearwards along each side of the engine.

At the rear ends the rods are held against turning by levers which react against adjustable stops. The adjustment is provided to enable the front end of the car to be levelled up after assembly.

The rear suspension is worked out in a different way so as to provide a measure of independent action, together with stability against rolling. A light tubular axle connects the wheels and is supported by radius arms coupled to transverse torsion bars. These arms also take the torque reaction produced by braking. They are made of deep yet narrow section so as to provide adequate strength in a vertical plane, together with an ability to twist to some extent when one rear wheel only is raised by an obstruction.

The Transmission System

Turning to the front-drive arrangement, it will be noticed that the engine is located further forward than usual so as to provide adequate weight for wheel adhesion. As a result, the whole body is moved forwards, bringing the rear seat ahead of the axle and so enhancing riding comfort.

The drive is conveyed from the clutch to a three-speed gearbox and is then taken back to a final drive, from which it is distributed to two transverse shafts. These are fitted with special universal joints and carry the power outwards to the wheels. Incidentally, synchromesh is used to facilitate gear changing, the controls being taken to a gear lever projecting through a gate in the facia. We found that after a few moments' practice no difficulty was experienced in handling this device.

Engine Details

Contrary to previous Citroën practice, the engine of this front-drive model is fitted with push-rod-operated overhead valves. Great care has been taken to ensure adequate lubrication and quiet operation in the valve mechanism. Another interesting point is that the cylinder bores are formed in detachable wet liners. Each liner is made from a durable iron-alloy, and is fitted separately to the block, the outer surface being in direct contact with the cooling water. Other engine features include fan and pump cooling, coil ignition and a pump-type petrol feed from a rear tank.

The double-walled steel shell, which combines the functions of chassis and body.

THE NEW FRONT-DRIVE CITROEN—Contd.

The engine is rubber-mounted on the floating-power principle which the Citroën Co. adopted, under Chrysler licence, some time ago. The cylinder dimensions are 72 mm. by 80 mm., giving a capacity of 1,302 c.c. and a rating of 12.8 h.p.

One other mechanical feature requires special mention, this being the steering gear. The rod system is divided in such a way that the steering is not affected by up-and-down movements of front wheels—a point which we have verified by noting the steadiness of the wheel when driving over extremely rough roads. Furthermore, in order to bring the steering wheel into a comfortable position, the driver's effort is transmitted by bevel gears fitted at the foot of a short column. Lockheed hydraulic brakes are standardized.

Many excellent features of the saloon body have already been mentioned, such as the good appearance, low build and roomy interior. It is equipped throughout in a thoroughly practical manner and is fitted with a sliding roof. Great care has been taken over the shape and springing of the seat cushions, which are upholstered in leather. A central armrest of the folding type is fitted to the rear seat, and

(Above) The Light Twenty Sports Coupe; a six-cylinder model. (Left) The new front-drive car in roadster form, listed at £270.

the whole squab can readily be lifted to give access to a luggage compartment in the tail.

A point of considerable importance is that, owing to the front-drive design and the forward position of the body, an outswept tail can be employed with a minimum of overhang.

From the driver's seat forward vision is very much better than in many

(Above) The Citroen Ten saloon which is continued for 1935 with various improvements. (Left) The engine of the new front-drive model is accessibly arranged.

modern cars, it being possible to see the near-side wing lamp by moving the head slightly. As the gear lever projects through the facia and the hand brake is of the pull-out type, the front compartment is entirely unobstructed. The electrical equipment includes duplicated stop lights and horns and a battery master switch. Triplex glass is fitted all round.

In addition to the saloon model, priced at £250, there is a fixed-head coupé with occasional seats at the back, and an open two-seater (called the Roadster), the width of which is actually sufficient to enable three people to be seated abreast.

MODELS AND PRICES

The full range of prices is as follows:—Super Modern Twelve: Saloon, £250; fixed-head coupé, £275; roadster, £270. Ten Saloon: £225. Big Twelve: Saloon, £245; seven-seater saloon, £255. Light Twelve: Saloon, £235; sports four-seater, £275; sports coupé, £305. Light Twenty: Saloon, £285; sports four-seater, £305; sports coupé, £340.

(Right) Front compartment of the Super Modern Twelve, showing the gear lever (projecting through the facia) and the pull-out hand brake. Note the easy entrance provided by the forward slope of the door opening.

2 YEARS AHEAD

- FRONT WHEEL DRIVE
- INDEPENDENT WHEEL SUSPENSION
- AERODYNAMIC BODY
- UNIMPEDED FLOOR AREA
- FRAMELESS CONSTRUCTION
- TORSION BAR SPRINGING
- AMAZING STABILITY
- REMARKABLE ROAD VISIBILITY

Here is the car of the future....built for you to-day! Beautiful in appearance, the "Super Modern Twelve" marks a new era in design, strength, safety and riding comfort. Smooth and vibrationless over the roughest roads, with a stability and immunity from skidding which is amazing, it is fully two years ahead in conception and performance. Breaking away from old traditions, it is new from radiator to tail, and is far ahead of any other design.
There is more room, more luxury, greater security and better value than in any other car at its price....send for Catalogue 19 and find out more about the most amazing car in history.

SUPER MODERN
Twelve

 DE LUXE £250 EQUIPMENT

PRODUCED AT
CITROEN WORKS · SLOUGH, BUCKS.
CITROEN CARS LTD.,
Crook Green, Hammersmith, W.6.   Showrooms: Devonshire House, Piccadilly, London, W.1

Without parallel for **SECURITY & STABILITY**

THE FRONT-DRIVE CITROËN

Four-cylinder Super Modern Twelve Tested Under Severe Conditions. Road Holding a Strong Feature

(Above) A view which gives a good impression of the low build; the overall height is slightly less than 5 ft.

(Left) The Super Modern Twelve saloon, as tested, which is priced at £250 complete. Note the streamlined shape, disc wheels and large tyres.

"THE MOTOR" RATIONALIZED ROAD TESTS

GENERALLY considered to be one of the most interesting of the new cars shown at Olympia last October, the new Citroën Twelve has since been selling in considerable numbers. In putting it through a thorough test we were particularly anxious to note its behaviour as effected by the front-drive principle and the novel torsion-bar system of suspension employed.

It will be recalled that this car differs from all others in current production by virtue of its "Monoshell" type of all-steel body, which is so rigid as to make a separate chassis frame unnecessary. The engine and transmission unit is simply fitted to a reinforced scuttle structure at the front, and the rear axle trails from radius arms connected to a stout cross-tube at the back.

At the front and rear long rods of circular section are employed in place of springs, the rise and fall of each wheel being absorbed by the twist of the rod to which it is coupled. The rear wheels are connected by a light tubular axle, while the front wheels are independently mounted on radius arms.

The absence of any transmission mechanism under the body has enabled the designers to provide a low, flat floor and plenty of headroom in a car with an overall height of only 4 ft. 10½ ins. Furthermore, the masses are all low, so that the centre of gravity is only 22¾ ins. from the road surface. As explained in a recent article in *The Motor*, a low centre of gravity reduces to a minimum the loss of front wheel adhesion which inevitably occurs when climbing a steep hill.

Tested under severe conditions this front-drive arrangement showed up very favourably. On a dry surface it proved easily possible to restart the car on a severe slope of 1 in 4.

Another interesting test was made on a slippery cart-track at the back of Box Hill, Surrey, where the chalk surface under the trees was in a treacher-

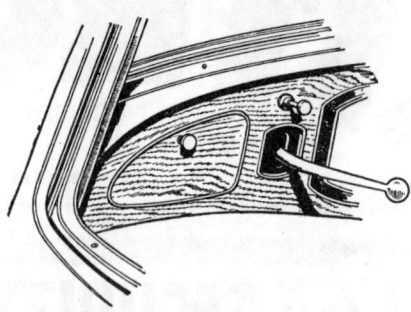

The gear lever projects through a slot in the facia where it is easily reached with the left hand.

ous condition. The steepness of the ascent gradually increases, and the car climbed steadily until it was finally stopped by wheel-spin on a gradient of 1 in 5. It was then allowed to roll back to the bottom and was driven up the hill in reverse, the driving wheels being then, naturally, at the lower end of the car.

The amount of wheel-spin experienced in reverse was slightly less than on the previous climb, but the car finally came to a standstill at almost exactly the same point as before. Our general conclusion was that even in an exceptionally hilly district no trouble is likely to be experienced with lack of adhesion.

An important point is that if wheel-spin does occur on a slippery surface the front wheels tend to keep to a straight course, whereas in a rear-drive car wheel-spin is always likely to be accompanied by sideways sliding of the tail. When negotiating sharp corners the front drive undoubtedly scores, this feature (coupled with the low centre of gravity) conferring an exceptional steadiness when open bends are taken at speed.

The springing is very flexible, and is seen at its best on a really rough road. There is some tendency towards pitching when main-road ripples are taken

THE FRONT-DRIVE CITROEN

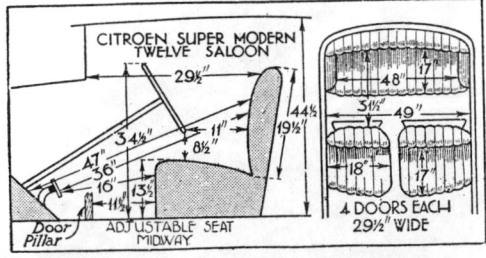

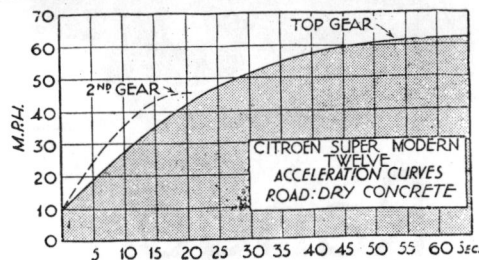

(Above) The interior of the body, showing ease of entry, wide seats and low floor; alongside are reproduced the body dimension diagram and acceleration graphs. (Left) An under-bonnet view of the engine and accumulators.

fast, but this is eliminated if passengers are carried at the back. Incidentally, the rear seat is placed just in front of the back axle, and there is a large space for luggage enclosed in the tail.

The driving position is fairly upright, and the pedals project horizontally, instead of being placed at an angle, as shown in a photograph reproduced. Once accustomed to this arrangement the driver finds no difficulty in control, but it is questionable whether this position affords quite so much comfort as does a more conventional layout. The wide range of forward vision is particularly commendable, as despite the low seating position both side lamps are within view.

Easy Gear Changing

The gear lever projects through a small gate in the facia to the left of the instruments, and is slightly angled to bring the knob within easy reach. Gear changing is made easy by the use of a synchromesh mechanism for the engagement of the top and middle gears of the three-speed box. The car is unusually lively for a 12 h.p. model, this being largely due to the weight reduction made possible by the unique design. For example, Brockley Hill, Middlesex, in which a long gradient of 1 in 15 leads to a final slope of 1 in 10, was climbed comfortably in top gear, with a maximum speed of 45 m.p.h. on the early stretch and a minimum speed of 30 m.p.h. at the crest.

The car is extremely handy in heavy traffic, as it steers well, is lively, and has powerful Lockheed hydraulic brakes. The hand control, incidentally, takes the form of a pull-out handle just beneath the facia, and as the gear-lever is also carried in a high position the front compartment is entirely free from obstructions.

The engine, in contrast to previous Citroën power units, is of the overhead-valve type, and is fitted with detachable cylinder bores made from a long-wearing iron. It pulls effectively at low speeds and affords a maximum speed of 62 m.p.h. The car cruises happily at 50-55 m.p.h., and on the middle gear of the three-speed box 45 m.p.h. can be reached. The transmission is a trifle noisy when the engine is pulling hard at low speeds.

The passenger accommodation is roomy and the seats are comfortably upholstered in leather of good quality. There are four wide doors, and the floor is so low that running-boards are unnecessary. Triplex glass is fitted.

The front-drive Citroën can, in short, be classified as a highly practical motorcar with many excellent features. The front-drive principle and torsion-bar suspension are now widely employed on the Continent and would appear to be mechanically sound and free from all forms of service trouble.

TABULATED DATA—CITROEN TWELVE

CHASSIS DETAILS

Engine: Four cylinders, push-rod operated overhead valves, coil ignition with automatic timing; 72 mm. by 100 mm. (1,628 c.c.). Rating 12.8 h.p., tax £9 15s.

Gearbox: Three forward speeds with synchromesh for second and top. Ratios, 4.9, 8.3 and 14.8 to 1.

PERFORMANCE

Speeds on Gears: Top, 62 m.p.h.; second, 45 m.p.h.; minimum speed on top gear, 6 m.p.h. approx.

Acceleration: From standstill through the gears to 50 m.p.h., 26 secs. Standing ¼-mile, 27 secs. (Average speed, 33.3 m.p.h.)

Tapley Performance Figures: Maximum pull in lb. per ton on gradient: top, 190 lb.; second, 340 lb. Corresponding gradients climbable at a steady speed are 1 in 11.8 and 1 in 6.6 respectively.

Petrol Consumption: Driven hard, 25 m.p.g.

Braking Efficiency: Measured by Tapley meter, using the pedal only: from 30 m.p.h., 85 per cent.; from 50 m.p.h., 80 per cent. Corresponding stopping distances are 35 ft. and 105 ft. respectively.

DIMENSIONS, ETC.

Leading Measurements: Wheelbase, 9 ft. 6½ ins.; track, 4 ft. 4 ins.; overall length, 14 ft. 2 ins.; width, 5 ft. 2 ins.; height, 4 ft. 10½ ins.

Wheels and Tyres: Bolt-on disc wheels with 140 by 40 ultra-low-pressure tyres.

Turning Circles: Left and right, 36 ft. dia.

Weight: As tested, with two up, 23¼ cwt.; unladen, 20½ cwt.

Price: Four-door saloon, as tested, £250.

The Fifteen seven-seater front wheel drive saloon.

The Citroën Range

Front Wheel Drive Models Go On Unaltered in Principle and With Two Engine Sizes : The Rear Wheel Drive British-built Ten Retained

THE front wheel drive Citroën models have had a successful reception in this country, and the cars now being delivered incorporate a number of improvements and modifications which have been put into effect during the course of the past twelve months, and not necessarily at any given time. The f.w.d. models are accordingly continued for next year, together with the Ten, the only remaining rear wheel drive Citroën.

There are three front wheel drive models, of which the Twelve was the original example of the striking breakaway in Citroën design represented by the change over to f.w.d. made last year. During this year, and now continued for 1936, there have been introduced two additional styles, the Sports Twelve and the Fifteen. These have a larger engine, rated at 15.08 as against 12.8 h.p. in the case of the Twelve.

Briefly may be repeated the outstanding features of the design of these most interesting cars. To start with, there is the front wheel drive, the clutch, gear box, and final drive itself being, of course, ahead of the engine and transmitting power to the front wheels through short universally jointed shafts; the engines are four-cylinder push-rod operated overhead-valve types, which are lively and give a good road performance.

The suspension is one of the most notable points, being, it will be remembered, independent for the front wheels by torsion bars instead of the conventional leaf springs, and by a torsion bar also for the rear wheels. The springing effect is obtained from the elasticity of steel rods to which the wheels are coupled, and besides remarkable comfort of suspension a steady-riding car results. It is stated that a microscopically small amount of trouble has been experienced with the torsion bars on the considerable number of cars of this type now in service.

There is no frame in the ordinary sense. A steel shell consisting of units welded together to form a strong whole is both frame and the basis of the body, besides carrying the engine and transmission units, and taking the attachments for the torsion bar springs.

In addition to points of improvement concerning the drive shafts, there have been changes in regard to shock absorbers, which now, in front, are of telescopic hydraulic type, instead of frictional as before, and these are mounted in such a position as to do more useful work than formerly. At the rear hydraulic shock absorbers are now used, but are of the more usual pattern. Refinements in the steering have also been made. Not only has it been made lighter, but also definite caster action has been introduced.

Three F.w.d. Models

The broad features of the design apply alike to the Twelve, the Sports Twelve, and the Fifteen. The Sports Twelve was introduced to give, with the larger size of engine, that extra performance that some people want, whilst with this size of engine and a longer wheelbase, as found in the Fifteen, more commodious bodywork can be provided, and even a seven-seater style offered, as is standardised among the range of models.

A great advantage as regards bodywork derived from the special construction of these cars is that perfectly flat and unobstructed floors are possible for the back as well as the front compartment. Also, there are no control levers in the way, for the gear lever protrudes from the instrument board—where, however, it is quite convenient—and the hand brake lever is of pull-and-twist type mounted just below the instrument board. Ease of entry and exit is another advantage of the construction, for the car is low in relation to road level, and there is nothing to obstruct the door space. Running boards are not fitted.

The whole finish, the upholstery, and the detail appointments are in keeping with our own style in these matters, for this work is carried out at the Citroën factory at Slough with British materials. Thus one finds leather upholstery, and that attention to small points which we as motorists appreciate here—the provision of a control for the rear blind, and accommodation in the shape of roof net, cubby holes, and pockets for oddments.

A modified instrument panel is now being used, still grouping all the gauges in one large dial, but in a more easily read form and better illuminated at night, with a rheostat control for varying the intensity of illumination according to taste. Also, the incidental controls have been grouped together, the horn button, the traffic signal switch, and the anti-dazzle control being carried on an arm extending above the steering wheel.

There is also a change in regard to the luggage compartment formed in the tail

The steel shell which is both body and frame.

(Above) Fixed-head coupé, which is available with either the Twelve or the Fifteen engine.

The Fifteen saloon is priced at £315, the seven-seater saloon at £345.

Other features of the design which may be mentioned are the flexible mounting of the engine on rubber at two points, the use of detachable cylinder barrels of a material which is particularly resisting to wear, a three-bearing crankshaft, 12-volt starting and lighting set, three-speed gear box, dry single-plate clutch, and Lockheed hydraulically operated brakes. Steel disc wheels are fitted.

The dynamo is of the latest automatically controlled output type, there is a concealed battery master switch, and twin stop lights are fitted at the rear. A petrol tank filler cap at each side is a convenience.

of the saloon, access to which is now given by a hinged external panel, which also carries the spare wheel. The latter is enclosed in a cellulosed metal cover.

The broad specification of the Twelve is : 12.8 h.p., four cylinders, 72 by 100 mm. (1,628 c.c.), tax £9 15s. Wheelbase 9ft. 6½in., track 4ft. 4in.

There are three body styles, of which the prices are: saloon £265, roadster £270, fixed-head coupé £275. The saloon is a four-door style; the coupé and roadster, which are similar in lines, allowing for the absence of a permanent head in the one case, of course, are particularly smart among standardised bodywork. They are two-seaters, with enclosed luggage accommodation in the dickey. The overall length of the saloon is 14ft. 3in., overall width 5ft. 3½in., overall height 5ft.

Slight changes in appearance have been made by an alteration to the upper part of the radiator shell, whilst the Citroën chevron design worked into the inclined slats in front of the radiator proper is more pronounced. The shell surrounds are now cellulosed to match the colour scheme, and the shape of the front wings has been altered slightly, an impression of greater depth being given to them. Also, the twin horns, built into the wing side valances and visible from the front, are now actually mounted on brackets behind the wings, but the plated grilles are retained by way of ornament. The front bumper, formerly divided, is now in one piece, being of distinct V formation.

The Sports Twelve is of similar dimensions as to the wheelbase and track and overall measurements, but has certain points of its own, among them the use of higher gear ratios, a 4.3 to 1 final

(Below) The rear wheel drive Ten saloon.

drive being employed as against 4.9 to 1 on the Twelve, with correspondingly higher second and first ratios. Also bigger tyres are fitted.

The engine of the Sports Twelve is rated at 15.08 h.p., four cylinders, 78 by 100 mm. (1,911 c.c.), tax £11 5s. The prices, for similar styles of bodies as apply to the Twelve, are saloon £285, roadster £305, fixed-head coupé £310.

Then the Fifteen, with the same size of engine as the Sports Twelve, has a wheelbase of 10ft. 1½in. for the five-seater saloon, and a wheelbase of 10ft. 9in. for the seven-seater, which has two occasional seats in addition to the main seats. The track in both cases is 4ft. 8in., the overall length of the saloon is 15ft. 3in., and of the seven-seater saloon 15ft. 11in., the overall width in both cases being 5ft. 8in., and the height 5ft. 1in. The bodywork of both the Twelves and the Fifteen is similar in general arrangement, differing in the accommodation provided.

In the case of the Fifteen, advantage is taken of the possibility of seating three in front, due to the absence of controls to get in the way, and a single-piece adjustable front seat is used instead of the separate seats of the other models.

The Ten, which is of conventional construction, with a four-cylinder engine of 69 by 100 mm. bore and stroke (1,495 c.c.), taxed at £9, is continued unaltered. This is a roomy, dependable kind of car which has a definite following, and it is built throughout at Slough from British parts. It has a four-speed gear box with synchromesh on top and third, as well as a free wheel, a cruciform-braced frame, normal half-elliptic springs, and provides unusual seating accommodation in a car of its size and price.

The wheelbase is 8ft. 10¼in., the track 4ft. 4¾in., and the price of the saloon with sliding roof and full equipment, including bumpers, luggage trunk, traffic signals, and so forth is maintained at £198.

(Left) Occasional seats in the Fifteen saloon, making it a seven-seater.

Engine and transmission unit.

The Autocar Road Tests

No. 999 (Post-War Series)

CITROËN SUPER MODERN TWELVE SALOON

An Interesting Car to Drive, and a Very Comfortable One

The Citroën on the London-Eastbourne road, in Ashdown Forest.

IT is not far short of two years since the front-wheel-drive Citroën first appeared in this country. In the meantime, much important development has been going on with regard to various points of the original design, as was natural where so revolutionary a car was concerned. Particularly, it has been found possible to incorporate improvements and refinements in both the front and the rear suspension layout, as well as in details of the transmission system, and in the steering.

The main features of the car will be remembered, principally the torsion bar suspension of all four wheels, independently in front, the drive being, of course, through the front wheels, and the construction of the body in an all-steel unit of restrained streamline shape to form a rigid shell which is both frame and body, no ordinary chassis being used.

From earliest acquaintance with this model it was clear that a greatly different type of Citroën from that formerly known had been produced. The 1936 version of it does undoubtedly represent a much advanced form of this always interesting car.

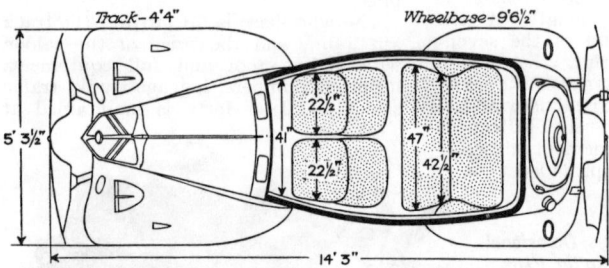

Its appeal is two-fold, even to a driver who is enthusiastic above the average about cars as cars, and who is acquainted with many makes and types. On the one side, it looks smart, being low-built but not freakish, and is extremely comfortable; on the other, it runs easily and well, and handles most satisfactorily, in no way tiring the driver on a long journey.

This Twelve has good acceleration—it can get away from a standstill briskly, it picks up usefully on top gear, being best over a range from about 20 to 50 m.p.h., and it can run very satisfactorily indeed at 50 as a maintained speed, with little sound of mechanism working. It provides, in fact, a useful and a usable performance out of proportion to what might be expected, due largely to an efficient ratio of power to weight, the complete car weighing little more than 21 cwt.

In the same way, the climbing on top gear is good, the engine having a creditable ability to hold on up long slopes, or on fairly appreciable gradients for that matter, whilst by use of the gears there are often occasions when an improved performance can be obtained, even though to employ them is not often actually necessary. The maximum speed of this Citroën Twelve is amply high for most requirements: the speedometer was slightly *slow* at a reading of 30, and virtually accurate up to 50, and *slow* throughout on another similar car driven; the highest reading was 68, dropping back to 65, when the car was being timed at 63.38 m.p.h.

There is a happy feeling of it swinging along without working hard, the engine running smoothly and quietly, in a manner above the normal for touring vehicles of 12 h.p. or so. Definitely it is an interesting car to handle, partly because of the performance, partly because of good brakes, and, above all, for the feeling of being "in one piece" it imparts to the occupants, which soon inspires confidence in, and a distinct liking for, the car.

A question always liable to be asked is how, if at all, this car, with its front wheel drive, differs from the normal? In general, it can be said that an average motorist, put into this Citroën for the first time, would not be likely to tell from any peculiarity of the running or of the actual handling that it differs in basic features from the majority of other cars on the road.

He or she could hardly fail to notice the extremely comfortable riding—soft, yet sufficiently firm for a strong feeling of safety—and the willingness of the engine to do what the driver wants.

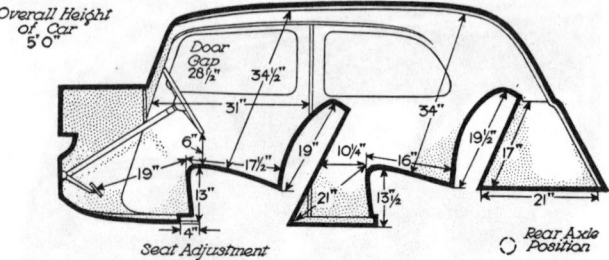

Also, the unusual arrangement of the gear lever might be a prominent impression, this projecting from the instrument board towards the left-hand side, with the "business" end curved over towards the driver. At first sight this might appear awkward as well as unusual. However, its placing has obviously been carefully considered, for it comes most readily to hand, and gives a very nice action, the lever working in a visible "gate" and with the gear positions

"The Autocar" Road Tests

clearly marked. There is an excellent synchromesh mechanism for the changes between top and second, which go through smoothly, quietly, and quickly, with only a single depression of the clutch pedal. Second gear is quiet.

As to other points, both now and in extended experience of the earlier model a year ago, no disadvantage has been found to result from the front wheel drive when cornering, on either a dry or a wet surface. Due to the design, it is possible, indeed, to take even appreciable corners much faster than is the general practice if a driver be so inclined; but whether bends are rounded with the engine pulling or on the overrun, or whether the driver happens to lift his foot on the throttle pedal midway round a corner, no disability as regards safe control of the car is discovered.

One point is that with the fast cornering possible and the low-pressure tyres employed, there is "screaming" from the tyre treads if enterprising methods are adopted, but this is at a minimum or non-existent in ordinary quiet driving.

If one is looking for points of difference, it is possible to detect the reversal of the drive when the throttle pedal is let back and the car goes on to the overrun, but that effect is not violent or disturbing. Again, there is at the same time some slight chatter from the front, where, of course, are the transmission units, but that, apparently, varies with individual machines, and is one of the points that has lately received attention.

As to the car's behaviour on severe gradients, a year ago tests were made with a similar car by *The Autocar* on well-known West Country hills, and not only were they climbed really satisfactorily in touring conditions, but also restarts could be effected on the worst gradient of each. Additionally, the present car could be restarted on the 1 in 4 section of the Brooklands Test Hill, it being best not to over-rev in doing so.

Starting from rest in the ordinary way is carried out quite smoothly without any particular care; just now and again a mild judder may result, possibly if the engine is speeded up too much, but this is not a specially sensitive car in that way.

The brakes, which are hydraulically operated, give good smooth power both for bringing down the speed surely in the ordinary course of driving and for a quick emergency stop.

The Citroën Super Modern Twelve has a smart, well-balanced appearance.

and down motion, nor sideway or roll, and over a "colonial" section comprising potholes and ruts the car can be driven as fast as 25 to 30 m.p.h. without anything going solid, without the steering being affected, and without even back seat passengers being thrown about. The steering is firm and accurate, and the firmness does not amount to heaviness at any time, though rather more effort has to be exerted when manœuvring than in straightforward driving.

One of the latest improvements consists of the incorporation of caster action in the steering, there being a definite tendency for the front wheels to come back of their own accord after a corner; it is reasonably high-geared steering, approximately 2¼ turns of the wheel giving full lock to full lock.

Very comfortable seats are provided, separate and adjustable in front, shaped in the back rests to afford adequate support, providing also a natural angle to the legs, and resilient without being so soft as to allow roll. An excellent outcome of the car's design is that the floor in both compartments is entirely flat and unobstructed; it is equally easy to use either door.

The wheel is brought well down to give an excellent driving position, vision being very good. The bonnet is not too high, and the pedals of a hinged type to which downward rather than forward pressure is applied, are actually quite convenient. The view behind in the driving mirror is useful without being comprehensive. The head lamp beam is extremely good.

The upholstery is in leather, and there are found all those important equipment details that motorists in this country are now accustomed to. The luggage compartment in the tail of this 1936 model is reached through an external lid, the handles of which embody locks.

The engine is neat and has a good big oil filler; the sparking plugs, the distributor, and the fuses are accessible, also the fluid reservoir for the hydraulic brakes. At each side of the car, towards the back, and at the front also, are pads to take the jack head.

The hand lever is of the pull-and-twist type, placed out of the way under the scuttle, though fairly conveniently, and this, provided it is in proper adjustment, will hold securely on a 1 in 4 gradient.

By ordinary main road standards the torsion bar suspension is particularly comfortable. It does not permit any marked up

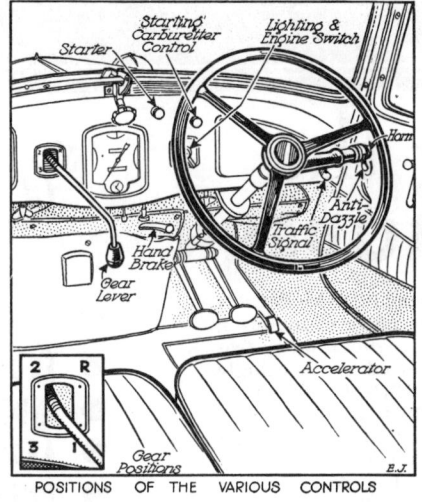

POSITIONS OF THE VARIOUS CONTROLS

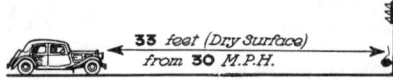

DATA FOR THE DRIVER

CITROËN SUPER MODERN TWELVE SALOON.

PRICE, with four-door four-light saloon body, £265. Tax, £9 15s.
RATING: 12.8 h.p., four cylinders, o.h.v., 72 × 100 mm., 1,628 c.c.
WEIGHT, without passengers, 21 cwt. 16 lb.
TYRE SIZE: 140 × 40 on bolt-on steel disc wheels.
LIGHTING SET: 12-volt; three-rate charging; 8 amps. at 30 m.p.h.
TANK CAPACITY: 8½ gallons; fuel consumption, 26–28 m.p.g.
TURNING CIRCLE (L. and R.): 44ft. **GROUND CLEARANCE**: 7in.

ACCELERATION			SPEED	
Overall gear ratios.	From steady m.p.h. of			m.p.h.
	10 to 30	20 to 40 30 to 50	Mean maximum timed speed over ¼ mile	61.43
4.90 to 1	13⅘ sec.	13⅕ sec. 16⅗ sec.	Best timed speed over ¼ mile	63.38
8.30 to 1	7⅕ sec.	9 sec. —	Speeds attainable on indirect gears:—	
14.80 to 1	—	— —	1st	21
From rest to 50 m.p.h. through gears, 23½ sec.			2nd	38–45
25 yards of 1 in 5 gradient from rest, 7¾ sec.			Speed from rest up 1 in 5 Test Hill (on 1st gear)	15.40

Performance figures of acceleration and maximum speed are the means of several runs in opposite directions.

Introducing the SUPER MODERN

With the advent of the Citroën SUPER-MODERN "FIFTEEN" models, the whole range of motoring requirements is covered. All the revolutionary features of design that have proved so amazingly successful in the SUPER-MODERN "TWELVE" have been retained. The "FIFTEEN," however, is available in two sizes, designed to seat five and seven people respectively. Providing far more accommodation and roominess than any other car in their class, they retain the stability, safety and comfort that has made the SUPER-MODERN range the greatest success in the history of motoring.

The SUPER MODERN

SALOON DE LUXE	**£315**
Seven-Seater	£345
Super-Modern "Twelve"	£250

with
- FRONT WHEEL DRIVE
- INDEPENDENT WHEEL SUSPENSION
- AERODYNAMIC BODY
- UNIMPEDED FLOOR AREA
- FRAMELESS CONSTRUCTION
- TORSION BAR SPRINGING
- AMAZING STABILITY
- REMARKABLE ROAD VISIBILITY

TWO YEARS AHEAD in *Construction, Appearance & Design*

Write for catalogue No. 11a.
CITROËN CARS LTD., Brook Green, Hammersmith, W.6
Showrooms: Devonshire House, Piccadilly, W.1

CITROËN

PRODUCED AT CITROËN WORKS, SLOUGH, BUCKS

The Front-drive Citroën Twelve Saloon

An Unconventional Car, Notable for Many Practical Features

NOT only is the Citroën Super Modern Twelve sports saloon an outstanding car from the point of view of appearance—it has graceful, yet essentially practical, lines—but it also offers the interest which must attach to a car of such unorthodox design. Virtually frameless, it employs both independent (torsion bar) suspension of the front wheels and front wheel drive; whilst the price of £250 is very low for such an interesting specification.

The practical nature of the design is revealed so soon as one enters the car to take the wheel. The wide doors allow plenty of toe-room when climbing in, so that one enters gracefully without contortion and, once in, the seat is comfortable and all-round visibility of a high order.

The controls—including the unorthodox gear lever situated on the left-hand side of the facia board—are well placed for instant accessibility. The only criticism which could be levelled in this respect concerns the "organ-type" pedals which, moving up and down rather than backwards and forwards, throw a certain strain on the ankles until one is accustomed to them.

Once under way, the gear-change proves very simple to handle. Very quick upward changes can be made from first to second and second to top, the synchromesh engagement of the two higher gears obviating all need for skill or sense of "timing." When requiring the best performance, over 20 m.p.h. can be reached on first gear and 44 m.p.h. on second, and the car very quickly reaches an easy cruising speed of 50-55 m.p.h. on top.

The four-cylinder engine is reasonably smooth when accelerating, but seems inclined to a distinct roughness on the overrun. Driving along a straight road it was difficult to judge, by aural impression, whether the car was driven by the front wheels or from the back, and when cornering quickly the Citroën gains by comparison with most conventionally driven cars. There is no sensation of having to pull the car round: rather does it feel—as is actually the case—that the front of the car is pulling in the desired direction and the back following easily. This sensation gives an impression of easy running, when the road becomes sinuous, which gives the driver considerable satisfaction and confidence.

This sensation is strengthened by the smooth working of the steering gear, which, though not particularly light in action, has nevertheless a self-centring action which was lacking on the earlier models. No road shocks are transmitted to the driver through the wheel.

The suspension is the outstanding feature of this car, and in this respect it is streets ahead of most other 12 h.p. models. It can be driven fast over pot holes, level crossings and cobble stones without the passengers experiencing any discomfort, whilst this method of springing combined with a good weight distribution makes the road-holding impressively good.

The car could be pulled right down into the gutter on a steeply cambered road without diminishing speed, and when main roads were left for bumpy lanes and unmade tracks it was always discretion, not discomfort, which

The figure in this picture is of average height, a fact which gives an idea of the lowness of the Citroën's trim lines. It can be seen also that accessibility is good.

THE FRONT-DRIVE CITROEN TWELVE SALOON

On the left is a photograph giving a good idea of the firm support to the thighs afforded by the front seats; also the unusual "wells" for the feet.

The accessibility to all seats is excellent owing to the low floor level and the absence of running boards. The former is the direct result of front-wheel drive.

governed the available speed. With the rear compartment occupied by two heavy passengers the same conditions persisted.

At Brooklands, where the track is extremely rough at this time of the year, the car gave a most comfortable ride at all speeds, smoothing out the "bump" at 60 m.p.h. in an impressive manner.

The car climbed the Test Hill very capably in first gear, and could be restarted comfortably on the 1-in-5 section. In making restarts of this nature the well-placed hand-brake lever, pulling out from the dash, in a line with the steering column, proved a convenience.

Restarting on the steepest part of the hill (1 in 4) proved more of an effort, and in this respect both the high first gear and the front-wheel drive approached limit conditions.

The "pulling" qualities of the car, shown by the Tapley meter readings recorded, are very useful on all the gears, bearing out the impression gained in general driving conditions that it is only very rarely that one feels the lack of a four-speed gearbox. The quick change into top from second at

TABULATED DATA—CITROEN TWELVE

CHASSIS DETAILS

Engine: Four cylinders, push-rod-operated overhead valves, coil-ignition with automatic timing; 72 mm. by 100 mm. (1,628 c.c.). Rating 12.8 h.p., tax £9 15s.

Gearbox: Three forward speeds with synchromesh for second and top. Ratios, 4.9, 8.3 and 14.8 to 1.

PERFORMANCE

Speeds on Gears: Top, 63 m.p.h.; second, 44 m.p.h.; minimum speed on top gear, 9 m.p.h. approx.

Acceleration: From standstill through the gears to 50 m.p.h., 23 secs. Standing ¼-mile, 23⅕ secs. (Average speed, 38.8 m.p.h.)

Tapley Performance Figures: Maximum pull in lb. per ton on gradient; top, 195 lb.; second, 350 lb. Corresponding gradients climbable at a steady speed are 1 in 11.5 and 1 in 6.4.

Petrol Consumption: Driven hard, 26 m.p.g.

Braking Efficiency: Measured by Tapley meter, using the pedal only: from 30 m.p.h., 80 per cent.; from 40 m.p.h., 75 per cent. Corresponding stopping distances are 37 ft. and 72 ft.

DIMENSIONS, ETC.

Leading Measurements: Wheelbase, 9 ft. 6½ ins.; track, 4 ft. 4 ins.; overall length, 14 ft. 2 ins.; width, 5 ft. 2 ins.; height, 4 ft. 10½ ins.

Wheels and Tyres: Bolt-on disc wheels with 140 by 40 ultra-low-pressure tyres.

Turning Circles: Left and right, 36 ft. dia.

Weight: As tested, with one up, 21¾ cwt.; unladen, 20¼ cwt.

Price: Four-door saloon, as tested, £265

40 m.p.h. wastes no time, and, in top gear the car quickly gathers speed on such a gradient as Putney Hill.

As has been indicated, the general comfort and finish of the coachwork are excellent. There is plenty of headroom and a roomy luggage locker is built into the rear panel. A central folding armrest is fitted in the rear seat, and both front and back-seat passengers appreciate the fact that the doors open a few inches above floor level; implying warmth for the feet. Two scuttle ventilators attend to air conditioning, and both front windows wind fully down. The back light is large enough to make reversing easy, and all windows are glazed with Triplex Toughened glass. The windscreen opens wide enough for a tall driver to see ahead in thick fog. The tandem windscreen wiper works very efficiently but is rather noisy.

In addition to having all these good features the unit construction of the car and coachwork makes for great sturdiness. The whole front assembly is attached to two massive scuttle members which are virtually part of the pressed-steel body, so that there is a complete absence of rattles. The resistance to impact would be great.

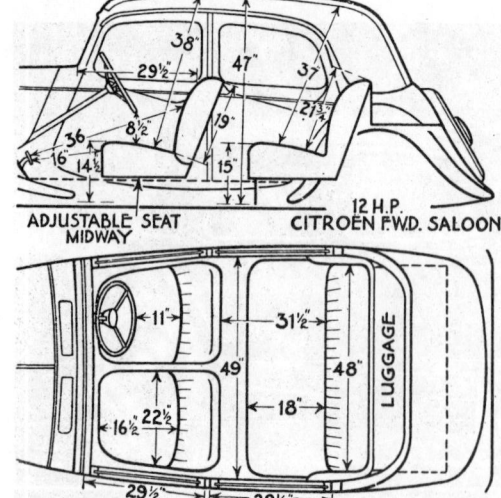

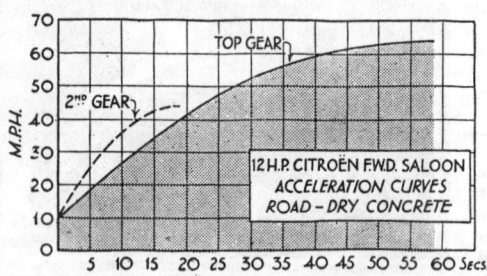

The Autocar Road Tests

On Beacon Hill, a loose-surfaced gradient over the Hampshire downs, in the West Meon district.

No. 1,040 CITROËN SUPER MODERN FIFTEEN SALOON

A Most Roomy and Comfortable Car, Offering Six-seater Accommodation, Which Gives an Interesting Performance

"SUPER MODERN" is the name given by the Citroën firm in this country to the now well-known front-wheel drive models, both 12 and 15 h.p., and there is considerable justification of the description. Undoubtedly this is one of the most advanced designs of the day developed to a practical stage and produced in quantity, with its front wheel drive, torsion bar suspension, and monopiece construction of the body in the form of an all-steel shell which acts as the frame as well, there being no ordinary chassis members.

On previous occasions reference has been made to the generally satisfactory performance of the 12 h.p. f.w.d. model. The same basic car can be obtained with a 15 h.p. engine of slightly less than 2-litre capacity, and is then known as the Sports Twelve, or this larger engine is available in a generally bigger car than the Twelve, with a lengthened wheelbase and wider track—the type now tested. In this form the f.w.d. Citroën becomes an extremely roomy and convenient type of car for family purposes, yet at the same time offers speedy transport with a minimum of fatigue.

One is immediately struck by the interior dimensions of this saloon body, which has a straight-across single-piece front seat and a very wide back seat. Owing to the front wheel drive there is, of course, no transmission shaft running beneath the floor of the car, which is made entirely flat and free from ridges and obstructions. Particularly is this apparent in the rear compartment, which provides a truly remarkable unrestricted area of floor space and seating width, together with good head room.

In front, because the gear lever is of a special kind, projecting from the instrument board, and the hand-brake lever is placed under the scuttle, again unusual freedom is given. It is fully possible to seat three in front, there being enough leg room for the centre passenger in spite of the fact that a projection caused by the engine housing comes into the driving compartment. Thus, this car is a really adequate six-seater, and, at its price—recently reduced—and considering the good performance recorded, is excellent value.

In spite of the big body of this model, the total weight is still not heavy, whilst the engine produces useful power, so that the top gear acceleration, especially from about 20 m.p.h. onwards, is notably good. In fact, there is a snap in the performance not often encountered in the kind of car usually referred to as being of touring type, still less in one that can be a six-seater. In particular reference to the performance, the speedometer was less than 1 m.p.h. fast at 30 and 2.8 m.p.h. fast at 50, and showed 79-80 as the highest reading when the car was being timed in the favourable direction at its creditable best speed of 72.58 m.p.h.

Including top gear, the ratios of the three-speed box are fairly high, which is a practical arrangement because of the unusually good power-to-weight ratio resulting from the special construction of the car, though first gear can conceivably prove over-high for a restart on a really severe gradient with a full load on board. A quick getaway can be made, the engine running rapidly up to its useful limits on the two indirect gears, and the change through from first to second, as well as from second to top, is rendered easy and quick by a good synchromesh. Furthermore, second gear is quiet, being altogether an eminently usable ratio.

This car's pulling powers on top gear are sound, the usual kind of 1 in 10 to 15 gradient being taken very well indeed when travelling in the normal stride at about 45 m.p.h., whilst on a hill approached more slowly for any reason second gear performs excellently, being able to deal,

(Left and right) Showing the unusually roomy seating and unobstructed floor space in the Citroën's front and rear compartments, respectively.

in fact, with a gradient of 1 in 6½. In close traffic or, for instance, when negotiating right-angle turns, it is usually best to engage the intermediate gear if the speed is to approach a crawl, otherwise engine "flutter" and a tendency to snatch may be experienced, no doubt partly because the mounting of the engine is quite flexible.

At any speed from about 20 m.p.h. onwards it is an exceptionally smooth-pulling engine, so much so that it is not in the least apparent as a four. At fast speeds there is a most pleasant feeling of swinging along in an easy way, and good times are possible from point to point. It is a car which very comfortably puts more than 40 miles into an hour even though the 60-minute spell may not be of a specially favourable nature.

As to the front wheel drive and its effect upon normal handling of the car, for all practical purposes the majority of people could not detect any difference from an ordinary rear-driven car. Certainly no instability on corners or need for extra care has come to light in prolonged handling of the f.w.d. Citroën as a type, and, dealing with a point sometimes raised, it does not seem to matter whether a corner be taken either with the engine pulling or on the overrun, or, again, with a combination of the two methods. If impressions are analysed, a slightly unusual interruption of smooth forward progress is noticeable on letting up the throttle pedal suddenly, or a similar opposite effect can be produced by opening the throttle suddenly, but these are matters rather of passing interest than of practical import. There is "scream" from the tyres if the excellent cornering capabilities are utilised, but it is hard to say whether this is in any way a result of front wheel drive and not largely a product of the real low pressure tyres employed.

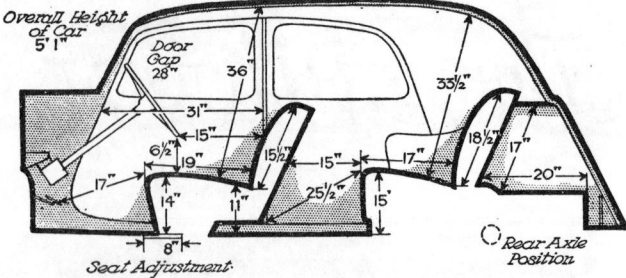

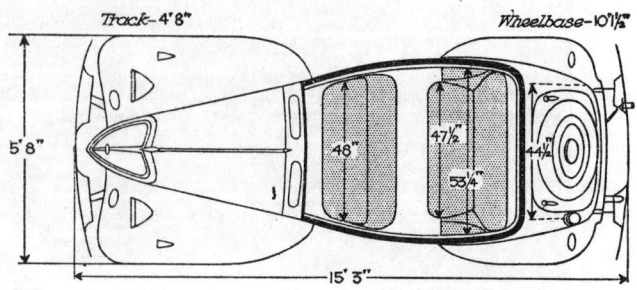

Seating dimensions are measured with cushions and squabs uncompressed

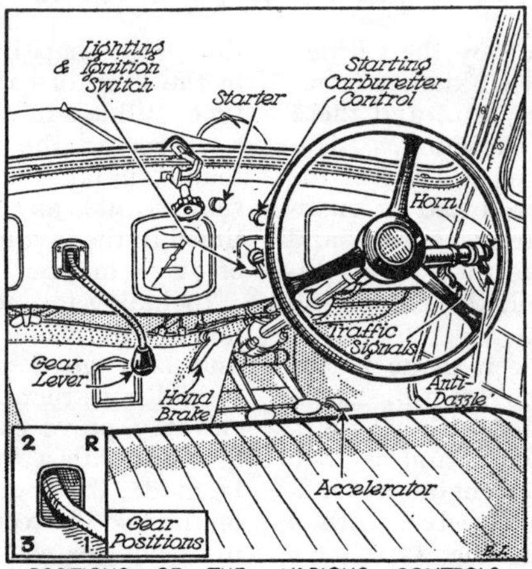

The suspension is quite out of the ordinary, both in design and in the riding qualities it affords. As is now generally known, leaf springs are not used, and for springing this car depends upon the torsional elasticity of steel rods, of which there is one for each wheel, independent action being allowed in front. At various times this f.w.d. model has been put through unusually severe tests over vile surfaces, and during the present test this particular Fifteen was taken over sections amounting almost to cross-country driving, on which it showed up outstandingly well. Although over such going, including large holes, there is fairly considerable deflection, chiefly noticed by any rear passengers, even then the suspension remains soft, and ground clearance is likely to be about the only difficulty where such extreme conditions are concerned. For general use it is exceptionally comfortable springing, seeming softer on this particular car than on previous examples of the type.

POSITIONS OF THE VARIOUS CONTROLS

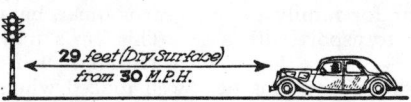

In general control the steering is light, and latterly it has been given distinct caster return action. It is not low geared, requiring 2¼ turns from lock to lock, and does not give back road shocks, becoming appreciably heavier when the car has to be manoeuvred or put round a sharp turn. There is a strong sense of safety and "one pieceness" about this car when travelling quickly, helped by powerful hydraulically operated brakes which do all that is asked of them without needing extreme pressure on any occasion.

The driving position is comfortable and confidence-inspiring, for the seating is well arranged and the view outward good. The width of body and front seat results in a driving position more to the right than in the Twelve. This causes the gear lever to be a little too far to the left.

The interior finish is well done and the equipment most complete. There is leather upholstery, and the rear seat has a centre folding armrest. Exceptional attention has been paid to accommodation for oddments, there being a shelf under the scuttle, two lockers with lids, pockets in all doors, a roof net, flap pockets behind the front seat, and even recesses in the rear quarters as well.

A tendency to warmth in the front compartment was noticeable during a recent hot spell, but there is a big area of sliding roof, ventilators are provided in the top of the scuttle, and there are opening panels in the bonnet sides to remove hot air at the source. Also, the windscreen can be opened for ventilation.

In the tail of the body is a luggage compartment of quite good size, where also the tools are housed. The battery is accessibly mounted under the bonnet.

DATA FOR THE DRIVER

CITROËN SUPER MODERN FIFTEEN SALOON.
PRICE, with four-door four-light saloon body, £289. Tax, £11 5s.
RATING : 15 h.p., four cylinders, o.h.v., 78 × 100 mm., 1,911 c.c.
WEIGHT, without passengers, 23 cwt. 2 qr. 12 lb.
LB. (WEIGHT) PER C.C. : 1.38.
TYRE SIZE : 150 × 40 on bolt-on steel disc wheels.
LIGHTING SET : 12-volt. Automatic voltage control.
TANK CAPACITY : 10 gallons ; fuel consumption, 21-25 m.p.g.
TURNING CIRCLE : (L. and R.) 45ft. GROUND CLEARANCE : 7in.

ACCELERATION.				SPEED.	
Overall gear ratios	From steady m.p.h. of				m.p.h.
	10 to 30	20 to 40	30 to 50	Mean maximum timed speed over ¼ mile	69.64
4.30 to 1	13.8 sec.	12.8 sec.	15.1 sec.	Best timed speed over ¼ mile	72.58
7.30 to 1	7.4 sec.	8.0 sec.	—	Speeds attainable on indirect gears—	
13.10 to 1	—	—	—	1st	19-23
From rest to 50 m.p.h. through gears, 23.2 sec.				2nd	38-43
From rest to 60 m.p.h. through gears, 37.7 sec.				Speed from rest up 1 In 5 Test Hill (on 1st gear)	15.80
25 yards of 1 in 5 gradient from rest, 9.1 sec.					

Performance figures for acceleration and maximum speed are the means of several runs in opposite directions.

STILL... THE MOST REMARKABLE CARS EVER MADE!

- FRONT WHEEL DRIVE
- FRAMELESS CONSTRUCTION
- INDEPENDENT WHEEL SUSPENSION
- TORSION BAR SPRINGING
- AERODYNAMIC BODY
- AMAZING STABILITY
- UNIMPEDED FLOOR AREA
- REMARKABLE ROAD VISIBILITY

Prices:
Super Modern "TWELVE" Saloon - £265
Super Modern "SPORTS TWELVE" Saloon - £285
Super Modern "FIFTEEN" Saloon £315
Super Modern "FIFTEEN" 7-Seater Saloon - £345
The "TEN" Saloon - £198

CITROËN CARS LTD.,
Citroën Building, Hammersmith, W.6.

Comparisons show the "Super-Modern" Citroën still to be "Two Years Ahead" in design, performance and comfort.

No other cars of similar size and price can compare with the "Super-Modern" range for safety, luxury, performance and beauty of line.

Write for catalogue "A" and the address of your nearest agent.... he will be delighted to arrange a trial run for you.

SUPER MODERN CITROËN
Two years ahead

PRODUCED AT CITROËN WORKS · SLOUGH · BUCKS

Improved Citroëns

Front Wheel Drive Models Still Further Developed, both Mechanically and in Bodywork: New Rear Drive Seven-seater Edition with 15 h.p. Engine

Latest Twelve f.w.d. saloon, with perforated steel disc wheels.

The Fifteen seven-seater saloon, which has folding occasional seats.

FRONT wheel drive passes into its third year of production on the 1937 Citroëns, which, whilst continued without fundamental changes, embody several mechanical improvements, and refinements in finish and equipment. The three types of f.w.d. models are retained, comprising the Twelve, the Sports Twelve, and the Fifteen. Of these, the Twelve is rated at 12.8 h.p., and the Sports Twelve and the Fifteen are rated at 15 h.p., all having four-cylinder engines of push-rod-operated overhead-valve design. There is also added to the range a rear wheel drive model, with a 15 h.p. four-cylinder o.h.v. engine, with conventional transmission and springing. This is offered as a seven-seater family saloon for those who want maximum accommodation in a saloon body of orthodox appearance.

Since this time last year considerable further development work has gone into the front wheel drive models, and recognised difficulties that were experienced at the outset with this form of transmission are now of the past. The general design may be briefly reviewed. In front of, instead of behind, an efficient o.h.v. engine are the single-plate clutch, three-speed gear box, and final drive, forming a flexibly mounted unit, and the power passes to the front wheels through short universally jointed shafts. Both front and rear the suspension is by means of steel torsion bars, the natural elasticity of which provides the springing effect against the action of the road wheels— independently at the front. There are no ordinary leaf springs. These Citroëns are also notable in possessing no separate chassis frame as ordinarily understood; pressed-steel units carry the engine and transmission, provide anchorage for the torsion bars, and form the basis of an all-steel body of very pleasing semi-streamlined shape. Remarkably comfortable riding, as well as a strong feeling of stability, is afforded by this special springing system.

Interesting mechanical modifications now announced include the adoption of an entirely new steering gear, a modern version of the rack and pinion principle. This, though essentially simple, achieves several advantages, including the provision of increased caster return action and a reduction in the number of arms and links formerly necessary in conjunction with the front wheel drive. A continuation of the steering column shaft

The fixed-head coupe, available with either a 12.8 or a 15 h.p. engine.

carries at its extremity a spiral pinion, which engages with a toothed rack, mounted in the same plane as the front "axle" and rising and falling with the latter under the influence of wheel deflections. Movement of this rack by normal turning of the steering wheel is imparted direct to the two halves of a divided push-and-pull rod, one for each wheel. The whole mechanism is enclosed, with provision against loss of lubricant and entry of water and dirt.

A different arrangement of the exhaust manifold and outlet pipe is now employed, the pipe being carried down more towards the front of the engine, and thus lessening any tendency for warmth to be transmitted to the car's interior. For carburation there is a new-type Solex instrument, to the air intake of which is attached a small and neat but effective cone-shaped filter-cum-silencer of Citroën design. Twelve-volt electrical equipment is fitted and, in accordance with a plan which is becoming more general, the positive and not the negative terminal of the battery is connected to earth.

Special points in the interiors of these f.w.d. cars are entirely unobstructed, flat floors, owing to the absence of an ordinary propeller-shaft, and also unusual freedom from controls in the front compartment, since the gear lever projects through the instrument board and the hand-brake lever is of the pistol-grip type, carried below the instrument board.

It will be remembered that these cars are assembled, trimmed, and finished off in the Slough works of Citroën Cars, Ltd., and the style of their interior appointments is thoroughly British.

This effect has been enhanced in the latest models by the use of a new polished wood instrument board in place of a metal facia. In this are now mounted separate instruments—a big speedometer directly in front of the driver, and individual ammeter and petrol gauge dials, replacing the previously used single big central dial embodying all the instruments on the one face. Clearness of reading and, even more important, ease of access in the event of derangement of an instrument, have been the deciding factors as regards this change. Also, a deep and really useful cubby hole can now be provided at the left-hand side.

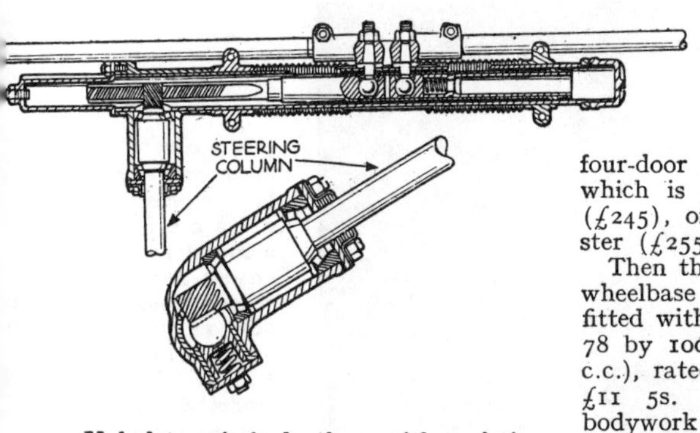

(Left) Details of the new rack and pinion steering gear on the f.w.d. models.

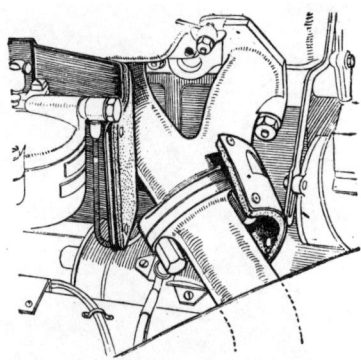

The exhaust pipe is now carried down at the front on the Twelve.

Upholstery is in leather, with a choice of single-piece or straight across front seats, and the trimming on the inside of the doors has been made more decorative. Also, as an option, "twin trim" upholstery is being offered in the front wheel drive models. An excellent quality wool cloth is employed for the main part of both cushion and squab, with leather at points of wear at the edges. Not only are advantages of both kinds of upholstery combined, but also the appearance of these "twin-trimmed" seats is decidedly attractive. The roof lamp has been moved to a position above the rear seat squab, the twin-blade electric screenwiper motor is mounted above the screen frame, where it does not obstruct vision, and there is a new type of "kick-plate" at the bottom of the door openings.

Externally, a new and neater mounting has been evolved for the rear number plate, now hinged from the metal cover enclosing the spare wheel, which is partially recessed into the lid of the internal luggage compartment. Use is made of the space in the thickness of this lid by providing an opening on the inside, forming a pocket for fairly bulky oddments. The front number plate is now flush-fitted into the bumper, and there is a rigid, permanently fixed abutment point on the bumper for the starting handle.

Part of the radiator shell has been chromium plated where before the metal was cellulosed; these Citroëns are finished all over—wheels, wings, radiator shell, lamps, and so on—in one colour, and a pleasing suggestion of "one-pieceness" is thus obtained. The perforated steel disc, or, as it may be regarded, short-spoked steel wheel has been adopted, and carries real low-pressure tyres.

These are well-turned-out cars that give a decidedly interesting performance, owing to a good power-to-weight ratio. As to specifications of the front wheel drive models, the Twelve four-cylinder o.h.v. engine is of 72 by 100 mm. bore and stroke (1,628 c.c.), the rating being 12.8 h.p. and the tax, therefore, £9 15s. The wheelbase is 9ft. 6½in., and the track 4ft 4in. This is available as a four-door saloon. the latest price of which is £228, as a fixed-head coupé (£245), or as an open two-seater roadster (£255).

Then the Sports Twelve has the same wheelbase and track dimensions, but is fitted with the larger 15 h.p. engine, of 78 by 100 mm. bore and stroke (1,911 c.c.), rated at 15.08 h.p., and taxed at £11 5s. This model carries similar bodywork to that of the Twelve, but runs on higher gear ratios in view of the greater engine power available. Latest prices of the Sports Twelve are: saloon £248; coupé £265; and roadster £275.

The Fifteen is of longer wheelbase (10ft. 1½in.) and wider track (4ft. 8in.) than the foregoing front wheel drive models, and thus takes a bigger body. From experience this can be described as a truly spacious type, rendered still more practical by the f.w.d. feature of flat, unobstructed low-level floors. The Fifteen four-door saloon costs £278, and there is also a seven-seater version, incorporating two folding occasional seats, at £298. This has a 10ft. 9in. wheelbase.

These engines, in addition to their push-rod-operated overhead valves, have a three-bearing crankshaft, detachable cylinder barrels cast in a special wear- and corrosion-resisting material, pump and fan cooling, and coil and distributor ignition. Lockheed hydraulically operated brakes are fitted on the f.w.d. models.

This pressed steel unit is the basis of the body, and also carries engine and suspension.

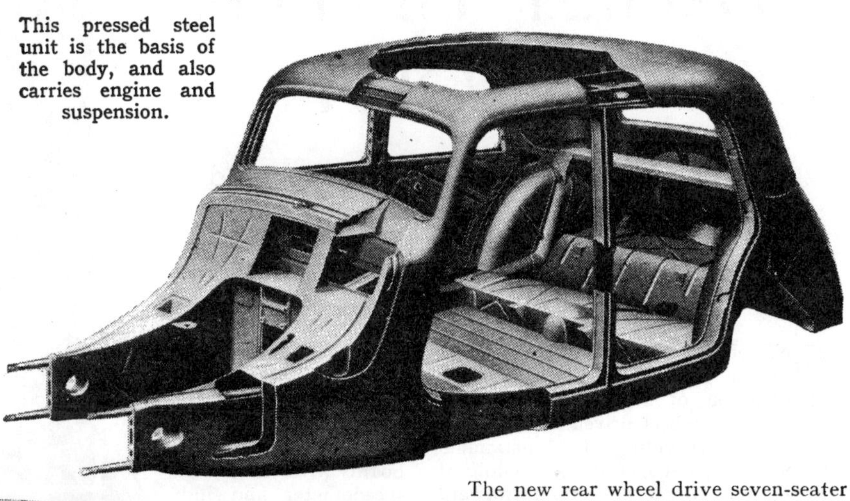

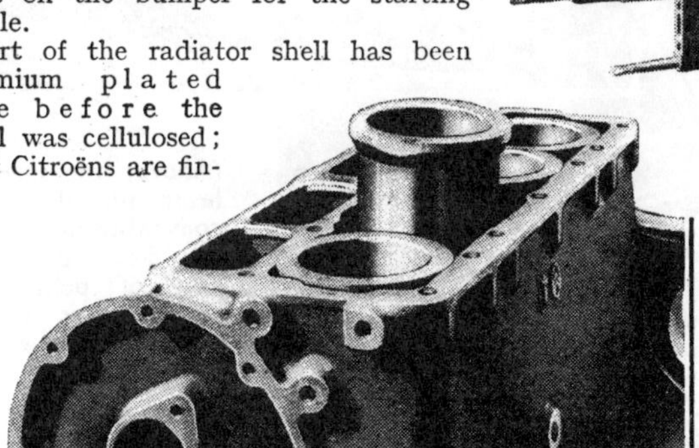

(Left) Detachable cylinder barrels are a feature of the present Citroen engines.

Built-in luggage compartment, as on the Twelve and Fifteen.

The new rear wheel drive seven-seater family saloon is fitted with a 15 h.p. four-cylinder o.h.v. engine of similar basic design and identical dimensions to that used in the Sports Twelve and Fifteen f.w.d. cars, but, of course, the clutch and three-speed gear box are mounted in unit in the conventional positions *behind* the engine. The gear box has an ordinary central lever, and there is synchromesh on top and second. In this car also the transmission passes normally through an open propeller-shaft to a spiral bevel final drive in the rear axle, and the suspension, at both front and back, is by conventional long half-elliptic springs. The frame, however, is built up of box-section side-members for rigidity, and Bendix duo-servo cable-operated brakes are employed, whilst the body is of steel construction.

This car's wheelbase is 9ft. 10in., and the track is of full 4ft. 8in. width; thus the four-door saloon body, with occasional folding seats in the rear compartment, is one of considerable spaciousness and comfort, as befits its purpose. There is a luggage grid at the rear, a sliding roof is provided, and the equipment includes flush-fitting traffic signals, safety glass all round, and 6.50 by 16in. real low-pressure tyres on wire wheels. The price of this seven-seater saloon is £285;

A roomy car but expressing graceful modernity: The seven-seater Fifteen saloon, now reduced from £315 to £298. This model has front-wheel drive.

A NEW "FAMILY FIFTEEN" CITROËN

Front-wheel-drive Range Continued with Better Steering, Improved Coachwork and Detail Price Changes

ONE of the most interesting features of the 1937 Citroën range is the introduction of a "Family Fifteen" saloon, rear wheel driven, a car which aims at providing the maximum amount of passenger room. Selling at £285, it is simple and straightforward in design and robust in build and finish and should supply the strong demand which exists for a really commodious car of moderate price.

The three front-wheel-driven, independently sprung cars continue for next season with detail changes only. All these models, the Twelve, the Sports Twelve (i.e., a 15 h.p. engine in the Twelve chassis) and the Fifteen, have four-cylinder engines, and torsion bar suspension is employed. The aerodynamically shaped steel bodies are integral with the frame, on the "Monoshell" principle.

Detail changes in this range of models consists of a new direct type of steering gear, giving lighter and more positive control, and "Twin-trim" upholstery, a combination of leather and cloth trimming, which is pleasant to look at and practical in use. The centre part of both cushions and squab is in cloth, the edges being of leather, to counteract wear, ruffling and soiling at the most vulnerable parts of a seat. This upholstery is optional.

Bumpers are of a new design, flat, and incorporating a slight central recess for the neat mounting of the front numberplates. An interior refinement is a new polished wood facia board with large dial speedometer and indi-

(Above) The gear-lever placed unconventionally in the facia board—now of polished wood—leaves the front compartment unobstructed.

(Left) A neat, but not far overhanging boot gives roomy accommodation on all the f.w.d. models. It is shown packed on the left.

vidual instruments accessibly grouped. There are also detail improvements to the front axle.

Voltage control dynamos, an electrical oil indicator with warning light, interior lights with self-contained switch have all been added, whilst a chromium-plated radiator shell, instead of being cellulosed, and a pocket for oddments on the inside of the luggage boot lid are appreciated modifications. Interesting also are the redesigned petrol pump unions, the positive earth for the electrical system and the new pedal gear.

On the Twelve only are new pressed steel spoked wheels of very clean appearance, and on the Twelve and Sports

THE CITROEN FAMILY FIFTEEN

The new model of the range: the rear-wheel-driven 15 h.p. Family saloon, in which roominess and simplicity of design have been the first considerations.

Twelve there is an improved exhaust layout, devolving from a new manifold which sweeps the pipe away from the front of the engine and not only enhances performance by freedom from back pressure but makes for a cool driving compartment. Before, it will be remembered, the exhaust pipe passed through the off-side body " horn," the extension of the scuttle which formed the " frame " at the front end and on which the engine and front drive and suspension were mounted.

Most interesting, from a mechanical point of view, of these detail improvements is undoubtedly the new steering gear which is perfectly simple and designed to be trouble free. The steering column leads directly on to a rack and pinion which actuates two ball joints. These are connected to each pushing and pulling section of a divided track rod, giving delightfully direct movement. Even engagement of the pinion is ensured by spring loading with easy adjustment.

The new Family Fifteen model has the same-sized o.h.v. engine as the Fifteen and the Sports Twelve (i.e., four-cylinder, 78 mm. by 100 mm., 1,911 c.c., 15.08 h.p.). A three-bearing crankshaft is used and detachable cylinder barrels are cast in wear-resisting steel. The three-bearing camshaft is driven by double silent-roller chain. Cooling is by pump and fan.

The wheelbase is 9 ft. 10 ins., the track 4 ft. 8 ins., and the suspension, quite orthodox, by semi-elliptic springs fore and aft. Duo-servo brakes actuate on large drums, and the wire wheels are shod with 6.50 ins. by 16 ins. super low-pressure tyres.

The body is of " Monopiece " steel and provides accommodation for seven —two in front, on individually adjustable seats, two on folding occasional seats, and three on the back, bench-type seat. A sliding roof is fitted, and a luggage grid at the back of the car.

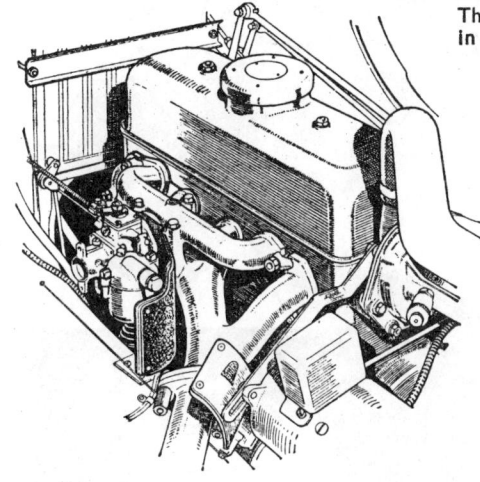

(Left) The new exhaust layout on the Twelve.

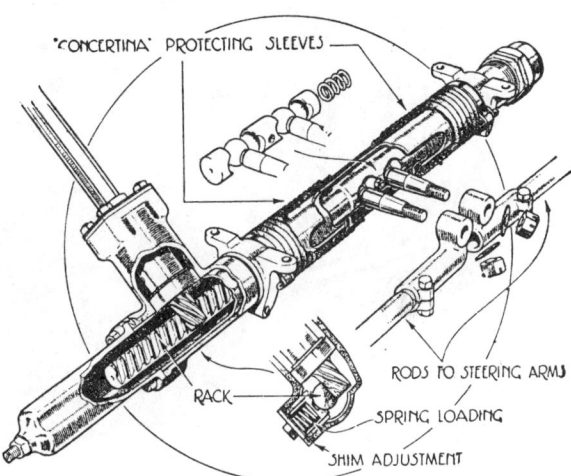

(Right) The simple rack and pinion steering gear, adopted on the Citroen models for 1937. The pinion is spring loaded and the concertina sleeves retain oil. Note the divided, "push and pull," track rod.

CITROEN PRICES

TWELVE
(4-cylinder o.h.v. engine, 72 mm. by 100 mm.; 1,628 c.c.; rating, 12.8 h.p.; tax £9 15s.; wheelbase 9 ft. 6½ ins.).
Saloon £228
Coupé £245
Roadster £255

SPORTS TWELVE
(4-cylinder o.h.v. engine, 78 mm. by 100 mm.; 1,911 c.c.; rating 15.08 h.p.; tax £11 5s.; wheelbase 9 ft. 6½ ins.).
Saloon £248
Coupé £265
Roadster £275

FIFTEEN
(Engine as in Sports Twelve; wheelbase 10 ft. 1½ ins.; long wheelbase 10 ft. 9 ins.).
Saloon £278
7-seater Saloon (long wheelbase) £298

FAMILY FIFTEEN
(Engine as in Fifteen; wheelbase 9 ft. 10 ins.).
7-seater Saloon £285

FROM ALL POINTS OF VIEW

Note torsion springing, front wheel drive and true aerodynamic lines completed by the flat undershield.

Beauty of line, unexcelled roominess, clear floor... change-speed lever is operated from the facia... and generous luggage space are features of this model.

The "Super-Modern" Citroën represents all the ideals of the owner-driver carried into practice. No other car has such road-adhesion... is so safe on corners... is so strongly constructed, yet so lively in action. It has raised the factor of safety fully 100%... augmented comfort immeasurably... made consistent high performance an effortless, everyday item... because it is fully two years ahead in design.

Write for catalogue "M" or, even better, ask for a demonstration run. A few minutes experience will bring realisation of the extraordinary superiority of the "Super-Modern" Citroën.

THE IDEAL OWNER DRIVER'S CAR

SUPER MODERN *Twelve*

CITROËN £265

PRODUCED AT CITROËN WORKS SLOUGH, BUCKS.

CITROËN CARS LTD., Brook Green, Hammersmith, W.6

PRACTICAL MOTORIST ROAD TESTS OF NEW CARS

The saloon model tested. It is listed at £228.

The Citroën "Twelve"

Unusually Large Seating Accommodation, Silent Engine And Transmission, High-Grade Finish and Pleasing Performance Are Additional To The Very Modern Mechanical Specification

FRONT-WHEEL drive, torsion-bar springing, absence of chassis frame, independent wheel suspension, aerodynamic body with flat undershield—these are a few of the very modern and unconventional features of a more-than-usually-interesting car. The Citroën "Twelve" does not rely solely on its technical specification as a means of attracting the driver, however, for it is a likeable car in many other respects. It has an air of refinement, and is particularly comfortable. In addition to its sound construction, the car is built throughout from high-grade materials. The design is equally attractive from either inside or outside; from the viewpoint of the passenger and driver, or of the casual observer.

The bodywork can truly be described as sumptuous, for there is ample room for three adults in the rear compartment, whilst three can also be accommodated in front if necessary. Extra space is available due to the absence of a transmission shaft running to the rear wheels, and also to the elimination of a chassis frame. In the front compartment very ample foot-room is provided by dispensing with the usual type of gear lever, and by placing the hand-brake grip underneath the scuttle. The gearbox is connected to a curved lever projecting from the nearside of the instrument panel and is mounted under the bonnet along with the engine.

Roomy Bodywork

Due to this form of construction the bodywork is appreciably more roomy than is that of many cars of almost double the horse-power rating. The two front seats are separately adjustable, and are upholstered in soft hide, as also is the wide rear seat. The upholstery is of the spring type, and really does provide arm-chair comfort at all speeds of the car.

Probably on account of the front-wheel drive and the absence of a propeller shaft, the car was particularly silent. In fact, it was possible for the driver to carry on a conversation with the rear-seat passengers at 60 m.p.h. without there being any need to raise the voice to a noticeable extent. The only occasion on which there was any sign of noise

The compact o.h.v.-engine-gearbox unit. Inset are shown the cylinder casting and detachable cylinder barrels.

which could be considered objectionable by even the most fastidious was when slowing down from speeds in excess of 50 m.p.h.; there was then a certain amount of "transmission rumble" which lasted until the speed fell to about 35 m.p.h.

Test for Road-Holding

Considered in conjunction with the size, the weight of the car is low (20½ cwt.), and this results in a performance which

Seating accommodation is generous, and there is ample room for three adults on the rear seat.

is well above that of the average 12-h.p. saloon. It is generally considered that a reduction in weight has a tendency to impair road-holding, especially on bad road surfaces. This certainly does not apply to the Citroën, for the excellent suspension system renders the car very steady on almost any road surface. One of our tests consisted of driving the car over a stretch of road which is well known to us as being very wavy and capable of giving most cars the "shivers" when traversed at any speed over 15 m.p.h. The Citroën could be driven over it at speeds up to 35 m.p.h. without it being evident that the surface was abnormal.

Performance Figures

On performance alone, this car has claim to high praise. The maximum speed registered by the speedometer was 67 m.p.h. (afterwards checked against a stop watch as 63 m.p.h.) despite the modest claim of a top speed of 60 which is made by the makers. In second gear, a speed of 46 m.p.h. could be reached, although the comfortable speed in this gear proved to be about 43 m.p.h. at which both engine and

There are separately-adjustable front seats. The gear-lever is mounted on the dash, and the handbrake is under the scuttle.

gear-box were quiet. We did not take a note of the highest speed which could be reached in bottom gear, because it was necessary to use this only when starting the car, due to the fact that the car could be driven at less than walking speed in second without any suggestion of "snatch" or roughness.

Acceleration times also were commendable. Thirty m.p.h. could be reached from a standing start in 7⅜ sec., using first and second gears; from 10 to 30 m.p.h. in second gear required 5⅜ secs.; whilst 50 m.p.h. could be reached from 30 m.p.h. in top gear in 14 secs. Times were taken with a Lougmes (Baume & Co.) Split-seconds first quality stop watch.

Brakes

The Lockheed hydraulic brakes acted very efficiently, and were found to stop the car from 30 m.p.h. in 29 ft. on a dry concrete road. Operating by servo action on the rear-wheels only, the handbrake is cable connected, and takes the form of a pistol-grip handle situated underneath the scuttle where it is easily reached; it is locked by turning the handle through 90 degrees.

It was at first thought that the dashboard-mounted gear lever would be rather difficult to use, but experience soon dispelled this idea. The lever is very accessible, and synchronised silent top and second gears could be engaged without taking any care to regulate road and engine speeds.

Engine Specification

There are so many interesting features on this car that one is apt to overlook the engine; it is certainly unobtrusive enough! It was found to start every time at the first touch of the starter switch, and the car could be driven away without waiting for the engine to warm up, provided that the throttle were kept fairly well open for the first few minutes. The engine is of the four-cylinder type with push-rod-operated overhead valves. Bore is 72 mm., stroke 100 mm.,

Section through the gearbox and spiral-bevel differential.

capacity 1,628 c.c., and R.A.C. rating 12.8 horse-power; it thus carries a £9 15s. annual tax. It is flexibly mounted and provided with an adjustable stabilising device, and has a three-bearing fully-balanced crankshaft. A point of importance is that the cylinder barrels are detachable, and cast in wear-resisting and corrosion-proof material.

Lubrication is by means of a gear-type pump, which feeds all bearings, timing chain and rocker arms. A horizontal Solex carburetter with easy-starting device is fitted, and this is fed by a mechanical pump which draws its supply from the 9-gallon rear tank. There is a 12-volt battery, charged by a voltage-control dynamo.

Principal dimensions are: overall length, 13 ft. 9 in.; overall width, 5 ft. 3½ in.; overall height, 5 ft.

The price of the car tested is £228, which in our opinion is very moderate for a high-grade car with such a fine performance and finish. Running costs also should be low, for the petrol consumption throughout our tests averaged 25 m.p.g., a figure which would probably be bettered on a long main-road run.

1938 CARS—THE CITROËNS

A DIESEL-ENGINED SALOON TO BE SOLD AT £375

Long range economy, smoothness and robustness features of the design

The six-light oil-engined saloon which will sell at £375.

Existing front and rear wheel drive models continued, with modifications

The compression-ignition engine is no larger than the normal petrol type. Four cylinders of 1,750 c.c. give 40 b.h.p. The engine is flexibly mounted in the frame.

A FEATURE of exceptional interest in the new 1938 Citroën programme is the announcement of a 1,750 c.c. compression-ignition-engined car priced at £375. Before proceeding with the technical description of the engine it may be as well to instance some of the reasons which have led the Citroën company to market it.

It is well known that the fuel tax of 8d. per gallon operates in this country equally on both petrol and oil fuels. Even so, however, the normal purchase price of the latter is one shilling, which is approximately two-thirds the price of petrol. In addition the high efficiency of the heavy oil engine in respect of fuel results in the quantity used being again only two-thirds, or rather less, of the amount of petrol needed for a similar-sized engine.

The result is that fuel costs per mile are reduced to under half by using an oil-engined car. Put another way, a 15 h.p. car with orthodox engine will consume £3 worth of fuel per thousand miles; an oil-engined car will consume only £1 4s. worth of fuel for the same mileage. There is a saving of £2 16s. per thousand miles or £100 per 35,000 miles.

Model and Specification
THE CITROEN DIESEL

Engine: Four cylinders, 75 by 100 mm., 1,750 c.c., maximum r.p.m. 3,500, minimum 350; Lavalette injection pump and Beru heater plugs for cold starting; three-bearing crankshaft, o.h.v., alloy pistons and steel con. rods. Weight, 450 lb. approximately.
Transmission: Three speeds, giving ratios of 5.3, 10.2 and 16.6 to 1. Synchromesh on top and second.
Chassis: Box-section frame, semi-elliptic springs.
Brakes: Lockheed hydraulic.
Equipment: 12-volt electric system, wire-spoke wheels, twin windscreen wipers, Triplex glass.
Price: Six-light all-steel saloon, £375.

The price of the compression-ignition model is £90 more than the same car with the petrol engine, so that once a mileage of 32,000 miles has been exceeded the Diesel engine begins to save money.

Many motorists, particularly those who use their cars for commercial purposes, exceed this mileage by a very considerable margin and if, for instance, 60,000 miles were run the eventual saving in cost would amount to nearly £80. As many of the larger firms in this country have travellers using perhaps 10 or 12 cars, it will be seen that very large savings are offered by the use of heavy oil fuel.

All this presupposes that maintenance costs will not be greater than those of the present types of engine, and that the general running characteristics are such that the car is suitable for everyday use.

There seems no reason to doubt that such is the case. The four-cylinder engine has a bore of 75 mm. and a stroke of 100 mm.; it develops 40 b.h.p. at a maximum speed of 3,500 r.p.m. The minimum running speed is 350 r.p.m., and this range of 10 to 1 is in itself an indication that many of the earlier problems of oil-engine design have now been solved. So also is the fact that the weight is only 11 lb. per horse-power, which is not very much greater than normal.

The overall dimensions of the unit are exactly the same as the petrol type which would be used in the same chassis, and in quietness the types are

THE CITROËN PROGRAMME

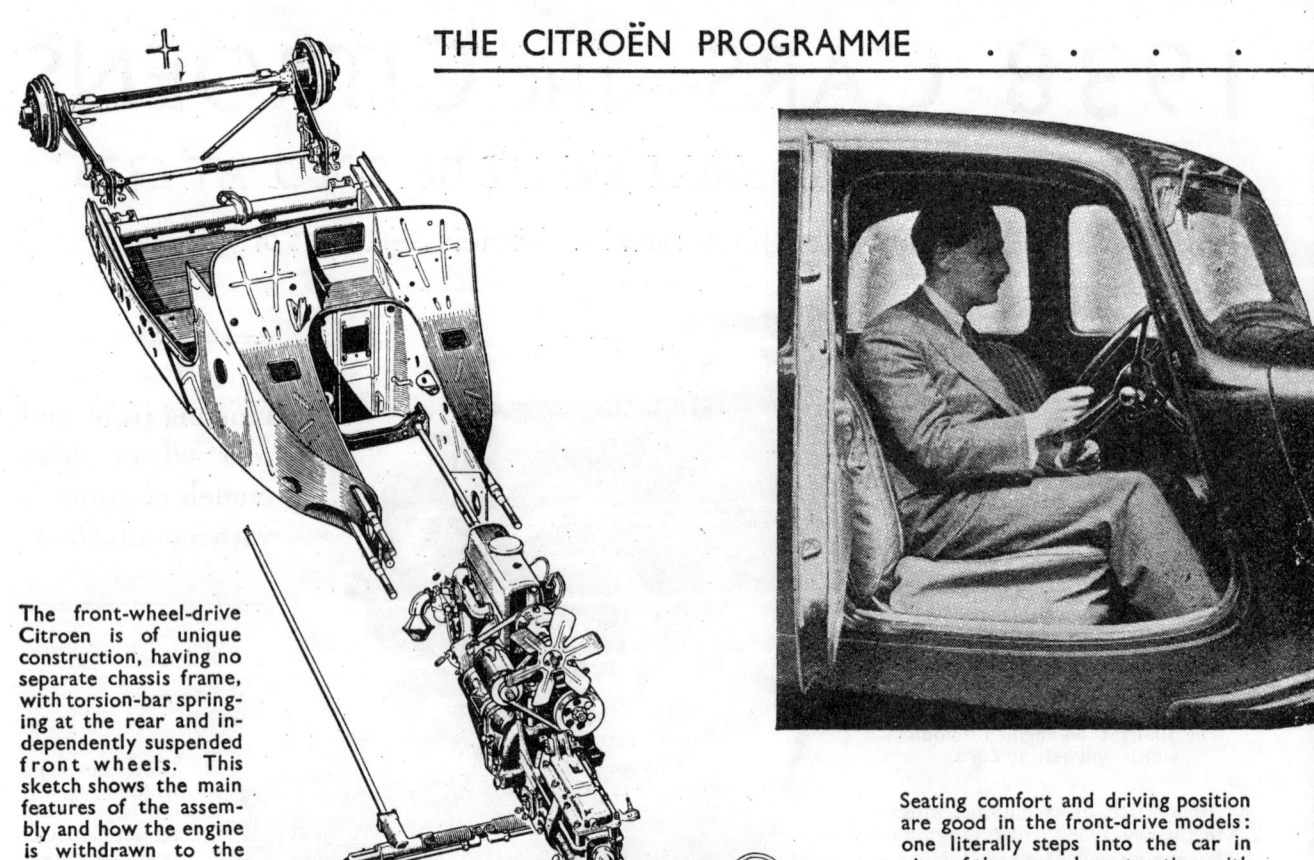

The front-wheel-drive Citroen is of unique construction, having no separate chassis frame, with torsion-bar springing at the rear and independently suspended front wheels. This sketch shows the main features of the assembly and how the engine is withdrawn to the front.

Seating comfort and driving position are good in the front-drive models: one literally steps into the car in view of the unusual construction; with the brake and gear levers mounted at dash level the front compartment is free of all obstruction.

(Right) Special iron liners are inserted into the block with direct full bore contact with the cooling water.

strictly comparable. Consequently, no special chassis arrangements are necessary, a feature which contributes to the low selling price.

The crankshaft is of large diameter and stiff section and runs in three bearings. So also does the camshaft, which is gear driven. The valves have been the subject of special study, and are made in an alloy steel which has a high resistance to pitting at elevated temperatures.

As is normal with compression ignition engines, the valves are situated in the cylinder head and operate through rockers and push-rods from the camshaft. The cylinder head itself is of the Ricardo Comet type which is being widely used all over the world; it gives very high efficiency and clean running with absence of smoke from the exhaust. Fitted into the head are four Beru heater plugs. Before starting, these are turned on and in all temperatures instant response follows, the engine being turned by the electric starter, as is normal petrol engine practice.

A feature of the compression-ignition engine is that it will develop full power without warming up, the heater plugs being turned off instantly after the engine is fired.

The injector system consists of Lavalette Bosch injection pump with a suction regulator. The amount of fuel injected by these pumps is very small indeed and their construction is such that it is essential that they should only receive clean oil, to ensure which, between the mechanical fuel pump and the injector, there is a special filter to eliminate the smallest impurities from the fuel.

The pistons are of heat-treated aluminium alloy and have two compression rings and three oil scraper rings. They are fitted into heat-treated steel connecting rods and the big-ends are lined with lead bronze. Lubrication is a full-pressure system from a gear-type pump, and the operating pressure can be varied readily to meet changes in oil or oil temperature.

Cooling is assisted by a centrifugal water pump, and especial care has been given to the cylinder head cooling so that very long life of the valves can be expected.

One of the fundamental problems of the Diesel engine is that of smoothness. On the petrol engine the internal pressures vary in accordance with the power which is being developed; this is not so on the compression-ignition engine, where the compression is always the same at all speeds and loads, and power is varied only by changing the amount of fuel delivered to the cylinder head.

It is for this reason that the type is so efficient. It is for this reason also that it is apt to be rough, especially at low engine speeds. The problem has been tackled by Citroën in two ways. First, by making the engine stiff so that it does not develop any undesirable internal roughness; secondly, by rubber-mounting the engine in the

. . . . Contd.

The characteristic, pleasing lines of the front-wheel-drive saloons. A road test report of the "Twelve" appears on the next page.

frame so that the reactions from the high explosion pressures are not unpleasantly felt by the driver and passengers.

This rubber mounting is a special Citroën design which has been called Pausadyne, from a Greek derivation. The engine is supported by a rubber block on the centre line of the crankshaft and steadied at the other end by two coil springs mounted relatively high up on the engine. At the far end of the engine, and also on the crankshaft centre line, there is a rubber buffer to prevent the rock becoming too great under abnormal conditions. The mathematics of this layout have been very carefully investigated and the result is certainly praiseworthy.

This interesting power unit is fitted into an orthodox rear-drive chassis with a track of 4 ft. 8 ins. and a wheelbase of 9 ft. 10 ins. The gearbox has three speeds, giving ratios of 5.3, 10.2 and 16.6 to 1, the gearbox change being synchronized between top and second. The frame is of box section and the suspension by orthodox semi-elliptic springs. Upon this chassis is mounted a single-piece all-steel body which has ample interior accommodation for up to seven passengers. Two occasional seats are, in fact, fitted, folding up into the floor when not required.

A Practical Proposition

This car is essentially one for practical usage, and the interior equipment is of a high standard in quality.

It can alternatively be obtained with a four-cylinder overhead petrol engine of 78 mm. by 100 mm. bore and stroke at a price of £285. It has the unusual feature of detachable cylinder barrels. The cylinder block is cast with a view to these barrels being inserted in it during the course of engine assembly. The barrels themselves are so machined that they fit one against the other and into recesses in the cylinder block. They can, therefore, be made of special wear-resisting iron, and in conjunction with the excellent and even cooling obtained have shown themselves to have an exceptional life without the need for reboring.

The Petrol-engined f.w.d. and Rear-wheel-drive Cars

Similar engines are also used in the highly successful front-wheel-drive models which are being continued with little change for the ensuing year. Of these, the Light Fifteen also has an engine of 78 mm. by 100 mm., and there is a smaller type of 12.8 h.p. R.A.C. rating, 72 mm. by 100 mm. As is generally known, the chassis con-

PRICES OF PETROL-ENGINED MODELS

"TWELVE" F.W.D.
Four-cylinder—12.8 h.p. (tax £9 15s.)
Saloon £238
Fixed-head coupé £255
Roadster £265

"LIGHT FIFTEEN" F.W.D.
Four-cylinder—15 h.p. (tax £11 5s.)
Saloon £248
Fixed-head coupé £265
Roadster £275

"FIFTEEN" F.W.D.
Four-cylinder—15 h.p. (tax £11 5s.)
Saloon £278
Seven-seater saloon £298

"FAMILY FIFTEEN" (REAR-WHEEL DRIVE)
Four-cylinder—15 h.p. (tax £11 8s.)
Seven-seater saloon £285

struction of these cars is unusual in that there is no separate chassis frame. Instead, there is a welded assembly of body panels joined to a flat punt-shaped base that constitutes the complete car unit, as shown in an illustration.

In the course of construction this unit is completely finished, trimmed, the seats placed in position and so on, after which the engine and transmission unit is attached.

The whole of the mechanical assembly is highly accessible and can be withdrawn complete with great ease should service attention be necessary. Independent suspension is provided at the front end of the car, torsion-bar springing being employed damped by very large piston-type shock absorbers.

Torsion-bar springing is also employed for the rear of the car, but here the wheels are not separately sprung, being connected together by a simple axle beam. This, of course, contains no drive mechanism, but it is interesting to note that for this year it has been made slightly cambered. So doing gives the rear wheels a slight inward inclination which, it is said, appreciably improves tyre wear.

As a result of the novel construction used, the floor level of the car is very low, yet at the same time ample headroom has been provided. In consequence highly stable cornering is achieved without the usual inconveniences of the low-built sports saloon type of body.

As is generally known, these cars are initially of French design, but they include considerable modifications that have made them thoroughly suitable for English use. Moreover, the whole of the constructional assembly of the car is carried out at the British works at Slough, and a very large proportion of British equipment is used.

A Variety of Improvements

A variety of detailed improvements have been included on the latest models. On the mechanical side these are not of great magnitude, but changes made in body equipment and finish are more noticeable. In the first place the facia board has been changed so as to provide grouped instruments of large diameter, and the trimming and construction of the seats have been improved. Another change concerns the location of the twin Lucas horns, which are now placed inside the radiator grille. The rear locker has been modified so that it opens flat and can be used as a luggage platform. The whole body is sprayed in the course of assembly with a sound damping material.

"THE MOTOR" ROAD TESTS
The Citroen Twelve (f.w.d.) Saloon

EXPERIENCE of driving the Citroën 12 saloon sets aside all bogies associated with front-wheel drive. The idea exists in the minds of many people that such form of transmission must of necessity be noisy and objectionable; must affect the steering and make it heavy; also adversely tell upon the lightness and pleasure of handling the car generally. While these may have been the early experiences with front-wheel drive, modern developments have surpassed all such criticisms, as many thousands of Citroën and other f.w.d. car owners have found to their satisfaction in the past few years.

To drive the Citroën one could, in the main, fully imagine that it is based on conventional layout, in so normal a way does it handle. The only difference which particularly struck us was the very strong self-centring action of the steering, which, however, has its advantage in that it pulls the car straightway back into the dead ahead position so soon as a change of direction has been completed. A good feature of the steering is its "directness," all normal directional changes as when in traffic, on moderate bends and corners, etc., involving little more than short forearm movement. With only 2¼ turns of the steering wheel, from full lock to lock, it will be appreciated that there is no winding and unwinding to be done, which can become so tedious.

Although the whole of the "works" are at the front, there is comparatively little noise or fuss emanating therefrom. Up to some 40 m.p.h. everything works quietly enough (except in full-throttle openings when accelerating from low speeds), becoming, however, more audible as cruising speeds increase to the 50-55 m.p.h. regions. These are comfortable cruising gaits.

The car settles down to its job, rides solidly and untiringly for mile after mile.

The farther you drive it the more its characteristics are enjoyed and an impression is gained that it is a soundly-built job which will tolerate uncomplainingly miles of hard driving.

The gearbox has synchromesh on second and top, giving a quick and easy change. Only three speeds are provided, but these suffice for all everyday purposes; our test took us into the by-

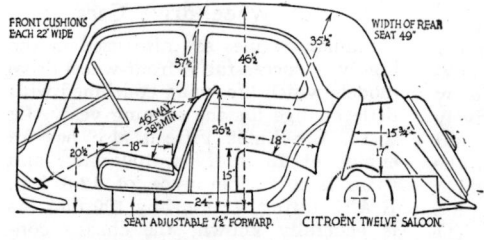

ways of Gloucestershire, where many of the lesser-used, hilly lanes over the Cotswolds were traversed and no difficulties were met on any gradients.

The gear lever extends through the dashboard: a rather uncommon position, and turns down at an angle which at first makes it awkward to use with aplomb, but, like so many things of this type, practice overcomes the difficulties. Similarly, on first acquaintance with the car one is struck by the odd angle at which the pedals are placed; instead of sloping in the usual way, so as to take a direct push, they lie almost horizontally, demanding a downward pressure. Thus, instead of leaving the heels on the floor, one has a tendency to lift the feet and tread down on the brake and clutch pedals.

The Citroën is unique among popular-priced cars in that it has independent suspension at the front and independent torsion bar springing at the rear. This arrangement proves extremely good, the car being practically free from bumping and pitching on all surfaces.

Brakes are particularly good and cornering calls forth a pæan of praise for the car can be pushed around bends with a feeling of security, so well does the Citroën hold the road.

The Motor TABULATED DATA—CITROEN 12 SALOON.

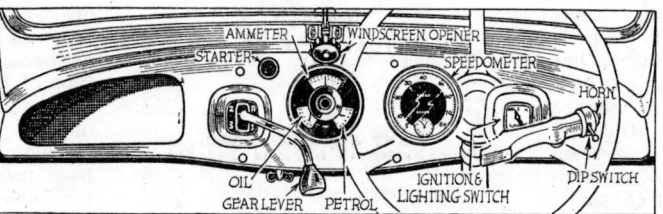

ENGINE

Four cylinders, o.h.v.; 12-volt coil ignition; 72 mm. by 100 mm. (1,628 c.c.); rating, 12.8 h.p. *Tax:* £9 15s.

MEASUREMENTS

W.b., 9 ft. 6½ ins.; t., 4 ft. 4¾ ins.; l., 13 ft. 9 ins.; w., 5 ft. 3½ ins.; g. clce., 7 ins.; turning circle, 40 ft. dia.

SPEEDS m.p.h.
- *Maximile mean timed speed .. 61.6
- Maximum mean timed speed .. 63.3
- Best timed speed 64.2
- Speed reached on second .. 40.0

ACCELERATIONS secs.
- 10-30 m.p.h., second gear .. 7.0
- 10-30 m.p.h., top gear .. 14.0
- 30-50 m.p.h., top gear .. 18.0
- 0-50 m.p.h., through gears .. 21.0
- 0-30 m.p.h., through gears .. 7.2
- Standing ¼-mile, through gears .. 25.0

METERED PERFORMANCE†

	Pull lb. per ton.	Gradient climbable.
Top (4.9 to 1)	180	1 in 12.5
Second (8.3 to 1)	310	1 in 7.2
First (14.8 to 1)	540	1 in 4.15‡

Petrol Consumption: 23.24 m.p.g., driven hard. Solex carburetter. 9-gallon rear tank.

Gearbox: Three forward speeds, facia control. Synchromesh on top and second.

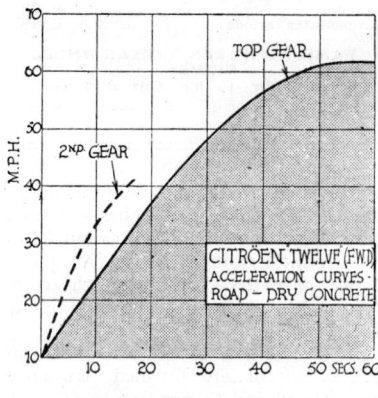

METERED BRAKE TESTS†

	Efficiency. %	Distance ft.
Pedal only, from 30 m.p.h.	80	37.7
Pedal only, from 50 m.p.h.	75	114
Hand only, from 30 m.p.h.	35	86

(Mechanical brakes; hand lever under dash.)

Wheels: Spoke, with 140 by 40 Michelin tyres.

Weight: Unladen, 20½ cwt.; with two up, as tested, 23 cwt.

Price: £238.

*Speed timed over ¼-mile after accelerating for one mile from rest through the gears.
†Pulling power and brake efficiency are recorded by Tapley and Ferodo-Tapley meters respectively.
‡Will climb South Harting, Edge, Kirkstone and Rest-and-be-Thankful Hills on this gear.

FRONT WHEEL DRIVE AND TORSIONAL SUSPENSION
EXTENSIVE EXPERIENCE OF THE CITROEN TWELVE

FRONT wheel drive and torsional suspension—features of specification of interest to all enthusiasts and students of design. It was to make better acquaintance with these features that we recently put a Citroen Twelve through its paces under a wide variety of conditions and over a quite extensive mileage. The Citroen Twelve saloon is certainly not a sports-car, but it is basically the same chassis as that used for the open two-seater Twelve and the higher performance Light Fifteen. It is a very interesting chassis indeed, for Citroen broke completely away from utility car design, after success since the early post-war years with the 7.5, 11.4 and 12 h.p. cars, and introduced, a few years ago, the existing torsionally suspended, front wheel drive chassis; one of the most successful cars of unorthodox layout ever marketed.

One's very early impression of the Citroen Twelve is that of being behind the wheel of a much bigger and more expensive car. For this the torsional suspension must undoubtedly take credit, for the car rides with rocklike solidity over all kinds of road surfaces, displaying hardly a trace of pitching or side sway. Indeed, no roll occurs even under really quick cornering, and the only time the suspension displays any trace of suppleness is after taking hump-back bridges at speed, although, from the degree of genuine comfort it affords, one would expect, and certainly excuse, some trace of flexibility. The special all-steel frameless construction must also contribute very materially to this steady riding of the Citroen and its rigidity does at times convey a slightly "dead" feel to the manner of riding.

It is not only luxurious running that suggests a larger car. The Citroen is a willing performer, as we soon discovered in hustling from the works at Slough to Brooklands, a journey for which we had craved a sports-car, until we found how easily the Citroen gets through traffic and settles down to an effortless cruising gait of 60 m.p.h. on derestricted roads. On second gear of the three-speed box excessive valve-bounce limits the maximum to 40 m.p.h., but up to that speed very brisk acceleration is commanded.

The most interesting feature of the Citroen is the front-drive layout. It makes a very considerable difference to control and has only very minor disadvantages to off-set this commendable characteristic. The Citroen Twelve is one of the most controllable cars we know. On wet roads, slimy with weeks of rainless weather, it will naturally slide if encouraged, but so effectively does it respond to the steering that after a little experience of its abilities in this direction one can literally sling it about on greasy roads without restraint. Here it is that the front drive is so valuable. Even on the over-run the car is very very stable, but if one has any doubts about where it is going one has only to turn on the power and the front wheels never fail to pull the car in the direction in which they are pointing. Particularly was this evident on London streets notorious for their slippery surfaces. Under conditions when the front wheels would spin furiously under acceleration, we certainly experienced loss of front wheel adhesion on the over-run on corners, but, on accelerating, directionability was at once restored. In consequence, one is able to do most remarkable things with the f.w.d. Citroen, on greasy roads and fast open bends alike. Round long lower-speed curves, there was a rather less pronounced feeling of stability, difficult to define, but always was the car under control. In cornering there is hardly any trace of severe roll, and the car goes round as on rails, with a fair amount of protest from the Michelin "Stop" Real Low-Pressure tyres, chiefly from those on the front wheels. What of the disadvantages of this front wheel drive? We have been warned of increased noise from having the gearbox and transmission under the bonnet instead of under the floor. In the case of the Citroen, there *was* a certain drumming or vibration on the over-run at times, but otherwise the transmission gave no clue to its whereabouts, and the gears were almost silent. We had heard that f.w.d. can adversely affect steering and cornering qualities. So far as the latter consideration is concerned, we have already commented on the effectiveness of the front drive in rendering the car stable under sticky going.

In the matter of steering, certainly under load the steering became stiffer than on the over-run, tending to vary from moderately light to fairly heavy, but it remained, even on the drive, smooth in action and was moderately high-geared so that undue work at the large diameter wheel was not necessary. There was sufficient castor action at all times and hardly a trace of return moment came back through the wheel, nor was there any column vibration. The lock is quite adequate.

By the time we had arrived at Brooklands we began to enjoy the sense of big-car comfort afforded by the Citroen and were impressed by its general solid construction. The interior finish is in leather, with carpeted rear compartment, and separate and adjustable bucket front seats of generous dimensions, and there is a sliding roof. Reverting for a moment to the front drive, it is possible to accommodate three persons on the front seats in view of the absence of a normal gear-lever. The only other point likely to be raised by those cautious of unorthodox layouts concerns adhesion on hills, and later in the test we tried some trials gradients with this in mind, with entirely favourable results, as detailed below. Outwardly, there is something very pleasing about the sleek clean lines of the aerodynamic saloon body, with its flush fitting bonnet with side inspection doors, inclined radiator grille, flowing wings, and artillery steel wheels. Incidentally, the floor is at a very low level so that, although no running boards are fitted, entry and egress are particularly pleasant to accomplish.

On Brooklands the Citroen rode in a dead steady manner, even to feeling a trifle "dead," and in consequence felt essentially stable. Lots of fast lappery did not appear to disturb it, and although no thermometer is fitted, the oil gauge showed a consistent 25 to 30 lb. per square inch at speed, as on the road—we dislike these modern oil gauges with spidery needles and dials as part of a main dial and often wonder whether they read true; but then they are common on modern utility cars and Americans. The lap came out at 60.5 m.p.h. and the flying half-mile at 65 m.p.h., at which speed the speedometer indicated 68 m.p.h.

The Citroen is thus not a fast car, but one judges it not so much by its 65 m.p.h. maximum as by its complete lack of fuss at all speeds on the road. It is especially complaisant at 40 m.p.h., but will cruise very happily at 55 to 60 m.p.h. The engine makes some carburetter noise when accelerating and there were a few body noises, otherwise the car does its work quite effortlessly, and covers the ground very fast by reason of its good cornering qualities.

Acceleration is of the order of standstill to 50 m.p.h. in 21 secs., and on the road the urge on the lower gears is distinctly usable. The gear-lever projects as a down-dropped fairly slender lever, from the centre of the facia, rather over to the near side. One very soon becomes accustomed to the action and thereafter only very occasionally does one feel a desire for the conventional sort of gearshift. Certainly on the car tried the lever moved stiffly, but with steady, as distinct from wild, movements the changes went through quite well, and reasonably rapidly, the synchro-mesh being fully usable, with no time lag. Bottom could always be located, albeit with a crunch under brutal engagement. Only once or twice did the driver become confused as to the sequence of lever movements. There is a remote possibility that a child, or a careless adult passenger, might get a finger caught in the dashboard gate—personally, we remembered to warn those privileged to sit on our left. The clutch was extremely smooth and progressive, required only moderate pressure to disengage, and felt positive.

The handlever is in the form of a cranked lever protruding by the driver's left leg. To hold the car in traffic it is pulled outwards and to lock it one turns the handle to the right. It was a trifle inconvenient to locate and rather vague at times in respect of locking on, but it functioned smoothly, one gradually became acclimatised, and, in company with the dashboard gear-shift, it permits of great spaciousness in the front compartment.

After having gained familiarity with the Citroen we left it in the drive over night, before starting off to meet the Antwerp boat at Harwich at 6.30 a.m. next morning. The engine commenced at once, and pulled away well, with the choke fully extended for the first mile or so. This night run fully endorsed our already high opinion of the car's ability to maintain a high cruising speed and of its extreme stability. The roads were awash and treacherous after the drought, but any slides we experienced were instantly brought under control and were

FRONT WHEEL DRIVE AND TORSIONAL SUSPENSION—continued

occasioned solely on account of an exuberance of enthusiasm. Taking a mileage reading at the grim hour of 4.50 a.m., while still in the Mile End Road, we covered 50.3 miles in the following sixty minutes. Even granting the absence of traffic, this indicates that the Citroen can get about very rapidly from place to place, and, moreover, in silence, riding smoothly and steadily.

The brakes are entirely adequate, and slow the car steadily with no fuss or detrimental effect on stability. For rapid stops a fairly heavy pressure was required and the action is a trifle spongy, but no loss of braking power was noticeable at the conclusion of the test, which extended over 520 miles. The lamps were good without being ultra-powerful, and the dimmer, operated by a lever on the steering column extension, very convenient to use and effective. This extension also carries the horn button (we prefer it on the wheel centre) and the direction indicator switch, well placed save that one was apt to knock it accidentally. The arms are of self-cancelling type operated from the steering wheel movement and at times tended to cancel rather early. The facia carries, from left to right, a usefully large cubby-hole with no lid, swivelling ash-tray beneath, gear-gate, combined fuel-gauge, oil-gauge and ammeter, speedometer with inset clock, and combined lighting and ignition switch, the latter with removable key. The concealed dashboard illumination was effective, but the left hand large dial was permanently lit, whether intentionally or because of partial failure of the dashlight switch we do not profess to know. It in no way resulted in dazzle, but, suffering in the early a.m. from liver, we would have given much to be able to douse it.

There are twin anti-dazzle visors and effective twin screen-wipers, and the interior illumination is excellent, controlled by a switch low down and within reach of the driver, or a passenger in the off side of the rear seat. The screen will open, by operation of a central knob, and was not entirely leakproof in a downpour. The sliding roof worked well. The screen pillars are rather heavy, but visibility is excellent, as the near-side lamp is visible. The high-set headlamps are brightly plated and in consequence it is easily possible to ascertain whether the side lamps are burning, quite apart from their indicator glasses, a reassuring point to those drivers who put out—as they should—their headlamps on entering well-lit streets. On the journey away from Harwich we appreciated the accommodation afforded by the rear luggage-locker, on the lid of which the spare wheel is carried within a metal case, and the spaciousness of the rear seat, which will comfortably accommodate three persons. Small ash-trays, inscribed with a tiny chevron, are provided in the rear compartment. The doors and luggage-locker are lockable and the front doors have elastic-topped pockets. A good point is the provision of jacking pads on the axles. The Citroen "Mono shell" all-steel coachwork must provide very valuable protection in the case of a serious accident. A useful driving mirror and rear blind are provided.

The dip-stick, distributor and plugs are accessible on the near side and the o.h. valves are enclosed in a neat, polished valve cover with oil-filler in the lid. The radiator filler cap is under the bonnet and the battery is accessibly placed on the engine side of the facia, while spare brake fluid and electrical fuses could hardly be more easily get-at-able. Over rough going the whole car retains its feeling of solidity, although there is some lamp movement and the bonnet top would ripple under clutch engagement in bottom gear.

Turning to the specification, the four-cylinder 72 × 100 mm. 1,628 c.c. engine is rated at 12.8 h.p. It has a balanced three-bearing crankshaft, detachable cylinder barrels, alloy four-ring pistons, and push-rod actuated o.h. valves. The larger engine of 78 × 100 mm. (1,911 c.c.), rated at 15.08 h.p., is obtainable in the same chassis. A Solex carburetter is used and cooling is by pump and fan. Coil ignition has automatic advance and retard.

The drive passes via a single dry plate clutch to the unit three-speed and reverse gearbox, differential and bevel front wheel drive. Front suspension is by torsion bars and parallelogram linkage and rear suspension by similar torsion bars. Lockheed hydraulic four-wheel brakes are used.

The pressed steel wheels carry 140 × 40 mm. Michelin Real Low-Pressure covers. The gear ratios are 4.9, 8.3 and 14.8 to 1, reverse 19.7 to 1. Wheelbase 9 ft. 6½ in., track 4 ft. 4¾ in., turning circle 40 ft., weight 20½ cwt. The fuel tank holds 9 gallons, giving a range of 180 miles approximately at a consumption of around 20 m.p.g.

In conclusion, the Citroen Twelve is a solid car possessing some remarkably interesting and practical technical features, and offering quite rapid transport with a high degree of comfort and convenience. It upholds the reputation enjoyed by Continental productions for safe sports-like road-holding, yet displays no flimsiness in its construction or manner of riding. On Allington hill, wet and very sticky, wheel spin defeated us only at the summit, and Coldharbour was a first and second gear ascent with full load, while Beechy Lees would have been a fast climb had not a gear jumped out. Our experiences of these and other Kent trials hill confirmed the belief that, correctly applied, f.w.d. is no bar to hill-storming. Priced at £238, taxed at £9 15s. 0d., and with a twelve months' guarantee, the Citroen is amongst the most interesting of Twelves. Full particulars may be had from Citroen Cars, Ltd., Trading Estate, Slough, Bucks.

THE CARS WITH THE FOUR YEARS LEAD

CITROËN
PROGRAMME FOR 1939

Four years have elapsed since the front-wheel-drive CITROËN was introduced into this country. Since then, continuous advancements in constructional details and general refinements have been made, and today it has attained a perfection of design that puts it far ahead for many years to come. Appreciation of the wonderful safety and stability of the CITROËN is now widespread. The 1939 models are the consummation of four years of successful production and offer the motorist a remarkable union of advanced design with established practice.

FULL LIST OF MODELS AND PRICES

Model	Price
"TWELVE" Saloon	£238
"TWELVE" Popular Saloon	£198
"TWELVE" Roadster	£265
"LIGHT FIFTEEN" Saloon	£248
"LIGHT FIFTEEN" Popular Saloon	£208
"LIGHT FIFTEEN" Roadster	£275
"BIG FIFTEEN" Saloon	£278
"BIG FIFTEEN" 7-seater Saloon	£298

CITROËN CARS LTD., SLOUGH, BUCKS.

1939 CARS
The Front-drive Citroëns

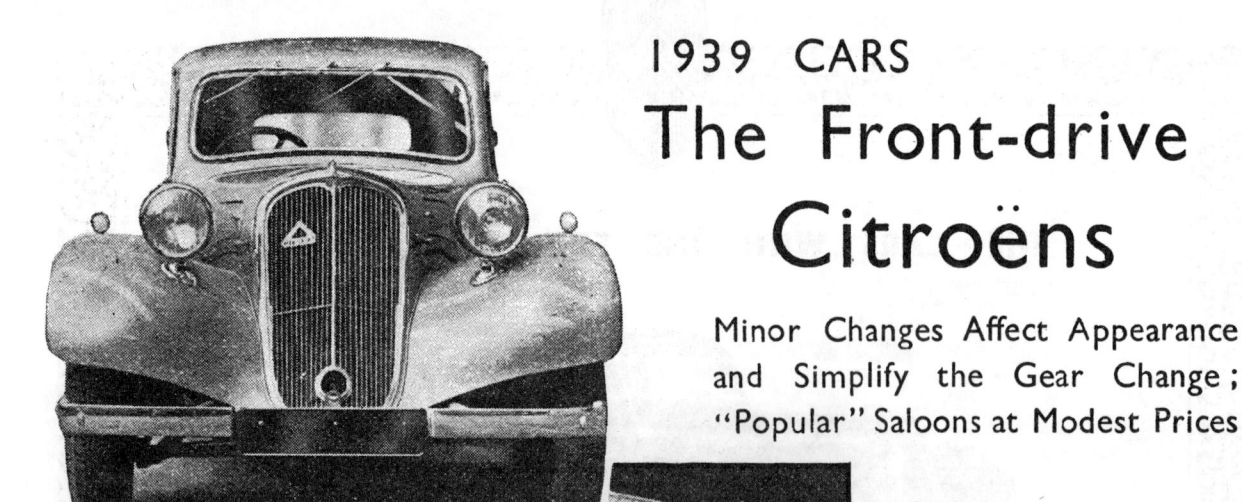

Minor Changes Affect Appearance and Simplify the Gear Change; "Popular" Saloons at Modest Prices

New radiator shell and grille (with a small badge replacing the big chevrons), and improved wings, are included in the 1939 changes. (Right) The wide door and unobstructed front compartment.

WHEN the current type of Citroën was first introduced in 1935 it created something of a furore by virtue of the advanced features of design incorporated. Now, with four years' experience, it is still very much ahead of a number of English and Continental layouts, and during that time has fully proved the soundness of the principles employed. It will be recalled that the main features common to all models are:—

Integral all-steel construction; front-wheel drive; independent torsion bar front wheel suspension; torsion bar rear springing; overhead valve engine with detachable

A low-built roadster 2-seater is available at £265 on the Twelve chassis and at £275 with the 15 h.p. engine.

wet cylinder barrels; and gear lever mounted on the facia board.

In the main there are no great changes in the cars for next year. The range will comprise the "Twelve," "Light Fifteen" and "Big Fifteen," so that the policy of progressive development and improvement on these models continues for the fifth year.

Advantages of the Torsion Bar Springing

Experience on the road has shown them to be capable of very attractive performances, with exceptional stability and road holding and an ease of cornering resulting from the low centre of gravity and the fact that the front-wheel-drive enables one to "pull" the car round a bend instead of driving it. The suspension also is outstanding and all seats are located within the wheelbase, being actually "slung" between the suspension points.

The Twelve and Light Fifteen comprise the same range of body styles with alternative engines. The Fifteen is more spacious and is available as a six-seater saloon or as a six-light seven-seater on a longer wheelbase chassis.

The New "Popular" Models

The range actually is augmented by a Popular Twelve saloon, costing £198, and a Popular Fifteen saloon, listed at £208, which have similar performances to their de luxe counterparts, but are equipped and finished in a simpler manner. For example, there is no sliding roof and the upholstery, instead of being in leather, is finished in cloth. The 12-volt equipment of the de luxe is replaced by a six-volt installation and the instruments are grouped in quite a small rectangular panel—about 9 ins. across—mounted above the steering column. The paintwork is all black, including the radiator shell, whilst the front of the car carries the chevron ornament across the grille.

The Light Fifteen Popular is the lowest priced 15 h.p. job at present available on the British market.

New Features to the Range

New features which cover the range generally are an improved gear shift with the selector locking device controlled by the clutch; this gives an exceptionally easy gear change but at the same time ensures a positive lock when once the change has been effected.

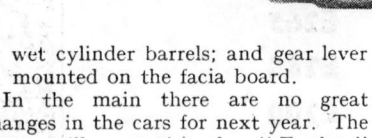

The new Michelin wheel with wider rim and unique "spoke" design.

THE FRONT DRIVE CITROËNS

A new type of Michelin wheel and tyre are used, notable for a reduction in weight (20 per cent. down on the old style) with a wider rim and tyre section. The wide metal "spokes" are laid out with a similar planning to that which is used for the old type of wire wheel. Balancing also has been improved and the tyre tread is of a very efficient pattern.

From the point of view of general appearance, the cars are outwardly the same as last year. The radiator shell has been modified, however, being simpler in appearance, yet it enhances the frontal aspect of the car and the familiar large chevrons—except in the case of the "Popular"—have been eliminated, a new small badge, which incorporates this, being employed instead.

The wings are wider and give better all-round protection against mud-slinging and the forward edge of the rear wings carry plated protectors to guard against damage to the paintwork, which might result from stones flung up by the wheels. Shock absorbers have been improved internally and the engine carries a new air filter and silencer.

These various changes, combined with the already high standard of finish and equipment which characterize the normal range of Citroëns, make them very attractive propositions.

Colour Schemes

Except for the Popular, a wide range of colour schemes is available with upholstery carried out in Connolly leather and the dashboard built up in a walnut panel. Jaeger instruments are fitted.

The saloons are of four-door construction, steel being used throughout, and the ease of entry is one of the important characteristics of all models. One sits well in the cars with a surprising amount of legroom, this applying particularly to the Big Fifteen, and good all-round vision.

In the case of the Big Fifteen six-light seven-seater, bucket seats replace the bench-type seat at the front, and behind them there are two occasional seats which fold into the floor when not required. Even when in use they do not hamper to any appreciable extent the legroom of those occupying the main back seat.

The gear lever, mounted on the dashboard, and the brake recessed under the panel on the right, leave the front compartments entirely unobstructed, so that entry from either side can be effected with ease.

Luggage accommodation is provided in the tail and when particularly big trunks are to be carried, the lid can be left partly open, a folding extension to the floor brought into use, and a surprising amount of impedimenta carried.

LINES AND ACCOMMODATION

The four-door saloon has modern lines and wide doors, all seats are located within the wheelbase so that the passengers enjoy good visibility although only four lights are provided. This shows also the luggage space.

Specifications in Brief

Brief specifications of the cars are as follow:—

Four-cylinder engines with push-rod-operated o.h.v. and detachable cylinder barrels cast in wear-resisting, corrosion-proof material. The units are flexibly mounted, but incorporate an adjustable stabilizing device to prevent excessive flex. The three-speed gearbox is mounted in unit with the engine and synchromesh is used on second and top. Power is taken to the front wheels, which are independently sprung, by means of torsion bars. The rear assembly forms one unit, comprising a

The "Popular" saloons on the 12 h.p. and Light 15 h.p. chassis have this compact instrument panel with finger-tip operation of the horn and light switches.

large tubular cross-member to which are pivoted two longitudinal links carrying the rear axle; torsion bars are attached to the forward end of the links and to the centre of the cross-member, an adjustment being incorporated. A diagonal radius rod completed the assembly.

Steering and Brakes

The latest type of steering is of a very efficient design, in which the spiral bevel drive operates direct on the push-and-pull rods, which form also the track rod.

Lockheed hydraulic brakes are fitted and the Lucas electrical equipment incorporates a 57 ampere-hour battery mounted under the bonnet. The general equipment is lavish and meets all normal requirements. Jacking at the rear is effected by placing the lifting device underneath pads mounted just forward of the rear axle. In this way jacking avoids soiling one's clothes by crawling about under the car, and becomes a comparatively easy operation.

On the polished wood dashboard, the speedometer, ammeter, petrol gauge clock and oil indicator are grouped in two circular dials—a pleasing and neat arrangement. The spare wheel is carried in a recess in the rear panel and is protected by a metal cover which blends harmoniously with the remainder of the coachwork.

The doors are fitted with remote controls and safety catches. A modified form of hub-cap embellishes the wheels and, apart from the distinction which it gives to the car, renders cleaning a simpler matter.

Citroën 1939 Models

Twelve
Four cylinders, 1,628 c.c., 12.8 h.p.; tax, £9 15s. Wheelbase, 9 ft. 6½ ins.; track, 4 ft. 4¾ ins.; overall length, 14 ft.; height, 5 ft.; width, 5 ft. 3 ins.
Saloon £238
Roadster £265
"Popular" saloon £198

Light Fifteen
Four cylinders, 1,911 c.c., 15.08 h.p.; tax, £11 5s. Wheelbase, 9 ft. 6½ ins.; track, 4 ft. 4¾ ins.; overall length, 14 ft.; width, 5 ft. 3 ins.
Saloon £248
Roadster £275
"Popular" saloon £208

Big Fifteen
Four cylinders, 1,911 c.c., 15.08 h.p.; tax, £11 5s. Wheelbase, 10 ft. 1½ ins.; track, 4 ft. 8 ins.; overall length, 15 ft.; width, 5 ft. 8 ins.; height, 5 ft. 1 in. (Seven-seater saloon: Wheelbase, 10 ft. 9 ins.; overall length, 15 ft. 8 ins.).
Saloon £278
Seven-seater saloon £298

New Cars Described —

Latest Popular Citroens

Low-priced Twelve and Fifteen Saloons with Simplified Equipment : Existing Range Continued with Detail Modifications

IT is a tribute to the advanced design of Citroen cars that the range of front-wheel-drive models should be continued for the fifth year with detail improvements only. The full range is now as follows:—

Twelve.
Popular saloon	£198
Saloon	£238
Roadster	£265

Light Fifteen.
Popular saloon	£208
Saloon	£248
Roadster	£275

Big Fifteen.
Saloon	£278
Seven-seater saloon	£298

The two Popular saloons, the Twelve and Light Fifteen, are newcomers. They are, of course, based upon the existing Twelve and Fifteen saloons, having the same body shell and major units, and, therefore, a similar performance to that of their de luxe counterparts. They are, however, equipped and finished in simpler style. The main points of difference are a fixed roof, cloth upholstery, six-volt electrical system, steel facia board with the instruments grouped in one rectangular panel above the steering column, lighter bumper bars, an absence of direction indicators, an all-black finish with chromium-plated fittings but with a black radiator shell, and smaller head lamps.

The radiator shell also carries the familiar double chevron running right across it, whereas all other models have a new style radiator shell of the same general shape as formerly, but simpler and more imposing and without the large chevrons. These have been replaced by a new small badge incorporating chevrons at the top of the off side of the grille.

Body Shell as Chassis

It may be well to recall the main features of Citroen design. These include integral all-steel construction (the body shell serving as the chassis), front wheel drive with independent torsion-bar suspension, torsion-bar springing for the rear axle, a four-cylinder overhead-valve engine with detachable cylinder barrels, in unit with a three-speed gear box having synchromesh second and top gears, controlled by a gear lever mounted on the facia board.

Amongst the advantages of the design are that it gives a low centre of gravity, which, combined with the front wheel drive, results in ease of cornering and also contributes towards exceptional stability and road-holding. Also, an unobstructed floor is secured in the body shell, which has a very roomy interior and true inter-axle seating.

The wheelbase and track of the Twelve and Light Fifteen models are 9ft. 6½in., and 4ft. 4¾in., the overall length being 14ft., while the Big Fifteen has a wheelbase of 10ft. 1½in. for the six-seater saloon, and 10ft. 9in. for the seven-seater saloon, the overall length being, respectively, 15ft. and 15ft. 8in. Both of the Big Fifteens have a track of 4ft. 8in.

Although the general design remains unaltered, certain minor modifications and improvements have been effected, apart from the new radiator shell. Thus, the gear change has been rendered lighter in operation by a simple device connecting the selector locking mechanism with the clutch.

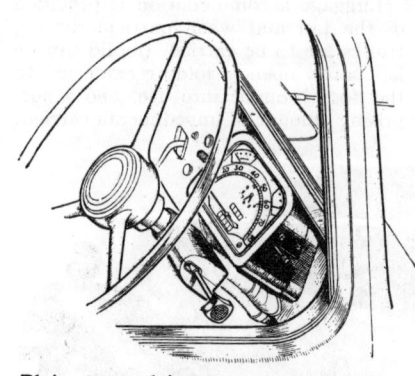

Plain grouped instruments are used on the Popular saloons. The control lever is for lights, dipping and horn.

Attached to the clutch withdrawal ring is a metal finger which connects with the T-shaped head of a spring-loaded rod running between the two selector rods. This rod has a tapered end which engages with the two locking balls of the selector mechanism, tending to force them apart into grooves in the selector rods. Thus, when the clutch pedal is depressed, the tapered end of this spring-loaded rod is withdrawn from the balls and the selector rods are free to move. It is impossible for more than one gear to be engaged at a time, since there is also between the selector rods a small ball-ended cylinder which engages with grooves in the rods and allows only one rod to be moved at a time.

Another important modification is the fitting of the new type Michelin wheels

A feature of both the Twelve and the Light Fifteen Roadster is a two-seater dickey which can be used for luggage.

Two folding seats, giving accommodation for seven in all, are a feature of the Big Fifteen 6-light saloon.

and tyres, which weigh 20 per cent. less than the previous design, and have a beneficial effect on acceleration and deceleration. These wheels are built up of steel pressings with broad flanged steel spokes and are distinctive in appearance. An important feature is the width of the rims and the shape of the tyre section, which is approximately that of a D. Indeed, these new tyres are 30mm. wider in the base than the corresponding Superconfort tyres and they are run at pressures of about 18 to 20lb. per sq. in. Despite these low pressures, the design of the tyre is such that there is a minimum amount of deformation when cornering, so that road-holding is improved.

Comfort and road-holding also gain from the fitting of improved shock absorbers of tubular type. These are hydraulic and double-acting. Wider wings have been adopted to give better protection to the body from mud splashes, and polished aluminium protectors are fitted to the front ends of the rear wings, enhancing the appearance and helping to maintain it in condition.

The engine is now fitted with a new type air filter and silencer, and in place of the oil pressure gauge formerly used there is now an oil tell-tale with a red warning light, giving a positive warning in the case of a failure of the oil pressure. As regards engine dimensions, the Twelve has a bore and stroke of 72 × 100mm. (1,628 c.c.), rating 12.8 h.p., and tax £9 15s., while the Light Fifteen and Big Fifteen have a bore and stroke of 78 × 100mm. (1,911 c.c.), rating 15.08 h.p., and tax £11 5s.

Features of the bodies are the wide doors, giving unrestricted entry in all models, the unobstructed floors, the roominess of the seating and its true inter-axle position, and the provision of a satisfactory amount of luggage space in the boot incorporated in the tail. The lid of this supports the spare wheel, which is provided with a metal cover, and on the

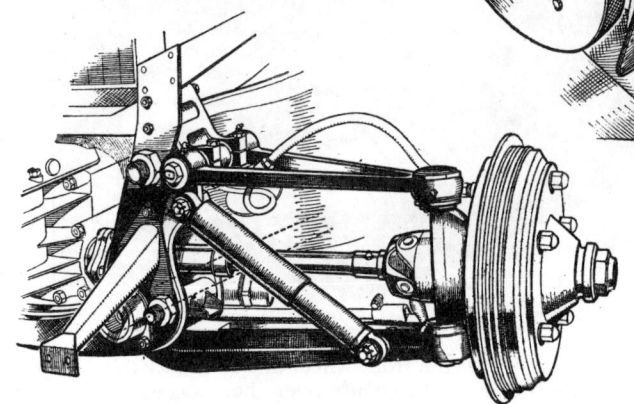

(Above) **New, the combined air cleaner and silencer, here shown partially cut away.**

(Above) **Improved tubular-type hydraulic shock absorbers are now fitted. Front wheel drive and independent torsion-bar suspension are features.**

(Left) **A lighter action of the gear lever is obtained by connecting the selector locking device to the clutch pedal.**

One of the roomiest cars of its h.p. rating is the Citroen Big Fifteen seven-seater.

Twelve and Light Fifteen saloons it may be used in the partly opened position since a hinged flap continues the flat floor of the boot, and the lid is rigidly supported by a stay at each side. On the Popular saloons a single-cable stay prevents the lid being opened too far.

The de luxe Twelve and Light Fifteen saloons have leather upholstery, the rear seat accommodating three in comfort, and the separate front seats having tubular frames and being independently adjustable. In the case of the Big Fifteen saloon the single full-width front seat is capable of accommodating three persons, while the seven-seater saloon has separate adjustable front seats, two folding seats, and a main seat to take three. The upholstery is in leather in both these larger cars.

Excellent visibility is given in all models owing to the wide screen and large windows, and also to the low build of the bonnet and radiator. The instruments are grouped in two circular dials in the centre of the polished wood facia board, which also provides a useful cubby hole in front of the passenger. Sliding Weathershields sunshine roofs are also standardised. Indeed, the equipment is very complete and includes driving mirror, twin screen wiper; ashtrays, door pockets, interior light, rear blind, scuttle ventilator, ventilating shutters in the bonnet sides, and built-in Trafficators. A wide range of colour schemes is also available.

A 4-light body is fitted on the Big Fifteen six-seater.

The Autocar Road Tests

No. 1,223.—CITROEN LIGHT FIFTEEN SALOON

ALWAYS an unusual design of car, the front-wheel drive Citroen in its latest form shows refinements over the model last tried. Main features, it will be remembered, additional to the transmission system, are the use of torsion bar suspension at front and rear—independent for the front wheels—and chassisless construction, the steel body forming the principal frame unit.

Since this Light Fifteen model has the same wheelbase and overall dimensions as the Citroen Twelve and weighs little more than 21 cwt. ready for the road, with an engine of 2-litre capacity the power-weight ratio is favourable. The benefits of this are demonstrated in the car's lively acceleration. It picks up very well on top gear, which is an important attribute in the eyes of the majority of drivers.

All-out maximum speed is not strikingly high in relation to the engine size, but top gear is by no means a low ratio, and, therefore, this car has a particularly easy way of getting along fast. The engine is comparatively big among four-cylinder units, but it is flexibly mounted, and is little noticed in the range between about 30 and 60 m.p.h.

When pulling away from low speeds, more especially if left on top gear, the engine can be felt to some extent, and the whole car seems to be quieter at medium and high speeds than in the low-speed range in town. In such conditions it is smooth and quiet enough, but one is conscious of the engine, and at times there is a slight jerkiness as the throttle is opened to accelerate.

But this car is singularly good in the way it swings along on the open road. It climbs the ordinary kind of hill without being pulled down appreciably in speed, and accelerates briskly after being seriously checked. Bottom gear is not likely to be wanted often.

Not the least pleasing point is the remarkable impression of tautness and solidity that the occupants receive. An entirely steady front-end is observed. The Citroen can be taken round bends as fast and steadily as a good sports type of car, and it sits down on the road as firmly and safely as could be wished.

Yet the torsion-bar suspension displays an outstanding capacity to absorb shocks. Scarcely ever, even over the worst type of third-class road surface, is appreciable movement transmitted to the passengers, and no pitching is experienced.

There is little to show a driver that he is not handling a car with orthodox transmission to the rear wheels. The steering is firm and accurate, being on the high-geared side, needing slightly less than 2½ turns of the wheel from full lock to full lock. It is light at ordinary speeds, tending towards slight heaviness when turning sharp corners at low speeds, whilst sometimes there is a mild suggestion of steering against the drive of the front wheels. Good caster action is present in the steering, and it is virtually unaffected by road-surface variations, no more than a slight rocking of the wheel occurring.

The ability of this front-drive machine to restart on steep gradients was shown by tests on the usual 1 in 4

DATA FOR THE DRIVER
9-9-38.

CITROEN LIGHT FIFTEEN SALOON.

PRICE, with four-door four-light saloon body, £248. Tax, £11 5s.
RATING: 15.08 h.p., four cylinders, o.h.v., 78 × 100 mm., 1,911 c.c.
WEIGHT, without passengers, 21 cwt. 0 qr. 17 lb. LB. PER C.C.: 1.24.
TYRE SIZE: 150 × 40 on bolt-on spoked steel wheels.
LIGHTING SET: 12-volt. Automatic voltage control.
TANK CAPACITY: 10 gallons; approx. normal fuel consumption, 25–27 m.p.g.
TURNING CIRCLE: (L. and R.): 41ft. GROUND CLEARANCE: 7in.

ACCELERATION				SPEED.	
Overall gear ratios.	From steady m.p.h. of				m.p.h.
	10 to 30	20 to 40	30 to 50	Mean maximum timed speed over ¼ mile	65.34
4.30 to 1	11.3 sec.	11.2 sec.	14.0 sec.	Best timed speed over ¼ mile	67.16
7.30 to 1	6.2 sec.	7.9 sec.	—	Speeds attainable on indirect gears (normal and maximum):—	
13.10 to 1	—	—	—*	1st	15—22
From rest to 30 m.p.h. through gears			7.5 sec.	2nd	38—42
To 50 m.p.h. through gears			20.6 sec.		
25 yards of 1 in 5 gradient from rest			—*	Speed from rest up 1 in 5 Test Hill	—*

Brooklands Test Hill not available.

BRAKE TEST: Mean stopping distance from 30 m.p.h. (dry concrete), 32ft.
WEATHER: Dry, cool; wind light, N.E. Barometer: 29.90in.

Performance figures for acceleration and maximum speed are the means of several runs in opposite directions, with two up.

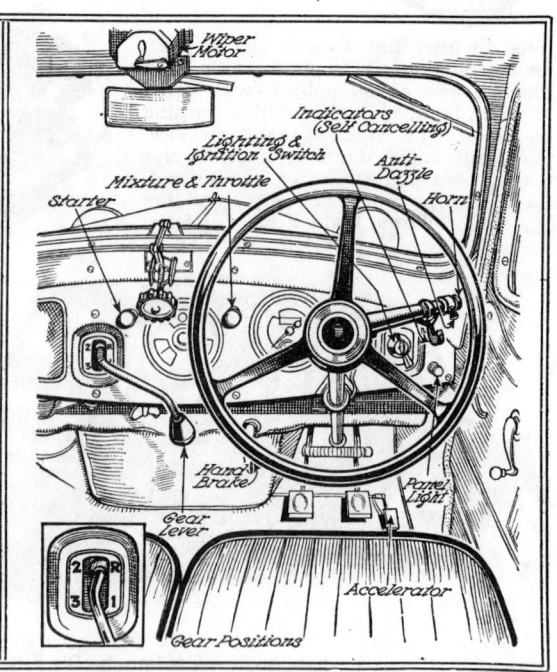

concrete-surfaced section, and, still more convincingly, on the appreciable gradient of a by-way hill used in trials, with a semi-freak rutted surface of damp chalk, which is shown in the heading illustration. This the car climbed without difficulty, not "bottoming" or scraping anywhere, and subsequently it was able to restart at two different points on the hill.

On the overrun on second gear a slight chatter is heard, but this is not now present on top gear, and the gears themselves are quiet. Tyre scream when cornering has been notably reduced, largely due to the use of a new and distinctive-looking type of steel wheel which gives a wide rim base for the tyre. It is only on smooth surfaces or with violent cornering methods that the tyres now produce any noise.

Hydraulically operated brakes give capital power, bringing the car down smoothly from the higher speeds with moderate pedal pressure, and being capable also of pulling it up decisively in a straight line under emergency conditions at town speeds. The hand-brake lever has a pull-and-twist action, moving horizontally. It is not specially convenient to reach, but holds securely on hills.

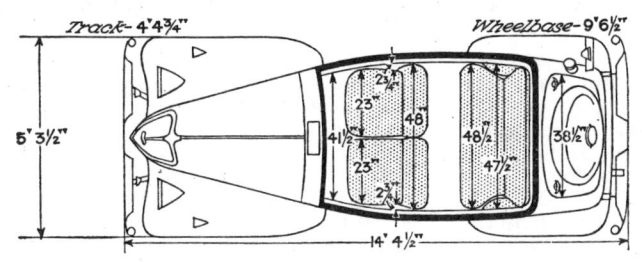

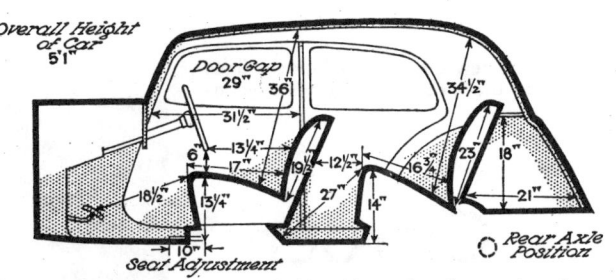

Seating dimensions are measured with cushions and squabs uncompressed.

As always on this model, the gear change is unusual, the lever projecting from the instrument board and having an up-and-down movement. The actual changing is convenient enough, and the synchromesh allows quiet and certain movements between top and second and upwards from first to second in the three-speed box, but the lever is a little far over to the left. The pedals are mounted horizontally, and have a downward action in contrast to the normal forward movement. The clutch pedal needs to be fully depressed to give quiet engagement of gear when starting.

So much for certain differences which the car exhibits from the conventional. Undeniably it gains by its design in a number of features—the stability and comfort already mentioned, for instance, in power-weight ratio, and notably in bodywork convenience.

For the size of wheelbase the four-door saloon is distinctly roomy in regard to head space, elbow width and leg room, and there is the considerable asset of entirely flat floors. In front no controls get in the way, and it is convenient to use either door. The floors are below the actual levels of the door openings, there being sills, as it were. The front-seat squabs give an upright position, tending to cause a slight leaning forward, but as a whole the seats are comfortable. At the back there is not a central arm-rest nor are upholstered elbow-rests provided.

General appearance is decidedly smart and clean-cut, and the finish both inside and outside is good, showing marked improvement over the standard of a year or two ago. Woodwork is used for the instrument board, there is pleasing leather upholstery, and the kind of detail fittings that British motorists expect is present, including an easily operated sliding roof and a rear-window blind control convenient to the driver. Citroen cars for this market are assembled and finished in England.

The instruments are neatly arranged, are clear to read and well lit at night. The speedometer was found to be considerably fast—by 3.7 m.p.h. at 30, 5.9 at 50, and showed a highest reading of 77 approximately, the needle floating.

Mirror, Head Lamps, Horn

A fair view is given by the mirror, the head-lamp beam is excellent, and the horn note effective as well as pleasant. Although not out of sight, the top-mounted windscreen wiper blades are not obtrusive; the windscreen can be opened sufficiently for ventilation. There is no special provision to this end in the arrangement of the door windows.

In the back of the body is a decidedly useful size of luggage compartment with an exterior lid which can be locked; the spare wheel is carried on the outside of this lid, but does not make it heavy to move. The bonnet opens easily and is of normal pattern; the oil filler is big and well placed, and the dipstick quite convenient to reach. The radiator filler is on the opposite side of the engine.

A Solex self-starting carburettor gives instant starting from cold, and the engine does not want much use of the mixture enricher before it settles down to work.

There is an excellent view to the near side and immediately over the bonnet; part of the near-side wing is just visible. The windscreen is not of exceptional width or depth, but its pillars are not specially thick. The steering wheel comes a trifle high.

The new car, modelled on the lines of the existing range, is a good-looker.

Citroën Introduce a Six-cylinder F.W.D. for French Market

AFTER a lapse of many years Citroëns have returned to the multi-cylinder engine, and a new 15 h.p. Citroën "six" will be one of the high spots at this year's Paris Salon. It is intended for the French market only at present, and will not be available here, nor will it be shown at Earls Court.

There is nothing sensational in the design of this model, which simply carries on tradition, but it is a first-class job and with its longer wheelbase and low, sporting lines has an attractive appearance.

The 1939 Citroën Six is in the 3-litre class, the bore and stroke being 78 mm. and 100 mm. respectively. The power unit is similar to that of existing Citroën models, but with two additional cylinders. Push-road operated o.h. valves and detachable cylinder liners are retained; the Citroën synchromesh three-speed gearbox, with a modification of the selectors that makes gear changing considerably smoother, is used.

So far as the chassis is concerned, torsion bars are retained, but these have been made more flexible than on former models, whilst cast-iron brake drums are employed with the object of preventing ovalization. A novelty in suspension arrangements is the position of the forward hydraulic shock absorbers, which now function vertically.

Down-draught Induction

The down-draught carburetter is fitted with an air filter and silencer, and the petrol filter has been taken to the rear and mounted on the fuel-tank outlet.

The standard saloon accommodates five people comfortably; even six can be taken without too much of a squeeze unless they are all of the 14-stone variety, and the new type of facia board has very few instruments on it. This simplification of the panel is becoming rather general in France and it may be that constructors have been shocked by the sight of certain modern aeroplane instrument boards when viewed from the cockpit.

One little point worthy of note is the arrangement of a floating oil-level indicator on the engine, which avoids the dirty and old-fashioned custom of wiping steel rods with a rag.

The standard saloon has a catalogue speed guarantee of 80 m.p.h. on the level, whilst fuel consumption when maintaining a high average, such as 50 m.p.h. on long runs, is said to be 15 litres per 100 kiloms. A fine range of standard coachwork is offered with this new model and a feature of the saloons is the arrangement of a central pull-out arm-rest for the rear seats.

Keeping Warm

A.C.-Sphinx Manufacturing a New Car Heater

CAR interior heaters of a very efficient type have been developed during the past two or three years and a new one, introduced to the British market, is the A.C., a product of the A.C.-Sphinx Sparking Plug Co., Ltd., and manufactured in their factory at Dunstable. It is of the small type of radiator mounted under the dashboard or in the rear compartment, utilizing a supply of hot water, by-passed from the radiator. The face of the heater is rotatable and is louvred, so that, in addition to warming the interior of the car, heat can be directed around the driver's or passengers' feet. Circulation is assisted by the incorporation of a small electric fan, there being models suitable for six or 12-volt installations.

The A.C. car heater is adaptable to a wide range of cars and can be fitted to any model having a pump or impeller in the water circulating system.

The provision of a supply of warm air inside the car not only adds to the comfort of motoring, but also prevents the formation of steam on the inside of the screen. An addition can be supplied to the heater by means of which warm air can be directed through a pipe to a nozzle which delivers a flow directly over the windscreen. This prevents formation of frost on the exterior of the glass, as well as condensation on the inside. The advantage of equipment of this type is that it involves no maintenance and costs nothing to run.

The heater is sold at £3 10s., the additional defroster selling at 10s. 6d. The dimensions of the radiator unit are 8¼ ins. by 6½ ins. by 6¼ ins.

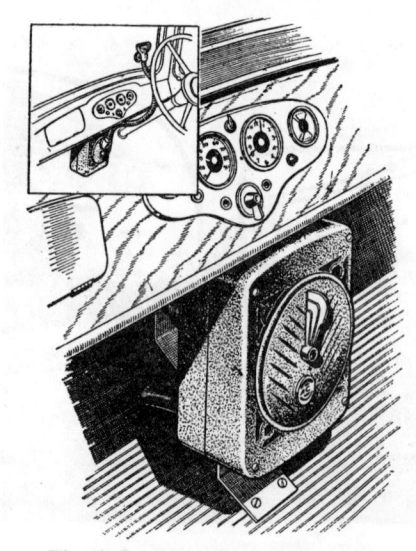

The A.C. car interior heater with the screen defroster shown inset.

GREAT DRIVE ON ALL FRONTS

The greatest drive of all is located at the front of every CITROËN. Great because it kills the motorists' bogey... skidding! Great because it makes high-speed cornering safe! And CITROËN front drive is allied to a lively o.h.v. engine that saves you money... because the replaceable cylinder barrels eliminate rebores, reduce oil consumption and improve cooling. There is a one-piece welded steel body to provide the safest and strongest of all designs, with torsion bar springing and independent wheel suspension combining to give the most luxurious riding of our time. Experience it yourself by asking your dealer for a trial run.

★ *In 1934 Citroën introduced* INTEGRAL ALL-STEEL BODY AND CHASSIS. TORSION BAR SPRINGING. FRONT-WHEEL DRIVE. INDEPENDENT FRONT WHEEL SUSPENSION. REPLACEABLE CYLINDER BARRELS. FACIA-BOARD GEAR-CHANGE. FLOOR DESIGN FREE OF CONTROLS, FOOTWELLS AND TUNNELS.

Prices and Models

"Twelve" Saloon	£238
Popular Saloon	£198
Roadster	£265
"Light Fifteen" Saloon	£248
Popular Saloon	£208
Roadster	£275
"Big Fifteen" Saloon	£278
7-seater Saloon	£298

CITROËN

CITROËN CARS · LTD · SLOUGH · BUCKS

The Motor Road Test

The Citroën Light Fifteen
FOUR-LIGHT SALOON

Safety and Stability are Outstanding Characteristics of a Car of Unconventional Design, Good Performance and Moderate Price

SELDOM do we drive cars which give such an outstanding sense of safety and security as that which is enjoyed when handling the Citroën. It is an unconventional car with what one might almost describe as unconventional stability. Drive hard over any sort of surface in the wet, even the despised wood blocks, or any other known slippery surface which you may encounter in the course of your everyday motoring, and it will be found that the Citroën goes just where it is steered—truly as if on rails. Corner a bit faster than usual, and round the car goes, with all the certainty one can desire. No front-wheel dither and no tail slide—stability such as is available with very few cars indeed. This fact in itself is helping to build up a very good reputation for the Citroën marque. Add to this a good performance, sureness of braking—of which we will write more anon—and excellent visibility for all occupants of the car, and you get the makings of a sound product.

Individuality of Design

The Citroën is quite different from all other cars in design, combining front-wheel drive with independent front-end transmission and torsion-bar suspension at the back, whilst the body and chassis are of integral construction, giving a very rigid assembly.

The result of this planning is to produce a low-built car (the overall height is a fraction under 5 ft.) without sacrificing headroom or passenger comfort. It further goes to prove how well the old bogy of noise from front-end transmission has been successfully overcome, for the car will cruise pleasantly with no more noise than one gets from cars of similar horse-power rating but with ordinary conventional transmission. Indeed, it is even quieter than some in this respect.

The only thing which makes one realize that the car is of the front-wheel drive type is the fact that the steering at low road speeds is a little heavier than one is accustomed to enjoy with the ordinary layout. There is, however, an absolute certainty of control, and above 12-15 m.p.h. steering control demands no great effort.

It has also a very marked self-centring action so that once the turn has been made, the front wheels fly straight back to the dead-ahead position and no reaction is felt in the

---

"The Motor" Data Panel (Citroen Light Fifteen)

Price, £248; 20 m.p.g.; tax, £11 5s.; weight (unladen), 20½ cwt.; turning circle, 40 ft. (2¼ turns of the steering wheel).

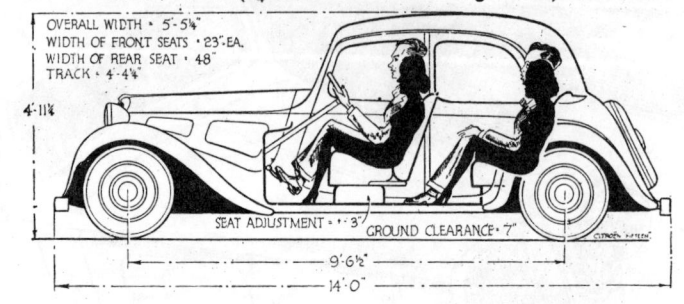

OVERALL WIDTH · 5′-5¼″
WIDTH OF FRONT SEATS · 23″ EA.
WIDTH OF REAR SEAT · 48″
TRACK · 4′-4¾″
4′-11¾″
SEAT ADJUSTMENT · 3″
GROUND CLEARANCE · 7″
9′-6½″
14′-0″

ENGINE
- No. of cyls. .. 4
- Bore and stroke .. 78 × 100 mm.
- Capacity .. 1,911 c.c.
- Valves .. O.h.v. push-rod
- Rating .. 15.08 h.p.
- B.H.P. .. 42 at 3,200 r.p.m.

CHASSIS
- Frame .. Integral construction
- Springs .. I.F.S. and torsion bar rear
- Brakes .. Lockheed hydraulic
- Tyres .. 165 × 400 super low pressure
- Tank .. 9 galls.
- Glass .. Lancegaye all round

PERFORMANCE

	Top secs.	2nd secs.
10-30	12.0	6.5
20-40	12.5	8.0
30-50	13.0	—
40-60	18.0	—
Max. m.p.h.	70	46

- 0-30 m.p.h. .. .. .. 7.0 secs.
- 0-50 m.p.h. .. .. .. 18.6 secs.
- 0-60 m.p.h. .. .. .. 28.0 secs.
- Standing ¼-mile .. .. 23.3 secs.

GEARS / HILLS
- Top .. 4.3 Max. grdnt. 1 in 12.0
- 2nd .. 7.3 Max. grdnt. 1 in 6.5
- 1st .. 13.1 Max. grdnt. 1 in 3.6

Engine speed, 2,500 r.p.m. at 50 m.p.h.
PULL, Tapley Q figure, 185.

BRAKES
30 m.p.h. to stop	lb. on pedal
120 ft.	38
60 ft.	70
Best 35 ft. (85%) ..	115

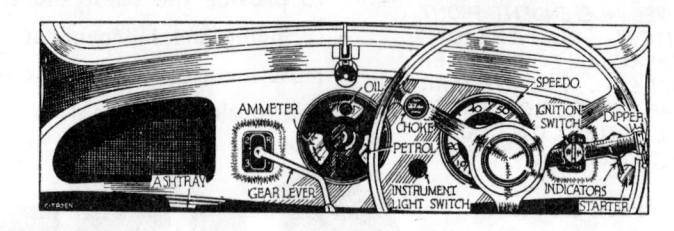

SEATING.—Black figure portrays woman 5 ft. 5 ins. high, 26 ins. from hips. White figure shows 6-ft. man, 30 ins. from hips. Scale of drawing 1/30 actual size.
HILL-CLIMBING.—Maximum gradients for each gear are shown. Where 1 in 6.5 is recorded the car will climb Edge, South Harting, Kirkstone, and Rest and Be Thankful Hills.
BRAKES.—Scale gives distance in feet from 30 m.p.h. as determined by a Ferodo-Tapley meter. Pressures needed to stop in shortest distance, in 60 ft. (normal short stop) and in 120 ft. or "slow up" are also shown. Average figures are 50 lb. for 60 ft., and about double for shortest; 100 lb. is the maximum pressure for average woman. If the 60-ft. and shortest-stop pressures are close together (e.g., 60 ft., 50 lb.—shortest, 72 lb.), the brake tends to fierceness.

THE CITROËN LIGHT FIFTEEN

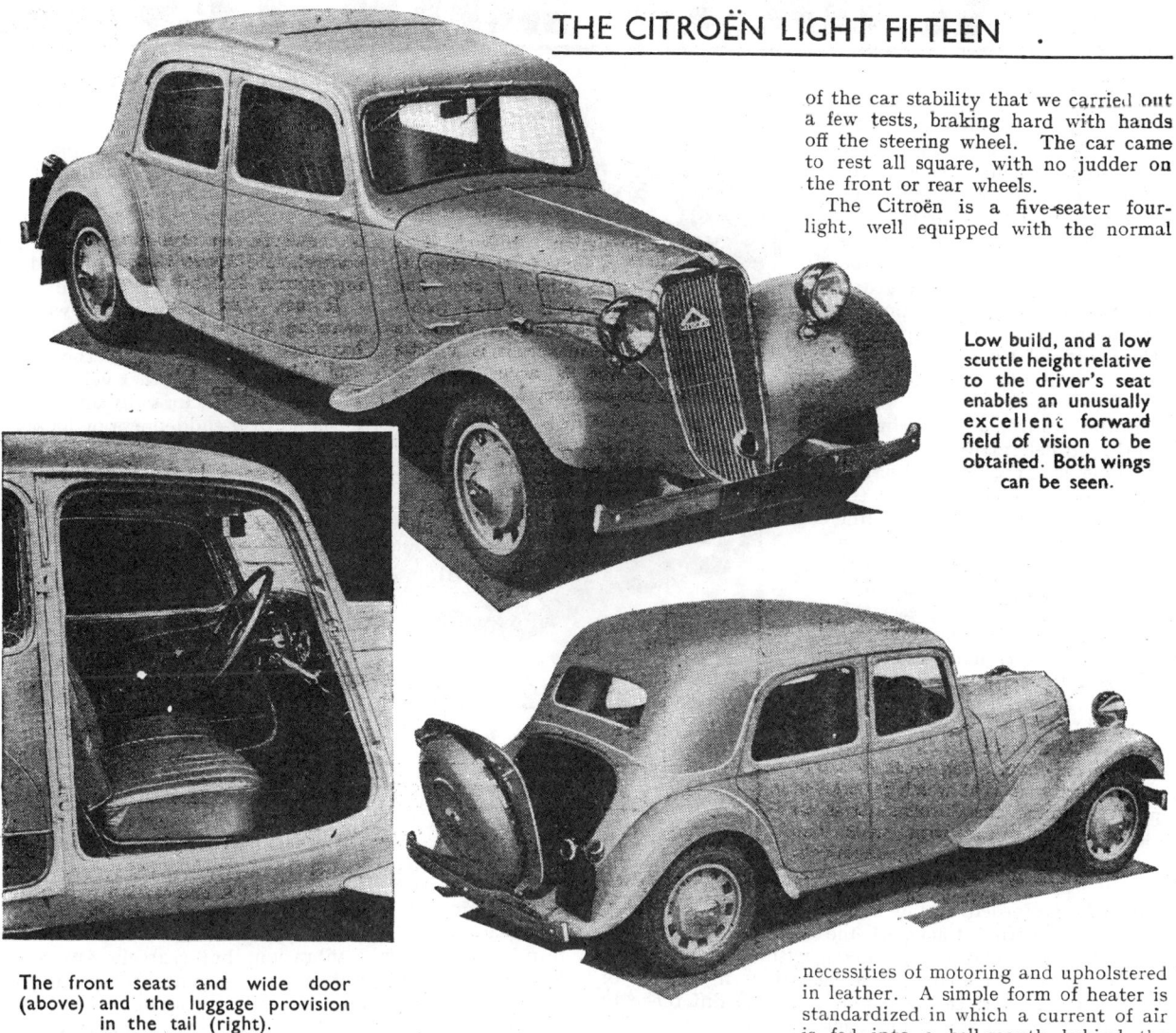

Low build, and a low scuttle height relative to the driver's seat enables an unusually excellent forward field of vision to be obtained. Both wings can be seen.

The front seats and wide door (above) and the luggage provision in the tail (right).

steering wheel no matter what nature of surface is being traversed.

Performance is pleasing. With the 15 h.p. engine in what is normally the 12 h.p. body and chassis, good acceleration and flexibility are available in spite of the fact that the top-gear ratio is 4.3 to 1, although for town work fairly frequent use of second is necessary. For fast journeys the car settles down to a healthy 60 m.p.h. cruising; above this the unit becomes a little more audible, but at no time does any harshness arise. This cruising gait, with the unusual stability on corners and good braking, enabled us to put up surprisingly good average speeds—higher than one might normally expect. The actual maximum speed proved to be 72 m.p.h., with a mean of 70 m.p.h. when tested over the quarter-mile at Brooklands.

All seats are somewhat of the chair type, giving fairly upright positions, but they are comfortable; this has the advantage of putting the occupants in such a plane that all-round visibility is enjoyed; the driver is able to see both wings—an almost unheard of thing these days—which helps the timorous when negotiating awkward turns or confined parking spaces. The seats are well within the wheelbase—a characteristic which adds considerably to riding comfort.

Liveliness becomes particularly evident above 20 m.p.h. Below this a change-down is really preferable, the actual change being quick, easy and quiet, so long as the clutch pedal is depressed to its full extent to ensure that the new selector locking gear is freed. The gear lever could, with advantage, be a little longer, whilst the pedals are of a rather unusual type, for instead of being of the normal design where one has to push them half forward with the toes, they lie in a more or less flat plane and involve lifting the whole foot to depress them. To this one becomes accustomed after a time, just as one does to the unusual position of the gear lever, which protrudes through the dashboard, and to the rock-solid, slightly heavy feel of the steering.

"Hands off" Braking.

Braking tests gave satisfactory stopping figures with light pedal pressures; in the ordinary course of motoring one could not help but be impressed by the certainty of retardation, whilst so confident did we feel of the car stability that we carried out a few tests, braking hard with hands off the steering wheel. The car came to rest all square, with no judder on the front or rear wheels.

The Citroën is a five-seater four-light, well equipped with the normal necessities of motoring and upholstered in leather. A simple form of heater is standardized in which a current of air is fed into a bell-mouth behind the radiator, heated as it passes down a pipe near the exhaust and delivered into the body of the car adjacent to the feet of the front passenger. This helps to maintain a reasonable temperature.

Mechanically the car is interesting, the torsion suspension, for example, being one of the most successful forms of independent springing yet devised. No attention whatever is needed in this system.

The engine has overhead valves with accessible tappet adjustment; the whole unit is flexibly mounted, is easy starting (a Solex self-starter carburetter being used) and quickly warms up to its job. An unusual feature is the use of separate detachable cylinder barrels cast in special wear-resisting and corrosion-proof material. It is claimed that cylinder life is practically indefinite, so that the need for a rebore is virtually abolished.

Body features include a flat floor, the front-wheel drive having eliminated the need for any transmission tunnel leading to the back.

Briefly, then, the Citroën Light Fifteen deserves high praise for its stability, performance, comfort and moderate price, which is £248 for the de luxe saloon.

The AUTOCAR ROAD TESTS

FRONT wheel drive, torsion bar suspension, monopiece or integral construction and consequent low weight render the Citroen one of the most advanced of existing cars, yet it is an interesting reflection that this design has been current on these same fundamental lines for some five years. Improvements and developments have naturally been incorporated in the course of time.

It is essentially an interesting kind of car, distinctive and pleasing in appearance, and proving in general road behaviour to be above the average of cars of approximately similar engine size. Because of the low weight, not much exceeding 21 cwt., fairly high gear ratios can be used, rendering it particularly easy-running nearly up to its maximum, yet acceleration is preserved. Indeed, it gives the immediate impression, which is maintained, of liveliness both in the way in which it gets going from rest—away from traffic lights, for instance—and in the sustained acceleration from around 30 m.p.h. up into the 50 to 60 m.p.h. range.

Then, again, it has the road holding and stability for cornering associated with a good sports car. The torsion bar suspension, using no leaf springs, of course, is extraordinary in the combination afforded of tautness and shock absorption. Really rough rutted tracks of the kind represented by dotted lines on a map can be taken without it being felt that the car is ill-treated. Wide variations in quality of the normal road surface make practically no difference to the comfort of riding in either back or front seats.

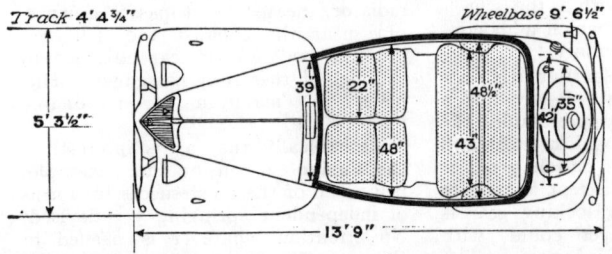

No. 1,284.—12.8 h.p. CITROEN DE LUXE SALOON

Even when on a given type of bend the speed is increased to what is regarded as reasonable by a comparative stranger to the car, it is still found to go round safely, virtually without "giving" an inch sideways unless the extreme is attempted. Although the tyres are of low-pressure type, tyre scream has now practically disappeared, apparently largely due to the use of a new type of wheel with a wide-base rim.

In addition, the Citroen feels entirely "in one piece," so that maximum confidence is experienced at all speeds. When the driver is in a hurry there is every possible encouragement to let it travel as quickly as a road permits, for no item connected with control or roadworthiness is lacking in response.

The steering is high-geared, needing less than 2½ turns of the wheel from lock to lock, and therefore "quick," giving the driver the feeling of being in accurate touch with the front wheels. It is light at low speed as well as normal driving rates, and is practically unaffected by the front wheels—that is, the driving wheels. The lock is good, so that the car is easily manœuvred.

Another important link in the sense of confidence imparted by this machine's behaviour is the braking, an hydraulically operated system. Particularly good emergency stopping results are obtained, and at higher speeds also the system shows up remarkably well. A rapid, smooth deceleration results when, for instance, the sudden emergence of another vehicle from a side road calls for swift action. There is no over-suddenness or locking of the wheels, and the pedal pressure does not have to be heavy.

Compared with earlier examples of the 12.8 h.p.-rated Citroen Twelve, an increase of engine power and consequent improvement of performance are evident in this latest model. Maximum speed is higher and acceleration improved. The impression is registered, as is often the case with French-designed cars, that a natural cruising rate of about 60 m.p.h. is intended for the purposes of Continental roads. Up to about that figure this Citroen is remarkably free from mechanical effort, whilst at 45 m.p.h. or so it is delightfully smooth and quiet, and the handling calls for singularly little concentration on the driver's part.

The four-cylinder overhead-valve engine pulls well on top gear. A smooth minimum on the level is 8 to 9 m.p.h.; some slight vibration, but scarcely any sign of flexible-mounting "flutter," is apparent when pulling away from low speed on top gear. When coming on to the overrun, more particularly on second gear, a slight vibratory transmission chatter is noticed.

Climbing Powers

Gradients of the 1 in 6 order can be taken on second, and it is often surprising how well the car continues to pull over an appreciable gradient on top gear, even though the approach speed is not high. The three-speed gear box has good synchromesh on second and top, first gear requiring double-declutching if it is to be engaged at appreciable road speed. The indirect gears are satisfactorily quiet.

As always on this car, due to the front wheel drive the gear lever is mounted towards the left of the instrument board in an open, visible gate, and has a vertical and across-the-gate movement. If the driver sits close up to

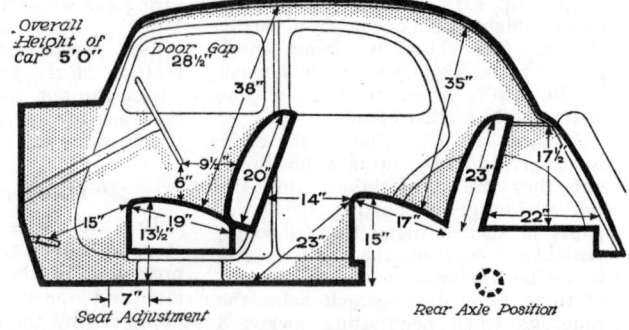

Seating dimensions are measured with cushions and squabs uncompressed.

48

DATA FOR THE DRIVER

2-6-39

12.8 H.P. CITROEN SALOON DE LUXE.

PRICE, with four-door four-light saloon de luxe body, £238. Tax £9 15s.
RATING: 12.8 h.p., four cylinders, o.h.v. 72 × 100 mm., 1,628 c.c.
WEIGHT, without passengers, 21 cwt. 1 qr. 14 lb. **LB. PER C.C.**: 1.47.
TYRE SIZE: 165 × 400 on bolt-on spoked pressed-steel wheels.
LIGHTING SET. 12-volt. Automatic voltage control.
TANK CAPACITY: 9 gallons; approx. normal fuel consumption, 24–32 m.p.g.
TURNING CIRCLE: (L. and R.): 40ft.   **GROUND CLEARANCE**: 7in.

ACCELERATION				SPEED.	
Overall gear ratios.	From steady m.p.h. of				m.p.h.
	10 to 30	20 to 40	30 to 50	Mean maximum timed speed over ¼ mile ...	65.34
4.90 to 1	10.9 sec.	11.2 sec.	13.6 sec.	Best timed speed over ¼ mile ...	67.67
8.30 to 1	6.3 sec.	7.9 sec.	—	Speeds attainable on indirect gears (normal and maximum):—	
14.80 to 1	—	—	—		
From rest to 30 m.p.h. through gears ... ... 7.5 sec.				1st	15—24
To 50 m.p.h. through gears 20.1 sec.				2nd	38—44
To 60 m.p.h. through gears 34.9 sec.					
25 yards of 1 in 5 gradient from rest ... ... —*				Speed from rest up 1 in 5 Test Hill	—*

* Brooklands Test Hill not fully available.

BRAKE TEST: Mean stopping distance from 30 m.p.h. (dry concrete), 31ft.
WEATHER: Dry, cool, dull; wind fresh, N. Barometer: 30.15in.

Performance figures for acceleration and maximum speed are the means of several runs in opposite directions, with two up.

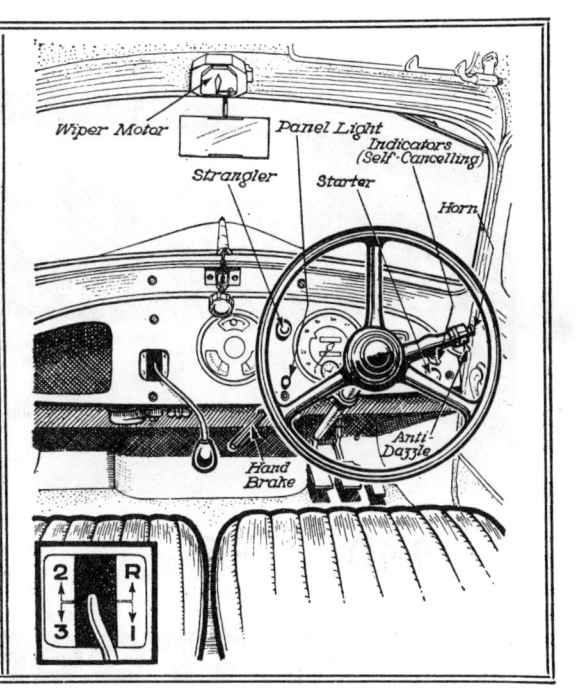

---

the wheel this lever is within convenient reach, and it is a good gear change of its kind.

The steering wheel is placed naturally, and outward vision is remarkably good, again helping to give confidence. The pedals are unusual, having a vertical downward movement, which at first feels strange. The decidedly important advantage of flat floors in both front and rear compartments is conferred by the front drive construction. The driver is free to use either door, and exceptional leg room and sense of freedom are provided for back seat passengers.

The seating is distinctly comfortable; the cushions of the separate front seats lead to a comfortable leg position, whilst the back rests provide an upright position which is altogether satisfactory. The upholstery is in good leather. It will be remembered that the Citroen is assembled and finished in England, and also incorporates a number of British components.

Consequently, the equipment is on lines that appeal to owners here, including the provision of a centrally locked sliding roof, a driver-operated rear window blind control, and instruments with British calibrations. There are two main dials, which are clear to read at night as well as in the daytime. Possible lack of oil pressure is indicated by a warning lamp, there being no gauge.

The speedometer read 1.9 m.p.h. fast at 30, 2.8 at 40, 3.6 at 50, and 4.4 at 60, a highest reading of 74-73 being shown during the timing of maximum speed over the quarter-mile.

There is a deep cubby hole for oddments, and a luggage compartment of useful capacity is provided. The mirror gives a useful view, though cut off rather close in certain circumstances. Although the windscreen wiper blades work from the top and do not go entirely out of sight when not in use, they are not obtrusive. Credit is due for the still strangely rare provision of an interior heating device. Warm air is led back from a scoop placed behind the radiator block, and the supply is controlled by a simple regulator inside the car.

An unusually effective and also pleasing note for the less expensive type of car is produced by twin horns, which, together with the traffic signals and anti-dazzle, are conveniently controlled by switches mounted on a steering column arm. The head lamp beam is very good for fast night driving. A 12-volt system is used.

The engine is neat and has an accessible oil filler, and such items as the dynamo, sparking plugs and electrical fuses are also placed conveniently. The engine starts at once from cold, requiring use of the mixture-enriching control for a little while before it pulls normally.

Good vision over the bonnet and of the wings—the near-side wing can just be seen from the driving seat—is an excellent feature. The windscreen is of fair width, tending towards shallowness in the other dimension; its pillars are not specially thick.

NEW—The six-cylinder 22.6 h.p. engine. In the foreground is the duct carrying warm air to the body (a feature of all models).

1940 CARS

Citroën introduce a new six-cylinder 22.6 h.p. model

Detail Improvements Also in the 12 and 15 h.p. Range; Development of the Moderately Priced Standard Models. A New Two-seater

WITH their all-steel chassisless construction, front-wheel drive and torsion bar springing, Citroëns may with some justification claim to have led the field in this combination of features. They have demonstrated, too, that front-wheel drive can be produced in large quantities as a quiet-running and efficient assembly, and have shown the outstanding stability which results from this form of traction combined with a suspension system in which the links are entirely free from driving or braking loads. Indeed the cornering power and stability of Citroën models is probably better than that of any other car of comparable price.

With all their mechanical advantages, they are listed at moderate prices. The range in the past has comprised the 12 h.p. and 15 h.p. four-cylinder models, but for the coming season these, which are being improved, will be augmented by an entirely new six-cylinder rated at 22.6 h.p.

This car will be available as a four-light, four-door saloon, seating six persons, selling at £328 or £370 in Grand Luxe form. The Six is the outcome of experience gained in the production of front-wheel-drive cars over a number of years. The transmission embodies detail refinements which make for quietness of running, the engine develops 75 b.h.p. and an attractive performance is available.

Further, the Citroën programme will include the addition of a Roadster model on the Big Fifteen (offering full three-seater accommodation on the front seat) and the introduction of equipment of a more elaborate nature; this will be known as Grand Luxe equipment. It will be offered at a small extra charge (£12) and has been designed to enhance considerably the appearance of the cars, also to add to driving comfort.

The Grand Luxe Equipment

Any of the normal de luxe models can be equipped and it will comprise the following items:—Lucas chromium-plated Bi-flex long range head lamps with 9⅝-in. fronts, and a pair of Lucas F.T.57 pass (or fog) lamps with 7-in. fronts mounted on the front bumpers

POPULAR — The four-door four-light saloon, a type of body which is available on all the 1940 chassis.

CUSHIONED—The front drive on the "Six" embodies big rubber joints which smooth out the transmission.

THE CITROEN 1940 PROGRAMME

"Twelve"
Models—Saloon, £198; de luxe with sunshine roof, £238, Grand luxe model, £250. Roadster, £268. Grand luxe, £280.
Four cyls., o.h.v., 12.8 h.p., three-speeds. Overall length, 14 ft., width 5 ft. 5¼ ins.

"Light Fifteen"
Models—Saloon, £208; de luxe with sunshine roof, £248, Grand luxe, £260. Roadster, £278, Grand luxe, £290.
Four cyls., o.h.v., 15 h.p., three-speeds. Overall length, 14 ft., width, 5 ft. 5¼ ins.

"Big Fifteen"
Models—Saloon de luxe, £278. Grand luxe, £290. Roadster, £298, Grand luxe, £310. Seven-seater, £298. Grand luxe, £310.
Four cyls., o.h.v., 15 h.p., three-speeds. Overall length, 14 ft. 9½ ins., width, 5 ft. 10½ ins.

The "Six"
Models—Saloon, £328; Grand luxe, £370.
Six cyls., o.h.v., 22.6 h.p., three-speeds. Overall length, 14 ft. 9½ ins.; width, 5 ft. 10½ ins.

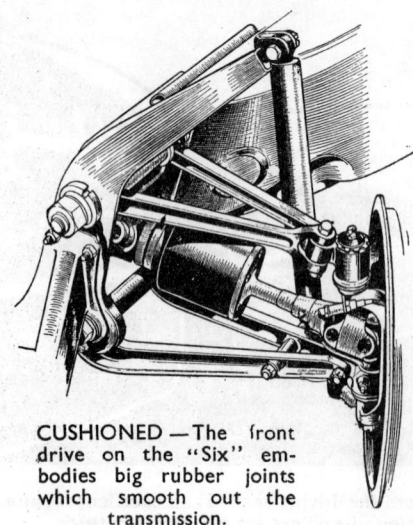

on each side of the radiator shell. Light switching arrangements are:—
(a) Two side and tail lamps, (b) two head and tail, (c) two side, one pass lamp and tail on dipper switch, and (d) a separate switch for the second pass light. There are also twin Lucas windtone horns, an Ashby spring steering wheel, chromium-plated hub covers, bumper overriders and an extra visor for the passenger. The six-cylinder Grand Luxe model will have even further refinements.

SEVEN SEATS—This Big Fifteen seven-seater at £298 is one of the lowest priced cars available offering such accommodation.

STANDARD — The black finish, cellulosed radiator and front end equipment of the lower-priced standard models.

The New Six

In overall dimensions the new car is similar to the existing Big Fifteen; it has the same wheelbase (10 ft. 1½ in.), with a very wide track, this being 4 ft. 10½ ins. at the front and 4 ft. 9¼ ins. at the rear.

The six-cylinder engine is of clean exterior design, and is built in unit with the double-plate clutch, an entirely new gearbox and differential for the front-wheel-drive mechanism.

The bore and stroke are 78 mm. and 100 mm. (capacity 2,866 c.c.). It has overhead valves operated by the push-rods, a compression ratio of 6.25 to 1, detachable cylinder liners (in common with other Citroën models)

QUIET—Sectional view (right) of the large air filter and silencer now fitted.

cast in special wear-resisting and corrosion-proof material and a four-bearing counterweighted crankshaft. Few cars have established such a reputation for prolonged cylinder life as the Citroën; 60,000 miles and more have been covered by users, with negligible wear in the cylinder walls—a high tribute to the corrosion-resisting nature of the materials used.

PLEASING—The walnut dashboard and neat instrument grouping of the de luxe types; the opening screen and large cubby. This actually is the Roadster model.

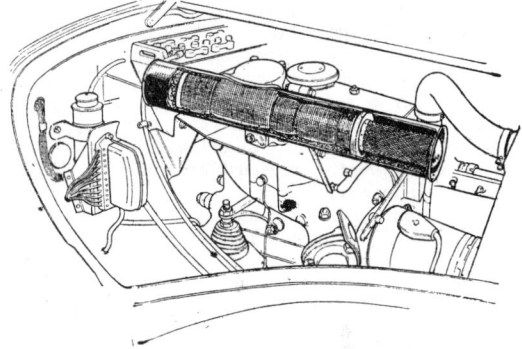

The whole unit has been designed for easy accessibility; all components can be removed readily and, indeed, anything which needs attention is located above the bonnet line.

Other features of the engine are the carburation by a Solex downdraught twin carburetter with accelerator pump and starter device; a large combined air filter and silencer which eliminates induction noise, and a mechanical fuel feed. Coil ignition operates on 12-volt equipment on the Grand Luxe type, the standard car having 6-volt. Exide batteries are standard equipment. The ignition advance and retard is fully automatic, but an additional hand control is provided for use when conditions or fuel demand a variation from the standard setting.

The compactness of the gearbox is a particularly commendable feature. It is only just over half the length of the unit employed on existing Citroëns, has a synchronized change on top and second, together with an automatic control for cancelling the gear selector locking devices. This is linked up with the clutch, and depression of the clutch pedal frees the selectors at the same time. The gearbox has also its own positively driven pump giving forced-feed lubrication to the whole of the box through suitable channels.

The transmission to the front wheels embodies a cushioned drive to each half-shaft, this being in the form of a rubber coupling, the elasticity of which takes up all transmission shocks. The result is improved sweetness and greater quietness in the front-wheel-drive mechanism.

The independent front wheel suspension system with torsion bars is based on that which has been used in the

CITROËN INTRODUCE A NEW SIX-CYLINDER MODEL. Contd.

past by Citroëns, but the swivels are connected by triangulated upper and lower links of heavy section; powerful hydraulic shock absorbers are mounted vertically. The front axle cradle is of very stiff construction with cross bracing and carries the integral front wing supports and shock-absorber mountings.

The rear suspension is by torsion bars with a trailing axle beam of cruciform section; vertical hydraulic shock absorbers are also fitted here. The brakes are Lockheed hydraulic with 12-in. drums and shoes operated by 1¼-in. twin cylinders on the front drive wheels, and a 1-in. single cylinder at the rear. The wheels are the latest Michelin pressed steel "Broadbase" type with 185 by 400 tyres.

Passenger Comfort

Consideration has been given to passenger comfort, the seat cushions and squabs being deeply sprung and upholstered in leather; equipment by Raco-Epeda is used. There is a central armrest front and rear.

The rear seat cushion is of considerable depth but it is so arranged that it can be moved backwards or forwards up to the base of the squab, so providing a range of adjustment to suit persons of different stature. Short persons, for example, enjoy a greater degree of comfort with what one would call a narrow seat cushion, whereas a tall person would not get the same degree of support under thighs, and for him, consequently, the cushion would be pulled forward.

The front seat is of the bench type on which three persons can be seated when occasion demands, although there is provided the central folding armrest for alternative use.

This model in Grand Luxe style has all the equipment described at the beginning of this article, with the addition of sunshine roof, a centrally controlled opening screen, safety glass all round, large luggage locker and built-in-car heater of simple and efficient design. Pile carpets are provided for the floor, there are two folding picnic tables of

TRANSMISSION. The compact gearbox (with its own built-in oil pump), front engine transverse mounting, gear selector lock (inset) and visible oil level indicator of the new "Six."

Rear end of the engine showing the accessible starter, exposed starter ring and vibration damper.

polished walnut which fold flush into the squab when not required, attractive walnut door fillets and dashboard panel, visors and Pyrene bumpers. The electric windscreen wiper provides twin blades and the motor is remotely mounted.

The Standard Model is designed to provide the same performance as the Grand Luxe, but it has simpler equipment, namely, steel facia with rectangular instrument board, glove pocket, no pass light, Rexine upholstery instead of leather, a fixed roof, and so on.

With its high standard of performance, quietness of running and smoothness of power unit, this model, although it will carry a £30 tax next year, will no doubt be popular.

ROADSTER. One of the most popular open cars is the Citroën Roadster, which offers good dickey seat accommodation. When folded the hood is fully concealed.

Improved Existing Models

The Light and Big Fifteen models now have a new engine. Improvements include a new cylinder head providing a higher compression ratio (6.2 instead of 5.9), an improved induction manifold working in conjunction with a downdraught carburetter; shorter and lighter valves with double springs; lighter connecting rods; new pistons; suction-controlled advance and retard; supplementary oil filter fitted to the oil pump; a new large air silencer and cleaner; heavier flywheel and new clutchplate. The effect of the last two has been to provide a considerable improvement in the quietness of transmission.

Improvements have been made in the Roadsters during the past season and these now represent a very popular type. One is offered also on the Big Fifteen now to sell at £298.

Additionally, more attention is being devoted to the low-priced standard models, which have the same quality and finish, performance, road-holding and safety as the de luxe versions, but carry simplified equipment. This comprises a fixed roof, leather cloth upholstery, 6-volt equipment, steel facia board, cellulosed radiator shell, twin filament head lamps, and other slight differences in detail, as compared with the de luxe. They do, however, offer excellent value for money.

THE NEW HIGH PERFORMANCE "SIX-CYLINDER"

CITROËN
Announce 1940 Programme

THE increasing popularity of the famous front-wheel-drive CITROËN cars has been a striking feature of the past few years. No greater contribution to the safety and comfort of the motorist has ever been presented and for 1940, these well-known models are being continued ... with many further advances. A Roadster model is added to the "Big Fifteen" range and an entirely new high-performance "SIX-CYLINDER" Saloon will be a striking feature of the programme. Every motorist or potential motorist should take the first opportunity of trying one of these CITROËN models, which will reveal the full pleasures of modern motoring.

FULL LIST OF MODELS AND PRICES

"TWELVE" Standard Saloon ... £198	★ "LIGHT FIFTEEN" Saloon de Luxe ... £248
★ "TWELVE" Saloon de Luxe ... £238	★ "LIGHT FIFTEEN" Roadster ... £278
★ "TWELVE" Roadster ... £268	★ "BIG FIFTEEN" Saloon de Luxe ... £278
"LIGHT FIFTEEN" Standard Saloon ... £208	★ "BIG FIFTEEN" Roadster ... £298
	★ "BIG FIFTEEN" 7-seater Saloon ... £298

and an
ENTIRELY NEW "SIX-CYLINDER" SALOON £370

Grand Luxe Equipment on models marked ★   £12 extra

CITROËN CARS LTD • SLOUGH • BUCKS

Citroen Six Now Here

New 2.8-litre Model, Announced Nearly a Year Ago in France, Offered in Slough 1940 Programme : Three Styles of Equipment on Four-cylinder Cars

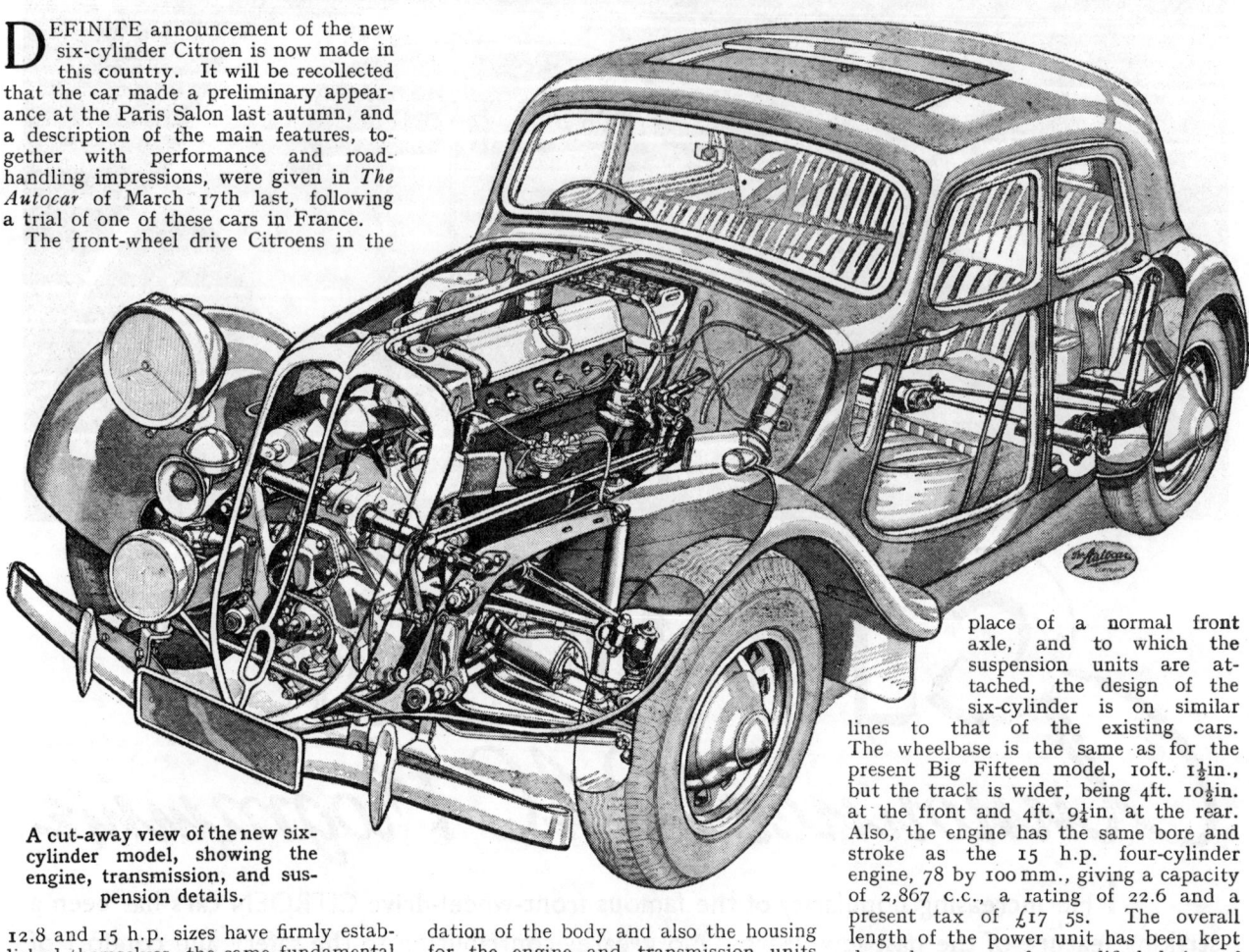

DEFINITE announcement of the new six-cylinder Citroen is now made in this country. It will be recollected that the car made a preliminary appearance at the Paris Salon last autumn, and a description of the main features, together with performance and road-handling impressions, were given in *The Autocar* of March 17th last, following a trial of one of these cars in France.

The front-wheel drive Citroens in the 12.8 and 15 h.p. sizes have firmly established themselves, the same fundamental design having been continued for five years. Improvements have naturally been incorporated from time to time, and, in particular, a distinct advance has been made in finish and equipment compared with the earlier examples. Assembly and finishing are carried out at Slough.

As to mechanical improvements, which are not necessarily seasonal, several items have been incorporated during the past year, taking their place in the production samples. The new six-cylinder should still further increase Citroen prestige, for it has a fine performance, is admirably equipped, especially in the more expensive form, and most comfortable as regards both seating and suspension.

The broad features of Citroen design, which is fundamentally the same for each of the models, may be outlined. First, of course, there is the front-wheel drive. Then the suspension is by means of torsion bars, not leaf springs, and it was really this car which first brought torsion bar springing into widespread notice on a production car. The front wheels are independently sprung. There is no separate chassis frame in the normal sense, steel pressings welded together forming a shell which is the foundation of the body and also the housing for the engine and transmission units and the mounting for the "axles." Low build is obtained, and a noteworthy advantage deriving from the front-wheel drive is the provision of flat floors in both compartments.

With certain differences necessitated by the heavier and larger unit, notably the cradle construction which takes the place of a normal front axle, and to which the suspension units are attached, the design of the six-cylinder is on similar lines to that of the existing cars. The wheelbase is the same as for the present Big Fifteen model, 10ft. 1½in., but the track is wider, being 4ft. 10½in. at the front and 4ft. 9¼in. at the rear. Also, the engine has the same bore and stroke as the 15 h.p. four-cylinder engine, 78 by 100 mm., giving a capacity of 2,867 c.c., a rating of 22.6 and a present tax of £17 5s. The overall length of the power unit has been kept down by means of a modified design of gear box, which is shorter from front to back. As opposed to rear-wheel drive practice, the clutch, gear box and differential and final drive casing are ahead of the engine itself, instead of behind it.

Before mentioning further details of the new design it is desirable to refer to the policy of Citroen Cars, Ltd., for the 1940 season in relation to the whole range of models. These cars are of two main types, the smaller being the Twelve, rated actually at 12.8 h.p., having a wheelbase of 9ft. 6½in. and a track of 4ft. 4¾in., the capacity of the four-cylinder overhead-valve engine being 1,628 c.c., tax £9 15s.; in this same size of car a 15 h.p. four-cylinder engine is available, the model thus provided being known as the Light Fifteen. A larger main type is the Big Fifteen of 1,911 c.c., with a normal wheelbase of 10ft. 1½in. and a track of 4ft. 8in., there also being a long-wheelbase chassis of 10ft. 9in., which carries a seven-seater saloon.

The following is the full list of models and prices for 1940 :—

Twelve.—Standard saloon, £198; De Luxe saloon, £238; roadster, £268. Grand Luxe: saloon, £250; roadster, £280.

Light Fifteen.—Standard saloon, £208; De Luxe saloon, £248; roadster, £278. Grand Luxe: saloon, £260; roadster, £290.

A cut-away view of the new six-cylinder model, showing the engine, transmission, and suspension details.

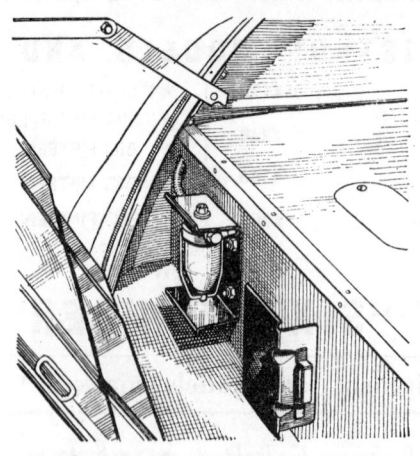

A "first-line" petrol filter on the Six, additional to the usual filtering.

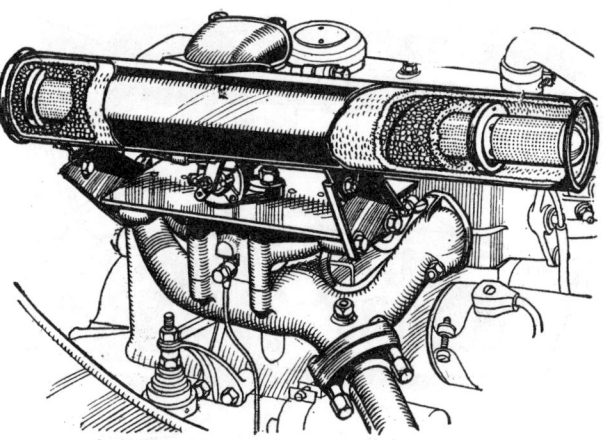

Citroen Six Now Here

(Left) New induction system and big air cleaner-silencer on the Light and Big Fifteens.

Big Fifteen.—De Luxe: saloon, £278; roadster, £298; seven-seater saloon, £298. Grand Luxe: saloon, £290; roadster, £310; seven-seater saloon, £310.
Six-cylinder.—Standard saloon, £328; Grand Luxe saloon, £370.

Particular attention is being paid to equipment, and, in addition to de luxe styles, equipment of a more elaborate nature will now be available, known as Grand Luxe. The de luxe models continue with detail body improvements, including a new polished walnut instrument board, improved trimmings to the doors and polished wood fillets, a chromium-plated radiator shell with the familiar double chevrons behind the grille, and new-type bumpers. An interior heater of simple but effective type is a commendable Citroen feature. Warm air from the radiator is led back through a controlled flap valve.

Grand Luxe equipment, now introduced, consists of bigger chromium-plated Lucas head lamps and a pair of pass lights mounted on the front bumper, also two wind-tone horns placed below the head lamps, a Brooklands spring-spoked steering wheel, chromium-plated hub covers to the special pressed-steel wheels with wide-base rims, over-riders for the bumpers, and a sun visor for the passenger in addition to that for the driver.

Also, Slough-produced standard models have been introduced possessing exactly the same mechanical specification as the more expensively equipped cars but simplified equipment, allowing a buyer of the Twelve and Light Fifteen, and now of the new six-cylinder, to obtain a Citroen at the lowest possible price.

The specification in this case includes a fixed roof, leather-cloth upholstery, six-volt electrical equipment, steel facia board, and cellulosed radiator shell, together with twin-filament head lamps.

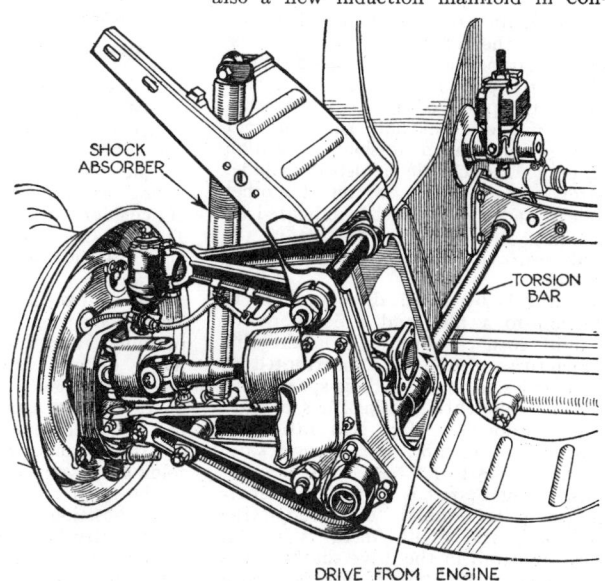

(Right) Front suspension details and flexible-joint drive shaft on the Six.

Thus there are three distinct ranges of the different models and, since many purchasers find the appeal of the car in its individuality, the introduction of the more elaborate style of equipment at extra cost is likely to be well received. The six-cylinder, however, is being offered in only two styles—standard and Grand Luxe.

The Light Fifteen and Big Fifteen engines have improvements which correspond to the features incorporated in the new six-cylinder engine, giving greater power and, therefore, an increased performance, without increasing the petrol consumption.

Among these are a new cylinder head used in conjunction with a higher compression ratio, 6.2 instead of 5.9 to 1, also a new induction manifold in conjunction with a downdraught carburettor, and a bigger air-cleaner and intake silencer. The valves are shorter and lighter than before and have double springs, connecting-rods have been lightened, there is a new type of piston, and so that the ignition timing shall still better conform with the conditions of load, a suction-operated advance and retard control mechanism is now used. Also, a point having important bearing upon the quietness of the drive and the smoothness of the engine at low speeds,

Produced at Slough, this Twelve is the lowest-priced of the range.

the use of a heavier flywheel, together with a new clutch member is noteworthy.

The roadster model, a smart convertible style with a bench-type three-seater front seat and roomy dickey seat, proved increasingly popular during the last season in Twelve and Light Fifteen form, and this style is now introduced also on the Big Fifteen.

This has more of the character of a drop-head coupé than of a normal two-seater with hood, forming a body which gives excellent protection and at the same time appeals to those who like fresh-air motoring when the weather

The Twelve and Light Fifteen are of similar appearance.

permits. It is available in de luxe and Grand Luxe forms. Improvements have been made in the head for ease of operation, and fixed windscreen pillars are now used, the glass being openable by a central winding control as in the saloon models.

Now again concerning the new six-cylinder; the engine is of business-like appearance and has a good finish, including a plated cover for the rocker gear of the push-rod-operated overhead-valves. The crankshaft runs in four bearings and has a torsional vibration damper. Detachable hardened cylinder barrels are employed, as on the other models, and the oil level is indicated by a pointer mounted on the side of the crankcase, avoiding the need for using a dipstick.

Cooling is by a belt-driven pump, in addition to a fan, and a Solex down-draught twin carburettor is used. The compactness of the unit contributed to by a redesign of the three-speed gear box and differential casings has already been mentioned. High gear ratios are used, top being 3.87 to 1, second 5.62, and first 13.25 to 1.

An unusual point is the use of a separate pump to circulate oil through the gear box, and again there is an interesting detail in the drive shafts to the front wheels. In these is incorporated a

New facia board and Grand Luxe style of interior, this range of equipment including a spring-spoked steering wheel.

semi-flexible joint of special pattern, utilising rubberised fabric bonded to the metal and acting as a form of cushion drive. The brakes are Lockheed hydraulic.

An excellent driving position is provided, and the flat doors are most convenient. A sliding roof is fitted on the Grand Luxe version. The finish is very well done. In the back of the one-piece front seat are picnic tables; a folding arm-rest is provided at the centre of the front seat, and comes in just the right position. The instrument board is attractively laid out, with the main grouped panel of dials, and, in Grand Luxe form, a supplementary panel comprising an oil-pressure gauge and water thermometer.

The luggage compartment is of useful size, and lined serviceably as well as attractively. Placed in the locker is an interesting detail, a petrol filter additional to the usual filter under the bonnet, designed to deal with any foreign matter before it has a chance of entering the main pipe-line.

ROAD IMPRESSIONS

Ample Power Allied to Smooth Running : Good Road-holding and Excellent Visibility

ON the road the instant impression given by the new six-cylinder model is one of solidity. This factor can, indeed, even be sensed from merely sitting in the driving seat. Although the car available for trial had not covered sufficient mileage to make it reasonable to use full throttle for any length of time, it is obvious that there is very good power indeed, with rapid acceleration up to that commonly usable range, 65-70 m.p.h., with plenty of power still in reserve.

Second gear suggests a maximum of about 60, and makes a capital high-speed accelerating ratio. There is a definite touch of quality in the running of the engine, the car by no means giving the suggestion that refinement has been sacrificed to performance.

Quite apart from the actual performance available, the handling is admirable, the steering being high-geared and quick without being heavy for low-speed turning, and there is just the response to the brake pedal that is so essential on a fast car. The steering wheel is excellently placed. The clutch action is rather lighter than has been the case on previous front-wheel-drive Citroen models, and also the gears engage with a pleasing smoothness, synchromesh being provided on top and second. The lever operates in an open gate placed vertically to the left of the centre of the instrument board. Cranked in shape, it is of a length that renders gear-changing a convenient movement in spite of the unusual position of the lever.

The engine is not perceptibly more noticeable when brought even fairly harshly on to the overrun than when pulling, and any sign of transmission chatter at that moment, as formerly noticeable in some of these cars, seems to have been removed. The torsion bar suspension is a striking combination of firmness for real stability of cornering in sports-car fashion, and of a remarkable suppleness that absorbs almost any form of shock without appreciable movement being transmitted to the car itself.

Vision is first-rate, the bonnet is not too high, and an average-height driver can see the near-side wing as well as the off-side one. The absence of any quiver or vibration in the front-end components visible from the driving seat—radiator, head lamps, wings and so forth—is particularly noticed.

There is every indication of this being one of the more remarkable and more promising cars of the 1940 season. A really long journey in it would be contemplated with pleasure in the knowledge, acquired even at short acquaintance, that it would "eat up" the miles. A full Road Test of this new model will be made at the earliest opportunity.

The six-cylinder Citroen is an imposing car with a good road performance.

The 1940 Citroën Cars

A New Six-Cylinder Model And A "Big Fifteen" Roadster; Special De Luxe Equipment Now Obtainable; Modifications To "Fifteen" Engine

FOR the 1940 season, Citroën have introduced an entirely new model. This is a four-light six-seater saloon with a 23 h.p. six-cylinder engine. The power unit has a capacity of 2,866 c.c., with a bore and stroke of 78 mm. by 100 mm. respectively. Short overhead valves are used, being operated by pushrods. Following usual Citroën practice, detachable cylinder barrels are used. There is a four-bearing crankshaft with torsional vibration damper. A Solex downdraught carburetter with acceleration pump and starter device is used. Ignition is by coil and distributer with suction-controlled advance and retard.

Front-Wheel Drive

The well-known front-wheel-drive system is employed. The gearbox and differential are in unit with the engine. There are three forward speeds, with synchromesh on top and second. Lubrication is carried out by a pump incorporated in the box. Ratios are: top, 3.87 to 1; second, 5.62 to 1; first, 13.25 to 1; reverse, 15.87 to 1. The gear lever is on the dash. The drive is transmitted to the wheels by universally jointed sliding cardan shafts, rubber couplings being incorporated in the shafts, a twin dry-plate clutch is employed. The front wheels are independently sprung with adjustable torsion bars and triangulated links with vertical hydraulic shock absorbers. Torsion bars and hydraulic shock absorbers are also used for the rear suspension.

"Clean" Interior

Steering is of the rack and pinion type with spiral gear teeth. Lockheed hydraulic brakes are used, the handbrake lever being mounted under the dash. As with all Citroens "clean" front and rear compartments, free from wells are thus obtained. Two models are listed, a standard fixed-head saloon at £328 and a "Grand Luxe" model with sliding head at £370. The "Grand Luxe" equipment includes special Lucas lighting system, Trippe pass lamps, a pair of "Windtone" horns, two visors, ash trays, thief-proof spare wheel cover, aluminium rear wing protectors, and other refinements. A 12-volt battery is used, whereas a 6 v. battery is used in the standard model.

The "Twelve" is retained as a fixed-head saloon, and with de luxe and "Grand Luxe" equipment: there is also the roadster, a two-seater with a roomy dickey. These models follow the usual Citroën practice of front-wheel drive, independent springing and integral body-chassis construction. The four-cylinder engine is rated at 12.8 h.p.

"Fifteen" Engine Modifications

Several modifications have been made to the engine of the "Light Fifteen" and "Big Fifteen" models. In each case a four-cylinder engine of 15.08 h.p. is employed. For 1940, however, the compression ratio has been increased from 5.9 to 6.2 by the fitting of a new cylinder head. The induction manifold has been re-designed and a downdraught carburetter fitted. New type pistons and lighter connecting rods and valves are used. The flywheel is heavier and the clutch plate is of an improved design; these have greatly assisted in quietening the transmission.

New Roadster

On the "Light Fifteen" are saloon and roadster models, while on the "Big Fifteen" are saloon, seven-seater models and a new roadster. The roadster follows the lines of the "Twelve" and "Light Fifteen" versions, but it has a bench type front seat, so that three persons can be carried. For the new season improvements have been made to the hoods of all roadster models, while the screen pillars are now fixed. The screen can be opened by a central control.

Prices

Prices of the cars in the Citroën range are :—

"Twelve"
Saloon	£198
De Luxe saloon	£238
"Grand Luxe" saloon	£250
Roadster de luxe	£268
Roadster "Grand Luxe"	£280

"Light Fifteen"
Saloon	£208
De luxe saloon	£248
"Grand Luxe" saloon	£260
Roadster de luxe	£278
Roadster "Grand Luxe"	£290

"Big Fifteen"
Saloon de luxe	£278
"Grand Luxe" saloon	£290
Roadster de luxe	£298
"Grand Luxe" Roadster	£310
Seven-seater de luxe	£298
"Grand Luxe" seven-seater	£310

"Six"
Saloon	£328
"Grand Luxe" saloon	£370

The new six-cylinder model which with 'Grand Luxe' equipment costs £370.

A new model—the "Big Fifteen" roadster with 'Grand Luxe' equipment at £310.

The 'Twelve' Standard saloon, priced at £198.

Anglo-French Cars for Export
Why Their Export is of Advantage to Great Britain

THIS issue of *The Motor* is only concerned with cars and accessories, the export of which is of particular advantage to this country by increasing its credit abroad.

In this category come two makes of cars of Anglo-French manufacture, the Renault and Citroën. Both originate in our Ally, France, which country is linked with Great Britain in its commerce and finance as well as in the prosecution of the war against Germany. In both, some 70 per cent. of certain models of the total value of the vehicles is of British manufacture, employing British labour and materials to that extent. Certain components are imported from France. Others are manufactured here and English coach finish adds to the British content.

In certain countries there is a preferential duty in favour of imports from Britain. Where a certain percentage is certified as of British origin such cars also enjoy a preferential tariff, and this applies to both Renault and Citroën models. We can, therefore, recommend the purchase of these cars by those anxious to assist the country.

The following are details of models exported from Great Britain:—

CITROEN

Models Available: Twelve, from £198; Light Fifteen, from £208. (Citroën Cars, Ltd., Trading Estate, Slough, Bucks.)

SOME years ago the Citroën Co. introduced a car which at that time was startlingly unconventional, but which embodied certain features which have now become accepted as first-class engineering practice. These include the use of an integral all-steel chassis and body, independent suspension by torsion bars and detachable cylinder barrels.

The first of these is now used by many companies, and the advantages of independent springing over really rough roads and under many other conditions are widely recognized.

The torsion-bar spring has been found to be the lightest type known, and has in practice proved 100 per cent. reliable. The advantages of detachable cylinder barrels are that the iron can be made very hard, giving virtual immunity from rebores, as the castings are very simple, the block, which also forms the crankcase, being made in an alloy devised specifically for this purpose.

The Citroen four-door saloon; this car, which has front-wheel drive and torsion-bar suspension, is proving itself under colonial conditions.

The 12 h.p. and 15 h.p. Citroën engines are four-cylinder overhead-valve types and, as is well known, drive through a three-speed transmission to the front wheels.

Front-wheel-drive cars have a steadily increasing number of adherents, and it is found that on greasy or sandy roads they have many advantages over the conventional type. In addition, they permit the use of a wide, low body, unobstructed by gearbox or propeller shaft.

The short-wheelbase chassis is similar whether powered by either 12 h.p. or Light Fifteen engines, and two body types are available, saloon and roadster, the latter a two-seater with folding head. The 15 h.p. engine is also available in a long-wheelbase chassis with either saloon or seven-seater saloon body.

Amongst detail points of interest on these cars are the use of Lockheed hydraulic brakes and a dashboard gear control, which, in addition to a scuttle-type handbrake, entirely clears the front compartment of any obstructions.

Gear changes work so that the gears are positively locked until full depression of the clutch pedal releases the lock and permits the ratio to be shifted. The rear axle is, of course, a simple beam, springing being by torsion bar, as in the case of the front member. Left-hand or right-hand drive is available, and de luxe models have 12-volt electrical equipment, sliding roofs, real leather upholstery and many other detail refinements.

Wide doors and liberal accommodation are features of the Citroens: the assembly, interior leather trim and finish, are all British work.

On active service – AT HOME AND OVERSEAS

No car compares with CITROËN for successful performance either at home or overseas. Rough roads make no impression on the immensely strong all-steel integral body and chassis ... a CITROËN design that is now acknowledged as the safest in the world. There are no road springs to break; CITROËN torsion bar suspension and independent wheel springing is infinitely superior to all other forms. The famous CITROËN front-wheel drive virtually eliminates skidding ... whatever the road conditions ... and gives perfect driving control. The brilliantly designed and lively o.h.v. engine is economical because the replaceable cylinder barrels do away with rebores, improve cooling and reduce oil consumption. And lastly, CITROËN Service is world-wide, and provides an organisation that is available in every corner of England and the globe. Wherever you are, there is a CITROËN agent handy ... glad to arrange a trial of the most luxurious riding of our time.

ONLY CITROËN offer ALL these features
INTEGRAL ALL-STEEL BODY AND CHASSIS. TORSION BAR SPRINGING. FRONT-WHEEL DRIVE. INDEPENDENT FRONT WHEEL SUSPENSION. REPLACEABLE CYLINDER BARRELS. FACIA-BOARD GEARCHANGE. FLOOR DESIGN FREE OF CONTROLS, FOOT-WELLS AND TUNNELS.

CITROËN
"AT HOME" on the roughest roads

CITROËN CARS LIMITED · SLOUGH · BUCKS

250,000 Miles in 369 Days

Lecot and his Citroën.

When Franoçis Lecot Did 12 Months Hard

By L. GRAHAM DAVIES

IN pre-war France we had two very famous exponents of long-distance driving, César Marchand and François Lecot. Marchand, whose exploits were the subject of an article in the issue of March 18, concentrated on track records of astonishing duration with teams of relay drivers. Lecot, on the other hand, went in for controlled performances on the road and invariably kept to the wheel himself the whole time.

Both men rendered extremely valuable service to automobile development, one by keeping cars running for very long periods at speeds which could only be maintained regularly on a track, and the other by covering even greater distances at normal touring speeds on ordinary roads, the cars in each case being subject to strict official observation and control.

Life-time in a Year

The technical value of these exploits lies in the fact that many years' normal use of a standard car are compressed into a relatively brief period. Examination of the car by research engineers after such an ordeal provides information which no conceivable amount of laboratory testing would give, for the car has been tested as a complete structure, not part by part, and under conditions unobtainable in any laboratory.

Marchand's team, for instance, drove "Petite Rosalie" at nearly 60 m.p.h. for 133 days, covering 185,353 miles on the smooth cement of Montlhéry track. Lecot's work, on the other hand, was done on the public highway and included the factors of traffic, mountain gradients and so forth.

Most of M. Lecot's enormously long drives were made with the Rosengart, a French copy of the Austin Seven, but for his last and most sensational one he used a Citroën saloon. With this, between July, 1935, and July, 1936, he managed to cover 250,000 miles under strict A.C.F. control.

Regarded merely as a feat of human endurance, the job looked impossible, especially as M. Lecot was then in his 59th year. Knowing something of the man and his previous exploits, however, I was convinced that he would stick it out if the car did.

At the conclusion of this astounding feat I had the opportunity of examining both car and driver in Paris. Lecot looked as fit and hearty as ever, but the car did surprise me. Although it needed a coat of paint, it appeared in excellent mechanical order after accomplishing the equivalent of about 20 years' ordinary use at 12,000-odd miles a year. In a short run I found the car quiet and still lively enough to behave well in Paris traffic.

In order to achieve this astonishing feat, he had to drive 19 hours out of every 24 and content himself with just 4½ hours' sleep. Carrying out such a programme for 365 days on end is not a simple matter in itself, but whilst we may be awed by the physical endurance of the driver, the performance of the car is certainly no less remarkable. Actually, he drove for 369 days. The vehicle used was a stock model 14 h.p. Citroën F.W.D. four-door saloon, and the technical organization of the attempt was carried out by M. Brisset, an engineer who had acted as M. Lecot's technical "manager" on many of his previous long-distance performances.

Car Outlasted the Observers

M. Lecot's headquarters were at Lyon, and he made alternate daily drives from Lyon to Paris and back and from Lyon to Monte Carlo and back. A commissaire of the A.C.F. travelled on the car to see that International Association rules were adhered to and that the special A.C.F. speed limit for such attempts of 65 k.p.h. (40 m.p.h.) was not exceeded. Eight A.C.F. officials took the duty in relays, but even so, they had quite enough of it. They were not all Lecots. Although the car did not wear out at all, Lecot certainly wore out the observers.

It is interesting to note that the fuel used was the ordinary French "poids lourd" spirit employed by industrial vehicle owners. This was a petrol alcohol mixture with a commercial alcohol content of 20 per cent. to 22 per cent. During the winter months, a little benzol was added to facilitate starting. Oil consumption proved very low indeed, lower even than that given in the Citroën catalogue for the model

CONTINUED ON PAGE 111

IN THEIR DAY . . .

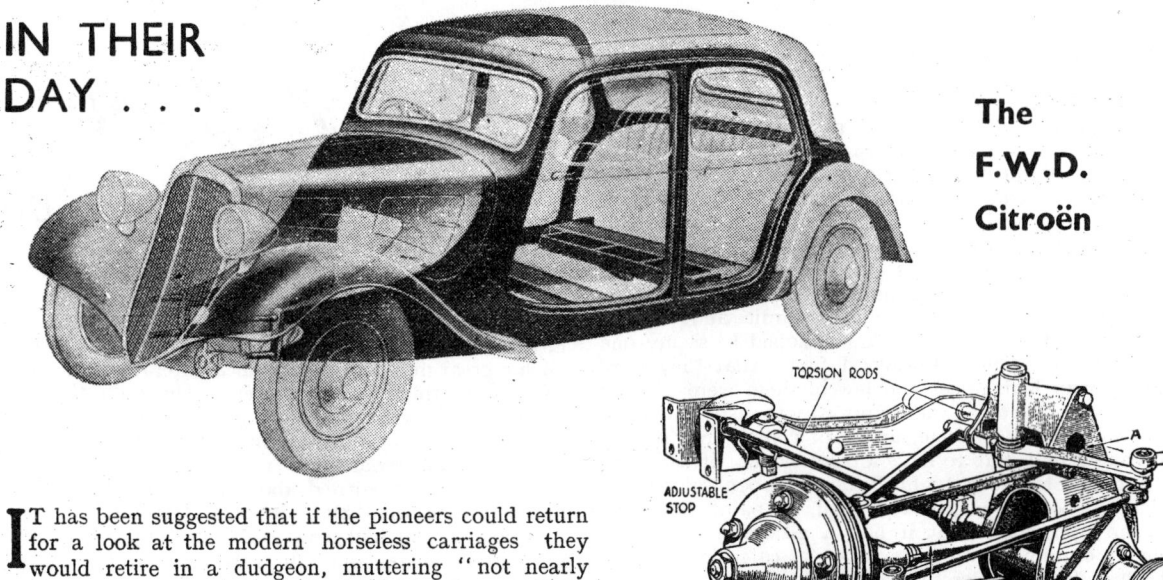

The F.W.D. Citroën

IT has been suggested that if the pioneers could return for a look at the modern horseless carriages they would retire in a dudgeon, muttering "not nearly horseless enough!" It is certainly true that until the last few years before the war popular motorcars were, to say the least, traditional in design, and no one can deny that engine development had got far ahead of the chassis and the conception of the car as a whole. By the beginning of the 1930s, however, people were realizing, especially those whose business took them down the long, straight French roads, up and down mountain passes and over rough by-ways in Europe, that something should be done to bring the chassis up to date.

The fortunes of the Citroën Company had been built upon the post-war "8 cv" and on the little "5 cv" car, described in No. 28 of this series (September 15); the larger cars which superseded it were a reasonable selling line, although not particularly noteworthy. The world slump hit France like everywhere else, and the company had its financial troubles. It certainly did not seem the moment to expect anything really exciting from that factory.

F.W.D. and Torsion Bars

Nonetheless, at the end of 1933, strange craft, heavily disguised, began to appear at Montlhéry, and for all the careful "security" which was observed, it was obvious that something very fresh was hidden behind the well-known swan trade mark. No amount of camouflage could conceal the low build of the car, however, and rumours began to get about. The new model was announced in May, 1934—so, on this, the 10th anniversary of one of the most popular Continental cars, it is appropriate to include it here.

Unlike most new models, this car was really new from end to end, and differed from its contemporaries in almost every possible way. Most striking departure for a best-selling car was front-wheel drive. Sweeping aside popular prejudice, Citroën secured the advantages of a forward-mounted engine, wide spring-track with consequent stability, extremely low frame, flat floor unobstructed by prop. shaft, a front compartment free of gear lever, and remarkable cornering power. But the innovations did not stop at F.W.D. The front suspension was, naturally, independent, but in place of the usual leaf springs, Citroën followed the late Parry Thomas and used torsion bars—longitudinal at the front, and transverse at the back.

The Super Modern Twelve and the Light Fifteen, with, respectively, 1,628 c.c. and 2-litre engines, were sheer unit construction. For example, body and chassis were in one piece, formed of large steel pressings welded to make a rigid box, in which the steel dashboard made a valuable cross-bracing. The body side-members extended forward of the dash, tapering down to form two arms. In this way a cradle or armchair was made for a rubber-mounted "power-egg" comprising the engine, three-speed synchromesh gearbox, bevel final-drive, and differential. At their tips the side-members each carried two prongs, which engaged with holes in sturdy, U-shaped cradles carrying the suspension wish-bones, torsion bars, frictional shock absorbers and steering connections.

The engine had push-rod overhead valves, an innovation for this factory, and had detachable steel wet liners. By placing the gearbox ahead of the final drive, the power unit could be mounted 12 ins. farther forward than was usual for F.W.D. cars, with consequently greater weight on the front wheels and better adhesion.

The new Citroëns at once made a name for smooth riding, high cross-country averages, and astonishing stability, especially on snow. The Frenchman loves controversy, and soon the country was split into little groups in bars, all arguing about "traction avant"; but the merits of the case for F.W.D. were soon proved by sales, which made the Twelve and the Light Fifteen two of the most popular cars in that country. In this country, too, it soon caught on, a saloon selling for £265. The English factory and assembly plant is at Slough.

250,000 Miles in a Year

By way of testing the car under private-owner, as opposed to track, conditions, a Frenchman, one François Lecot, set out to cover the greatest possible distance in a year. Based on Lyons, he would drive to Paris and back one day, and then to Monte Carlo and back the next, all the time under A.C.F. observation. To do this meant observing the arbitrary A.C.F. speed limit of 40 m.p.h. for such attempts and driving for 19 hours out of the 24. Although then in his 59th year, Lecot kept this programme up for 369 days and covered a distance of a quarter of a million miles, or a daily average of 680. Bearing in mind that much of the route is over mountains, a great proportion would be done in the dark, that snow is frequent on that run, and that ordinary traffic would be encountered all the time, it must be admitted that Lecot gave his Citroën as searching a test as any production car, front drive or otherwise, has ever had. It was still running satisfactorily after its year's hard labour.

With sundry small changes and refinements the cars continued in production up to the war.

Across A...

3,500 Miles in Seven Days i...

THE story begins last summer when I went out to California to work on a Navy film at Warner Bros. studios in Hollywood. As I needed some form of transport during my stay I looked upon arrival for an interesting but inexpensive sort of car. Friends had loaned me one of their extra cars, an old 1932 Dodge coupé, too old to be even reliable, much less interesting. Because of low taxes and garage costs nearly every family has one or two ancient and almost useless cars which help them to cope with the rationing problem, as each is entitled to an "A" card!

I knew that some three hundred front-wheel-drive Citroens had been sold in Southern California and, having wanted one of these cars for years, I hoped to secure one at a reasonable figure. However I found that they were selling for very high prices and few of them were in good condition. On the day I was on my way to inspect another kind of car I saw a strange, unkempt vehicle in a service station and I pulled up to have a better look. It had once been a handsome 1939 Light Fifteen Citroen saloon but now had no bonnet, no radiator shell, no paint and practically no upholstery—but it had a gleaming new Ford V8 engine crammed into the narrow space instead of the little 2-litre four-cylinder Citroen unit. It being no longer f.w.d., the gear box projected back into the driver's compartment, and the torque tube passed through a fairly high tunnel to a standard Ford rear end sprung by the usual transverse spring. All this was made possible by an incredible amount of cutting and welding, ingeniously executed.

The Work of Reconstruction

I asked the owner questions about the conversion and found that he was willing to sell for a reasonable figure. A cheque for 250 dollars changed hands, the owner's certificate was signed, and I drove off the proud owner of one of the most dilapidated cars in the Los Angeles area. I went to work evenings, week-ends, every spare moment during the next six weeks working on the reconstruction of "Barlow's Folly." Every shred of remaining upholstery and soundproofing was removed, the body was sandblasted and then repainted, re-upholstered and refitted. A new drive shaft was made up, minor repairs were carried out, a new Delage-like radiator grille made up and the Citroen made its triumphant reappearance, a handsome little car. It had been repainted grey and upholstered in two shades of green with pale green carpeting.

The Citroen seemed to be in fine shape; the rebuilt V8 engine had been carefully run-in on graphited oil, but with such a great number of mechanical alterations and improvisations anything *could* happen. The only thing that I could definitely suspect was overheating, as the natural operating temperature around Hollywood and Los Angeles was about 190 degrees, and I was somewhat worried about what would happen when we "hit" desert and mountain country. I had managed to pick up a few Citroen spares and now carried an extra torsion bar, some rubber suspension bushings, and a pair of slightly worn front wheel bearings.

My work completed the Navy Transportation Office gave me my petrol tickets and at three o'clock one Wednesday afternoon, the back seat loaded down with luggage and a spare tyre, my wife and I left Hollywood, heading north along the coast road towards San Francisco. As we rolled along the superb highway not more than fifty feet from the Pacific Ocean it seemed to us that we were off on a holiday rather than facing a stiff 3,500-mile drive which had to terminate in Washington, D.C., in exactly seven days.

Just before reaching Santa Barbara we drove off the main highway and down an

Eucalyptus trees lining a ranch road near Santa Barbara, California. The car is the Light Fifteen Citroen saloon, with Ford V8 engine, described in the accompanying article.

merica

8-Engined Citroen

By ROGER BARLOW

Moving sand dunes, like snow banks, along the coast road near Carmel, California.

Garages on the ground floor of San Francisco houses and apartments. By this means an effective effort has been made to deal with the problem of housing the city-dwellers' cars.

unbelievably beautiful ranch road lined with tremendous eucalyptus trees and behind, blue in the distance, the Santa Barbara Mountains We continued on to Santa Maria, which was hopelessly overcrowded, and no hotel or even auto-court accommodation was to be had. So, as it was still early, we set out again. Twenty minutes later, just as we topped a small hill and entered a village, the engine spluttered for a second and I realised that we were out of petrol. I wished I had put the gauge in working order before starting on such a long trip. The village was entirely dark and the three petrol stations were deserted, so I drove on, thinking that perhaps there was still enough fuel to take us four miles to the next town; but hardly a quarter of a mile further on the engine hesitated again—I saw an auto-court with the lights still on, pulled out the choke, pumped the accelerator and just managed to climb up the driveway into the courtyard as the last drop was sucked out of the Stromberg. Fortunately they still had a vacant cabin which looked warm and friendly to two travellers who had just had visions of spending a cold, damp night in a stalled car.

Early the next morning we were off again and in about 30 miles left the main highway, which now swung inland, and took the second-class road that followed the coast. This road, though well surfaced, was narrow and twisty and fifteen or twenty miles of hard driving along this stretch made it evident that the Citroen had suffered no loss whatever of its ability to get around curves rapidly without fuss and we were thoroughly satisfied with the handling of our latest acquisition. Now the road passed by the famous Hearst ranch and ten miles or so to the right could be seen the fabulous castle which Mr. Hearst had imported from Scotland and rebuilt stone by stone on the crest of a high ridge in the centre of his huge estate.

In a few more miles we reached the section of the coast road which made it famous—it is here the mountains of the coastal range extend to the sea and end in great cliffs. Here the highway continues near the sea but is forced to hug the face of the cliffs, winding from sea level up to 1,500ft. or more, then down to sea level again every few miles. Here, too, at one point the giant redwood trees of the Big Sur forest come down to the very edge of the sea. This section of the highway is most difficult to maintain and during the rainy season is often closed for days by severe landslides.

The rugged mountainous section dwindled rapidly and we soon reached an area where sand dunes, under the impetus of the strong prevailing wind off the Pacific, tended to drift slowly across the road—at times looking for all the world like huge snow banks. Now, too, began to appear the dramatic and spectacular looking Monterey pines, peculiar to the peninsula which gives them its name. A few miles farther on we passed through Carmel, a charming and popular artists' colony, and later reached the picturesque city of Monterey and had lunch while the oil was being changed in the Citroen. The rest of the drive into San Francisco was pleasant but uneventful and, though we reached the city after dark, we were fortunate enough to find a room at a good hotel.

Garages in San Francisco

The sun was just up and the city already stirring when we left the garage and threaded our way through unfamiliar streets in search of the route to the Golden Gate Bridge. Going through the residential section of the city we were most intrigued by the fact that every apartment house, every private dwelling, had garage space on the first floor or in the basement. We noticed that the street floor of every building in street after street was devoted to the garaging of private cars. No other city in America that I have visited has made such an effective effort to deal with the problem of housing the city-dwellers' cars.

Across America *Continued*

The view from the top of one of the ranges of the Blue Ridge Mountains, about 100 miles from Washington.

A superb four-lane highway led off through low hills in the direction of Sacramento and we made the most of this opportunity to put the miles behind us in the cool of the morning, for I knew that as we reached the Sacramento Valley during the heat of the day driving would cease to be a pleasure and become instead a real job of work. Then, too, we would find out how the Citroen would stand up to desert-like temperatures. We were now headed east, away from the cooling breezes off the Pacific that had made the trip up from Los Angeles so comfortable, and by ten o'clock the engine head temperature stood at 190 degrees at a steady cruising speed of about 50 m.p.h. By noon, when we arrived at Sacramento, it had crept up to 200. At a Shell station I was able to get a new oil filter cartridge and, as it was evident that the V8 engine was going to run extremely hot, I thought it best to add a quart of Shell motor-cycle oil of about S.A.E. 70 weight to the 40 weight oil in the sump—although it raised the level a quart over the "full" marker on the dipstick.

Before leaving Sacramento I made another search through all the garages and second-hand shops in the vain hope of finding a jack of some sort. I had been unable to procure one in Los Angeles and here I found nothing for sale save a 20-ton hydraulic truck jack for 18 dollars—which decided us to continue on without one. Actually our tyres were in good condition with plenty of tread left.

After Sacramento our course lay toward Lake Tahoe, in the High Sierras, and immediately upon leaving the city it was evident that we were going to have trouble with the Citroen's cooling system. It was impossible to detect any gradient visually or by any labouring of the engine, but we were climbing steadily and the engine temperature hovered on the boiling point. After an hour's driving we pulled into a petrol station and found that over a gallon of water had boiled away, though we had not yet reached even the foothills of the Sierras, and the attendant told us that we could not expect appreciably cooler air until we had climbed to the summit of the grade ahead of us.

As soon as we were again on the road we could definitely detect the gradient and in a few minutes the Citroen began to boil in earnest, and well it might, for the air that was supposed to cool the engine felt like

The large sign erected at a petrol station in Wendover, Utah, to commemorate the land speed record attained by John Cobb on August 23, 1939.

A skid mark caused by hitting an unseen bump on a bad stretch of road between Salt Lake City and Laramie.

64

the breath of a blast furnace. At every petrol station or roadside spring we encountered we stopped to top up the radiator and were somewhat consoled to find that a number of other cars were also boiling on this climb. The grades could be pulled in high gear at almost any speed by the Citroen and dropping to second did not help to prevent boiling as there was no fan to increase the flow of air. (The previous owner had removed the V8's fan because of space restrictions when fitting the large engine and had experienced no serious overheating around Los Angeles.) By the time we reached an altitude of 5,000ft. the air had become noticeably cooler but, of course, because of the reduced atmospheric pressure the boiling point had also dropped several degrees, so nothing much was gained.

After one of the hottest stretches I found that the battery, mounted above and slightly to one side of the engine, had been subjected to such intense heat that the fluid in the front cell had boiled and had eaten off some of the paintwork where it had run down the side of the bonnet. The case itself was so hot that it was plastic and could be easily bent with light pressure. I rigged a makeshift baffle under the battery to keep off some of the heat from the manifold and to direct more of the incoming air over it and it seemed to do the job, as from this point onward boiling was entirely restricted to the radiator.

At the summit of the 7,500ft. grade we stopped to cool off. As soon as we started down the Eastern slopes the engine temperature dropped to a satisfying 180 degrees—we had really been worried about the brutal treatment to which the V8 engine had been subjected during the last few hours and were most relieved that a cylinder head had not cracked or a gasket blown.

Now that the Citroen could take it easy we, too, settled down and relaxed and were able, for the first time, really to enjoy the simply magnificent mountain scenery. Later we reached Lake Tahoe itself, the water of which was much too cold for swimming. The road here follows the contour of the lake and pro-

Close-up showing the texture of the surface of the famous Salt Flats. The tyre has passed over the salt clumpings without crushing them.

Road building in Western Nebraska, where a section of the highway was under reconstruction. Ten-ton trucks, bulldozers, scrapers and other " monsters " raised clouds of dust.

vides a very beautiful drive before swinging east into the mountain again on the way to Carson City, Nevada, which we reached just as the sun was setting.

The evening air was so pleasantly cool that we decided to drive all night and sleep the next day so as not to subject the Citroen to another day of over-heating by crossing the Nevada desert during the heat of the day. I took the wheel for the first hour and only a few miles out of Carson City noticed the water temperature right on the boiling point again after only a very slight climb. Also I realised that the car had lost its snap acceleration. There was a softness and fluffiness to the engine that made me fear that the continual boiling had taken the best out of it. I also noticed that when very hot, and pulling hard on a grade, it was impossible to make the engine pink even by stepping the throttle hard down, a state of affairs most unusual with Ford V8 engines, which will usually pink at normal temperatures if accelerated hard on top gear.

Somewhere between this spot and the mountains in the background John Cobb made the world's record run on land.

Across America

3,500 Miles in Seven Days in a V8-engined Citroen

By Roger Barlow

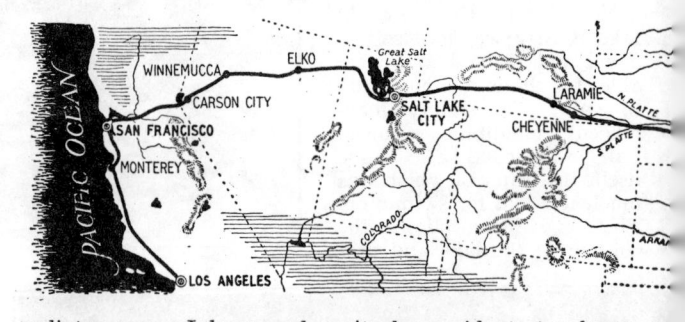

The start of a journey across America from the Pacific to the Atlantic Coast was described in the last issue. The car used by our contributor was a 1939 f.w.d. Citroen fitted with a Ford V8 engine. In this article the narrative is concluded with an account of the trip from Carson City to Washington.

AFTER Lovelock we put the next 72 miles of flat desert road behind us in just one hour and fifteen minutes, reaching Winnemucca only a little after 10.30. We left Winnemucca just as the sun was nosing over the horizon and crossed the Nevada desert wastes. As far as the eye could see the only evidence of man was the highway itself, lost in the morning haze twenty miles ahead of us, and disappearing over the last rise ten miles behind us. I was not comforted to note that on the first easy and hardly discernible grade the engine boiled and that the engine seemed fluffier than ever.

In Elko we found a Ford dealer who offered to sell me a new unit for the distributor. I secured the help of a rather heavy-handed youth who did most of the dirty work while we had lunch. In a very short time we were off again—this time with high hopes that all would be well. It was not. In fifteen minutes the temperature gauge was again at the boiling point.. At the next petrol station I took on more water and set the distributor advance as far forward as it would go. Still no pinking, and still the tendency to boil on every slight grade. Now, at the hottest part of the day we had a long steady hill to climb and after fifteen miles most of the water had boiled away. Ahead of us lay another long steady hill to the horizon—a climb that was useless for us to attempt without more water. Then we saw a signpost bearing a name we had seen on our map and up a dirt road, three-quarters of a mile away to our right, were a house and a railroad water station.

We took water from a hose when the boiling and steaming had subsided enough for me to remove safely the radiator cap. I happened, quite by accident, to glance down into the header tank and saw a piece of bent and battered metal lying on top of the core. I carefully probed into the inside of the radiator and finally with long-nosed pliers triumphantly withdrew the jagged remains of a baffle plate which, I was now certain, had been blocking the flow of water enough to account for the tendency to overheat.

The "patient" was slowly given a transfusion of a couple of gallons of cool water and then started up. No leaks were apparent anywhere, so we started off again, and began the long climb that had seemed so formidable a few minutes before. In spite of the 110 degree afternoon heat and the steady climb, the head indicator seemed content to sit right on the 200 mark—just short of boiling—proving that the reduction of circulation caused by the broken baffle had been just sufficient to make the difference between boiling and not boiling.

In less than half-an-hour we reached the barren cinder-like hills and topped the crest of the grade, and there, spread out before us, was the vast, dazzling, shimmering whiteness of the famous Salt Lake Flats.

As soon as petrol was aboard at Wendover we were off towards Salt Lake City, a hundred and twenty-five miles away. The road was of smooth black asphalt, raised a couple of feet above the surface of the salt. Five miles or so east of Wendover we were about opposite the area where the record runs were made, and I drove off the highway to sample the surface that had made possible these incredible speeds. I then experienced the surprise probably felt by all newcomers to the Flats, for the surface is not made up of fine granular table salt, but of clumps of quite large crystals which produce lumps and jagged projections half an inch above the solid surface of the salt bed, which at a hundred miles an hour would probably cut a tyre to shreds in no time. This accounts for the presence of the steam roller at the site of the record runs—the surface used is rolled and the larger crystalline projections are crushed, producing an almost perfect surface for high speed motoring because there are no bumps or undulations in the salt bed.

Salt Lake City

The pure white salt beds continued for some miles and then we approached and skirted a low range of hills. As darkness fell we could see in the distance the lights of Salt Lake City. The dry thin air of the desert rapidly gave up its heat and before we reached Salt Lake City we had all the windows up and the heat that found its way back from the engine was more than welcome.

The next morning, only a few miles out of the city, our road entered a steep-sided canyon and we began climbing the first grades of the Rocky Mountains. Our road, which up to now had been quite good, began to deteriorate and every now and then there would be a caution sign warning motorists of an especially rough section that should be taken carefully at about 25 m.p.h. Just after passing through one such section I had accelerated to 40 or 45 m.p.h., with an apparently smooth road ahead, when I hit an unseen bump with a frightful crash. With a horrible screeching of tyres, shuddering and shaking, the Citroen slid to a halt without my having even touched the brakes.

A signpost showing the maximum elevation reached in the Sierra Nevada mountains.

I could smell burnt rubber as I climbed out fully expecting to find a burst tyre, but as I came to the right front wing I saw that the tyre was still fully inflated and that a skid mark of black rubber led back for seventy feet. What had brought the car to such a violent stop? Half a dozen unpleasant possibilities flashed through my mind as I looked under the car.

Everything seemed intact, and then I noticed that the wing was clearing the tyre by only an inch or so, and that led to the immediate discovery that the accident had been caused by the bump forcing the wheel up six or seven inches so that the tyre had been driven hard against the wing bracket which pressed into the tread and was itself bent forward by the rotating wheel. The self-wrapping effect had pulled the bracket ever tighter against the tyre, almost instantly locking the wheel and bringing the car to a stop—whereupon the self-wrapping effect ceased and the bracket was able to spring up slightly, releasing the tyre. I crawled out from under, drove the car off the road, jacked up the front end and was able, by lying on my back and pushing with my feet, to bend the wing and bracket back into their proper positions. I was positive that in no circumstances should it have been possible to force the wheel up to where it fouled the wing, so I deduced that the added weight of the V8 engine had reduced the normal clearance by depressing the torsion bar springs to

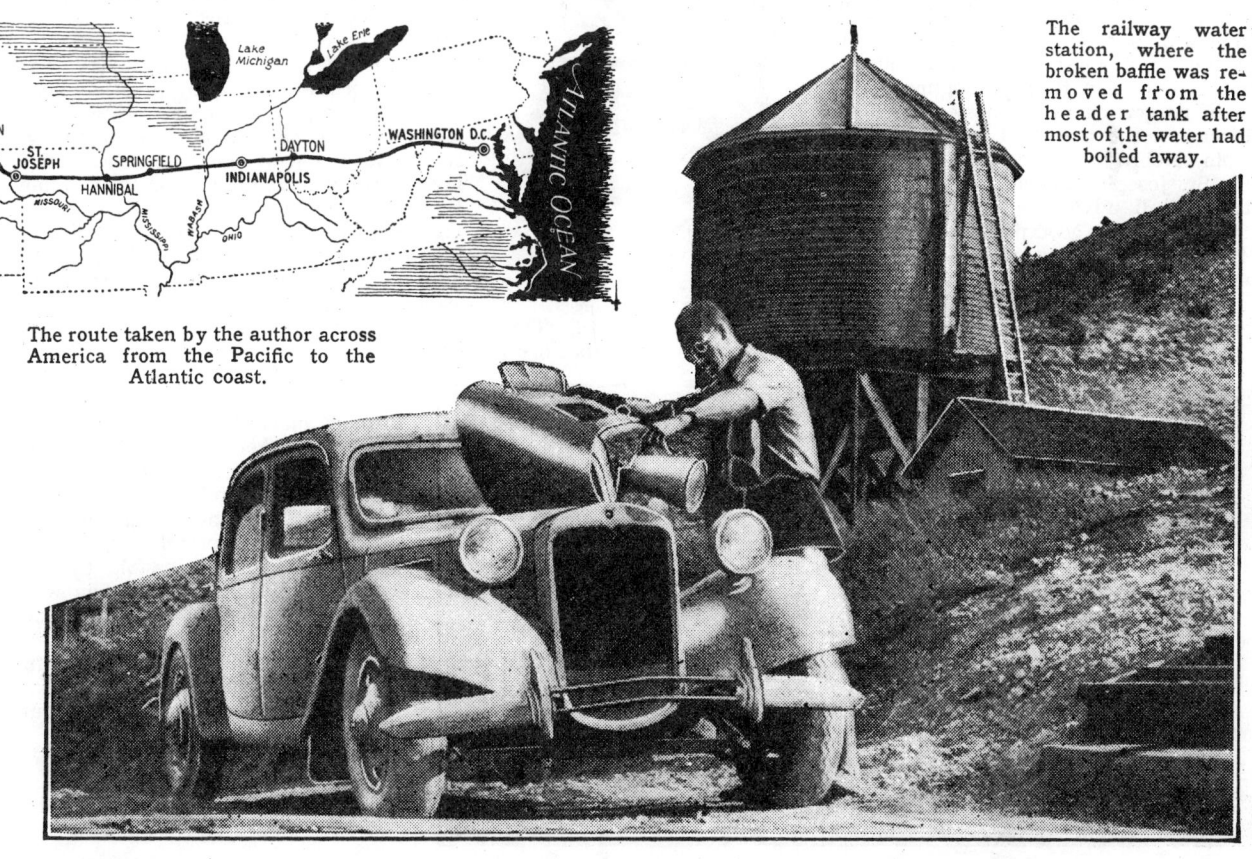

The route taken by the author across America from the Pacific to the Atlantic coast.

The railway water station, where the broken baffle was removed from the header tank after most of the water had boiled away.

Across America

an unusual degree. With the Citroen's system of suspension it was a simple matter to remedy this, only a couple of turns of the adjusting screws at the back mounting of the torsion bars being needed to increase the clearance.

After this excitement the hours and the miles rolled by monotonously and we reached Laramie in time for dinner. We decided to go on east to Cheyenne before stopping for the night. The next morning we made an early start in the frosty dawn and in less than an hour covered the 42 miles to the State line and crossed into Nebraska. The highway was smooth and straight and on one especially well-surfaced section I brought the car up to over 80 m.p.h. and held it for a couple of miles.

Shortly before arriving at North Plate I noticed an improvement in acceleration, and then I began to hear a faint pinging when accelerating . . . this was more like the car I had started out with from Los Angeles. Then I understood—the engine had not gone fluffy from the terrific beating it took climbing the Sierras; the apparent loss of tune was not because of the continued boiling in the desert but was only the natural loss from poor cylinder filling at high altitudes.

That evening we were barely half-way across the continent and we had already used five of our seven scheduled days of travelling. Obviously we had no alternative but to plan to drive continuously, each of us getting as much sleep as possible when not at the wheel. After leaving Lincoln, Nebraska, we headed south towards St. Joseph, Missouri.

Taking full advantage of the lack of traffic I kept up a good pace and put over half the State of Missouri behind by the time the first faint glow of dawn showed up ahead. We were nearing the Mississippi River and had breakfast at Hannibal (one-time residence of Mark Twain and the locale of the events chronicled in Tom Sawyer). We took turns sleeping and driving until just outside Indianapolis we came upon the first dual highway since California.

Eroded rock outcroppings in Wyoming. Left: A sign on the fine road in West Virginia.

With Indianapolis behind our route lay through Dayton, Ohio, and then through the West Virginia mountains, missing all other large cities. A beautiful new wide concrete road wound up the West Virginia mountains in easy sweeping curves that could be taken at 55-60 m.p.h., but, alas, this road soon degenerated into a narrow, twisty and exceedingly rough-surfaced secondary road that lasted most of the way across the State.

Our journey across the country was almost over and we had by this time been on the move for 52 hours. Now, with less than 150 miles to go, we ran into a steady downpour of rain which made driving somewhat more difficult on the winding Blue Ridge mountain roads. After crossing seemingly endless ridges we finally found level going and, eventually, swung into the twin-track divided highway that leads into Washington and soon we were there. Three thousand five hundred miles, the last 56 hours since leaving Cheyenne spent on the road!

The Citroen, with all its alterations and modifications, had come through marvellously. The V8 engine had taken no oil during the entire trip other than the additional quart of heavy oil I had added in Sacramento, and by now the oil level had dropped to normal. The added performance made possible by the 85 h.p. engine had not in any way proved too much for the suspension or chassis of the Citroen. In fact, it seemed to me that the general design and handling qualities were such that the added horse power brought out their best features.

The road down one of the ranges of the Allegheny Mountains.

Comment on the f.w.d. Citroen

HAVING been an f.w.d. enthusiast for many years, it gave me some pleasure to see space being devoted to the praise of this type of transmission.

I have personally been the owner of two Citroen f.w.d. cars and had a great deal of experience with many others. As what I might term an "ordinary" motor-car, that is to say, a car of average price available in large numbers to the general public, I think the Citroen is a truly remarkable product and one deserving much more attention from the enthusiasts.

The first model which I owned was a 12-h.p. saloon of 1938 vintage. This was bought second-hand in 1940, having done a very hard 22,000 miles, but after a little attention to universal joints the car seemed almost as good as new. The performance for a car of 12-h.p. was most amazing and she would climb Newlands Corner in top gear with six up and never drop below 40 m.p.h.

The cornering and roadholding were of the usual Citroen order, and this really has to be tried to be believed. I have never succeeded in skidding a Citroen, and I should be most interested to hear from someone who has. I have driven both this car and other ones on winding roads at really terrific speeds and they always appear to be glued to the road, so much so, in fact, that it leads me to wonder exactly what might happen if the car did finally skid.

The 12-h.p. model was sold early in 1942 and, after a number of other makes, was replaced early in 1945 by a "Light Fifteen" Citroen. This again was a car which had been driven very hard indeed for 30,000 miles and, for a 1939 car, it was in a remarkably rough condition. The previous owner, who is fairly well known in the motoring world, had used the car for carrying very heavy loads of steel in the course of his business.

The purchase of this car would probably amuse many enthusiasts. I had made up my mind to have another Citroen, but, after hunting all over London and the South of England, I finally resorted to the time-honoured method of chasing and stopping any Citroen I saw on the road and asking the driver if he or she would be open to an offer. Before I was successful, these tactics resulted in a number of rude and comic answers, and also in some very hard wear indeed upon my Renault Twelve, which, although good, was far from up to the Citroen standard.

The engine of my "Light 15" was in a very poor state indeed, and I am sure you could have dropped pebbles down between the pistons and cylinder walls. The oil consumption was rather fantastic and the rest of the car mechanically was in about the same state. In spite of this, however, the performance was still all that can be expected from a Citroen, 70 m.p.h. being easily obtainable at any time, coupled with some excellent acceleration. An interesting point, upon which I should like a little confirmation from other enthusiasts, is that I obtained a better petrol consumption from the "Light 15" than from the 12-h.p. model, and I assume this to be due to the difference in final drive ratios.

In the August issue we appealed for readers' experiences of the f.w.d. Citroen, a normal car which, like many other Continentals, is a worthy substitute for a sports car and which should be an object-lesson to the British industry. P. A. Whittet writes enthusiastically of this car, with particular reference to the "Light Fifteen," and Myles Wadham equally so in respect of the "Twelve."—Ed.

Most readers will probably know that the Citroen is fitted with wet liners and it is, therefore, the easiest matter to fit a new set of pistons and liners. I have actually fitted a friend's "Light 15" with new pistons and liners and had the engine running again in three hours single-handed. There is, of course, no need to remove the sump to fit these parts. In view of the excellent performance of my "Light 15," however, I was very loath to take the engine to pieces and, to overcome the great usage of oil, I drained and refilled the sump with Castrol R, and this not only resulted in a most pleasing smell, but the engine actually seemed to like it and ran many thousands of miles without any trouble whatever. Whether Castrol R has some remarkable re-building effect, or whether it merely gums up all the clearances, I should not like to say, but the oil consumption was certainly reduced to something almost approaching normal and I found no ill effects, although many friends had warned me that I should ruin the engine.

Recently, after having saved up much basic ration, I decided to set out on a tour of Devon and Cornwall with my wife and daughter, aged two, and I think that our week's running was probably as severe a test as one could give any motor-car. We ran down the north coast of Devon and Cornwall, choosing all the cliff roads. The mere sight of a notice suggesting that a road was impracticable for motors was enough to make us try it, and, touch wood, we always got through somehow.

At one point beyond Lynmouth we had taken an "impracticable" road and had descended about a couple of hundred yards, where the gradient was about 1 in 2½, and it would quite obviously have been impossible on a grass surface for any vehicle to go up the gradient again, so we just had to hope that there was a way out by going ahead. A little further on we came upon a very small humped-back bridge over a ditch, and this very nearly proved our undoing. The bridge was so humped that the car grounded in the middle long before we were over, so there was only one thing to be done. My wife and daughter got out, I backed the car a couple of hundred yards and made a terrific rush at the bridge. The Citroen sailed into the air and landed some yards on with a most frightful clatter, but upon inspection seemed none the worse for wear. Appropriately enough, the registration letters on this car were "FLY."

Further on during our tour, we were ascending a hill of about 1 in 4, or even more, along a road so narrow that we were touching the verge on either side, when, on rounding a bend, we came upon a Standard Twelve very stuck indeed, with the passengers hastily scratching lumps of stone out of the hedge to put under the wheels. Apparently the hill was so steep that the petrol from the float chamber was simply pouring straight down the air intake and had stopped the engine. After a small amount of fiddling, we got it running again and, with several people pushing, the Standard got away, much to my concern, taking with it all its passengers and leaving the Citroen perched upon a remarkably steep hill. However, the clutch proved better than I had expected and we got away without any trouble at all and, in fact, without any wheel spin.

For the whole of our tour we averaged 25 m.p.g. with the "Light 15," a truly remarkable performance when you consider how much first and second gear work we were doing. The car was never really driven at all with any view to getting good petrol consumption, and wherever there was a decent open road a steady 65 m.p.h. was maintained. Our last stage home to Bagshot from Hamworthy, a distance of 89 miles, was done in exactly two hours.

Much to my sorrow, my "Light 15" has now gone.

It is my good fortune to be able to own, within reason, almost any type of motor-car, and, of course, I have tried most makes. But I still repeat that at similar speeds on the average English roads I feel safer in the Citroen.

Finally, for the sake of those who have not had the pleasure of diving into the works of a Citroen, here are just a few interesting details. The engine is a very nice, large, 4-cylinder overhead-valve, push-rod affair, the "15" being of 2-litres capacity. The push-rods and rocker gear are nice and light and very quiet in operation. Clutch is ordinary single dry plate and a 3-speed gearbox is mounted right in front of everything underneath the radiator grill, the drive is thus taken through the clutch over the top of the differential into the gearbox and returned out of the back of the gearbox to the differential. The gear-change lever is mounted on the dashboard and operates two selector rods running from behind the scuttle down to the gearbox. The three-bearing crankshaft is pressure oil fed and runs in bearings of ample proportion, which never seem to give any trouble, however hard they are used. Oil is also supplied under pressure to the rocker shaft, which is drilled underneath each rocker arm. A point worthy of note is that when decoking a Citroen it is well worth while dismantling a rocker shaft and clearing the oil ways, as I find that these are rather inclined to become clogged, resulting in excessive rocker wear.

The suspension is by full torsion bar, coupled with good direct-acting hydraulic shock-absorbers. The torsion bars are all adjustable, and the height of the car can be re-set very easily in a matter of a few minutes. The previously mentioned shock-absorbers work better than any other type I have struck and never seem to require any attention. Steering is of the rack-and-pinion type and extremely accurate. At any speed the car can be placed beautifully to within a fraction of an inch.

The body is a pressed and welded steel affair, serving also as a chassis, the whole of the works being bolted on to the front end of the body. It is, in fact, quite an easy matter to unbolt all the works and wheel away the engine, gearbox, front suspension and, in fact, everything, leaving just a body, and, once you get the knack of wheeling everything away like this, it is a lovely easy method of getting to work on the car. Furthermore, for an enthusiast it means to say that practically a complete spare car can be kept and worked on for use at comparatively short notice without the expense of licensing two cars.

Of criticisms on the Citroen I have very few indeed, and these are only minor details. The gear-change lever might well be improved in operation, and, in fact, I think the whole system of linkage between gear lever and gearbox might be re-designed to advantage. The position of the front seats in many models can also be improved, and in each of the cars I have owned I have re-made the front seats and set them some 2 in. higher than standard. I do wish, too, that manufacturers would return to fitting an oil pressure indicator instead of merely a warning light. I know that for some people the disappearance of the red light is enough to make them feel happy about their engine, but personally I would much rather know exactly what pressure there is.

From this article readers will gather that I am a confirmed Citroen fan, but I still remind them again that this is in spite of the fact that I have the opportunity of trying other makes of car. Whether it be for the everyday motorist, who merely wishes to drive about the country in safety, or for the enthusiast who likes a car that feels " right " and gives some real performance, I really do not think there is anything else within reach of the average pocket which you can even begin to compare with the Citroen.—P. A. WHITTET.

* * *

In 1937 I began to reach the age when sports cars, which I had always previously owned, were beginning to lose a little of the appeal which they had had for my youthful eye and ear. Comfort had begun to seem more desirable than the hard springing then almost inseparable from the sportsman's ideal, and silence rather than a healthy bark insidiously began to state its claims. I still felt the necessity of avoiding a closed car at all costs, although I wanted one that could be rendered waterproof and draughtproof if necessary — also I wanted something " different " and progressive in design. After much searching of the motor papers, I came to the regretful conclusion that there was nothing British that met my requirements within the limitations of my purse. Thus I came upon the 12-h.p. Citroen, which, as a roadster priced at £255, had almost everything that I desired.

I have now had it for eight-and-a-half years and covered over 93,000 miles, and every one of those has been a real joy. Repair costs have been very, very low, though I did have new cylinder sleeves put in at 50,000 miles in 1942, thinking that if the war persisted it might not be possible to get them later on when they were really needed.

This car has been truly " designed "— not just grown up year by year from something pre-historic, like the average English car. Its general layout of chassisless construction, front-wheel drive and independent suspension by torsion bars all round are perhaps too well known to readers of MOTOR SPORT for me to more than mention, but the easy accessibility of all components, the provision for adjustment of even minor things, such as bonnet flaps to prevent rattle, have to be experienced by the fastidious owner-driver to be fully appreciated. Even the ground clearance can be easily adjusted if the suspension should "settle" with age, or the owner decide to traverse unusually rock-strewn byways.

The engine, a 1,628-c.c., 4-cylinder, push-rod o.h.v. job, flexibly mounted on the Chrysler principle, is very smooth and quiet running right up to maximum revolutions, whilst the drive, taken from a 3-speed gearbox mounted underneath the radiator through universally-jointed cross-shafts to the front wheels, is entirely silent and free from roughness.

In appearance the car never fails to attract admiring glances and comments, whether it be open or closed. With the hood erected and attractively rounded glass windows wound up into position, the car is completely weathertight and has the snug, sleek look of a rather dashing coupé. In fine weather, with the hood stowed in a recess out of sight and windscreen folded flat, it is transformed into a very well-bred looking sports car.

As regards roominess, three slim people can travel long distances in comfort in the wide front seat (the floor being unobstructed by gear or brake levers) and two broad ones can spread themselves in the dickey, where there is enough legroom for the tallest guardsman.

The suspension and roadholding are excellent, thus proving that by initiative on the designer's part comfort need not necessarily be divorced from ability to stick to the road like the proverbial leech.

I think it should be no criticism that the very direct steering is inclined to be on the heavy side, because it is accurate and steady to the highest degree, whilst in all the mileage I have covered no adjustment of any sort has been done to the rather unusual steering mechanism, yet the steering wheel cannot be turned a fraction of an inch without producing an answering movement of the road wheels.

The technique of cornering with a front-wheel-drive car needs a little mastering, for the beginner must remember always to have the engine pulling him round the corner if he wishes to get the maximum result. It is on wet roads, or better still on snow or ice, that the Citroen really comes into its own, for it is almost impossible to skid this delightful car, and one can safely drive at high speeds when other drivers have to crawl. I said " almost " impossible, for it can be done if one really tries hard enough and tyres are suitably smooth. I have practised it on ice on open bends and the technique of getting out of such a skid quickly needs a little courage, though the method is unfailing—simply remember to tread on the accelerator pedal instead of the brake, and the traction on the front wheels hauls the car into the direction in which one is pointing them.

My one criticism is the gearbox. Though the car is a top-gear performer, almost ranking with the Americans, and though the acceleration is good enough to cause one's passengers to comment admiringly, I sigh for a close-ratio 4-speed box instead of the 3-speed affair with a fool-proof, but " uninteresting," type of change as at present fitted. That's what comes of having graduated through sports cars and gained a liking for something one could play really snappy tunes on !

Petrol consumption is good and averages 28 m.p.g. on my daily 16½ miles to work, which I invariably cover in under 20 mins. (not straight or arterial roads either !), whilst on a recent 270-mile run the total petrol used was a fraction under 9 gallons.

Surprisingly enough, I see that I have not mentioned speed, although I have indicated in the previous paragraph that excellent averages can be maintained. The maximum by speedometer is still, as it always was, 74 m.p.h. I believe the instrument to be some 5/6 m.p.h. fast in these regions, for I once tried to cover 60 miles in 60 mins. on the dead straight, dead level Brescia to Milan autostrada by keeping the needle as steadily as I could on the 65 m.p.h. mark. The journey actually took 60½ mins.

Now that my article is almost at an end, readers will have realised that it is written by an enthusiast, but that is inevitable. I have never talked to a Citroen owner who didn't declare it the best car he had ever had. In case critics should say that I have painted too rosy a picture, I can only advise them to try one for themselves. It is a grand car and I only wish that England had produced it 100 per cent.—MYLES WADHAM.

The Citroen Light Fifteen

Unconventionality Stands the Test of Time

IT is true to say that if the Citroen Light Fifteen had been making its first public appearance to-day instead of continuing where it left off at the outbreak of war, it would have created a buzz of sensation as the latest in advanced automobile engineering thought. In point of fact, of course, the car arrived way back in 1934 exactly as it is to-day but for subsequent detail changes, and still features characteristics which are reflected in the latest designs of the post-war world.

The Citroen is rather unusual than unorthodox, but there is technical reason backed by years of success behind every departure from the conventional—the chassis-less construction, the front-wheel-drive, the proven success of the torsion-bar independent suspension, the detachable wet barrel-type cylinders—even a simple solution of the interior heating problem.

This 2-litre 5-seater saloon weighs only 21¼ cwt. and is the lightest car per foot of wheelbase of any car on the British market, yet no light metal enters into the design apart from the pistons! Those who still regard front drive as undesirable may also remind themselves that among pre-war Continental sales, one car in four was f.w.d

Placing of the gearbox in front of the engine brings the weight distribution as between front and rear axles to something closely approaching the recognized optimum (for f.w.d. designs) of 60-40 per cent. while retaining the known advantages of front drive. The suspension design has reached what is possibly the best compromise yet achieved between the conflicting demands of constant wheel track and parallel motion, while the carefully thought-out, torsionally sprung back axle preserves parallel motion and ensures that the wheels remain vertical during fast cornering. This last is, of course, of major importance with a front-driven vehicle in eliminating the tendency of independently sprung wheels to bank with the body on curves, resulting in oversteer which constitutes a particularly distressing vice with a front-wheel-driven car

CITROEN DATA

	Light Fifteen Saloon		Light Fifteen Saloon
Present tax	£18 15s. per annum	Suspension	Independent front, dead axle rear ; torsion bar springs
Cubic capacity	1,911		
Cylinders	4		
Valve position	o.h.v. in line at 10 degrees	Steering gear	Citroen rack and pinion
Bore	78 mm.	Steering wheel diameter	16¾ ins.
Stroke	100 mm	Wheelbase	9 ft. 6¼ ins.
Comp. ratio	6.2	Track, front	4 ft. 4¾ ins.
Max. power	55.9 b.h.p.	Track, rear	4 ft. 4¾ ins.
at (A)	4,250 r.p.m.	Overall length	14 ft.
Max. torque	88.2 lb./ft.	Overall width	5 ft. 5¼ ins.
at (A)	2,000 r.p.m.	Overall height	4 ft. 11¼ ins.
H.P.: Sq. in. piston area (A)	1.89	Ground clearance	7 ins.
		Turning circle	41 ft.
Wt.: Sq. in. piston area (B)	80.4 lb	Weight—dry	21¼ cwt.
		Tyre size	165 × 400 Broadbase
Ft./Min. Piston speed at max. h.p. (A)	2,788	Wheel type	Michelin pressed steel
		Fuel capacity	9 gals.
Carburetter	Solex 35 F.A.1.E down draught	Oil capacity	11 pints
		Water capacity	14 pints
Ignition	Lucas coil	Electrical system	Lucas 12 volt c.v.c.
Plugs: Make and type	Champion J8-B	Battery capacity	57 amp./hrs.
Fuel pump	A.C. mech.		
Clutch	Citroen 9 in. diam.		
1st gear	13.1		
2nd gear	7.3	**Top Gear Facts :**	
3rd gear (top)	4.3	Engine speed per 10 m.p.h.	575
Reverse	17.5		
Final drive	Spiral bevel (f.w.d.)	Piston speed per 10 m.p.h.	377
Brakes	Lockheed hydraulic		
Drums	10 ins.	Road speed at 2,500 ft./min. (piston)	66.25
Friction lining area	97.5 sq. ins.		
Car wt. per sq. in. (B)	24.4 lb.	Litres per ton-mile (B)	3.104

(A) With normal setting of carburetter, etc. (B) Dry weight.

As an approximation, 2,800 litres per ton-mile enables a car to climb a gradient of 1 in 10 in top gear. Similarly pro rata—a gradient of 1 in 9, for example, requires 3,100 litres per ton-mile.

INDEPENDENT FRONT SUSPENSION on the Citroen is by means of parallel torsion bars and swing arms, married to front wheel drive in a design which successfully masters the problems of parallel motion and constant track. The suspension is mounted in a massive pressed steel cradle.

With these features, the Citroen presents a modishly sleek, low vehicle capable of racing-car stability on corners, which will not slide under even abnormal surface conditions and which is capable of very high average speeds without a very high maximum.

The engine is specifically designed for limited output and long life. It peaks at 4,250 r.p.m., which cannot be exceeded in top gear, at which speed the piston is moving at only 2,788 ft. per minute. Engine performance is meant to be of that type that goes on going on without frequent maintenance.

Two-part Construction

The Citroen can be regarded as fabricated in two major component parts—(a) the front end, including engine, gearbox, transmission, front axle assembly and radiator, and (b) the back end, meaning the welded steel body and integral floor, together with the rear wheels and springing.

The chassis must be considered in one with the body and is based on a flat steel floor, crimped up at the back to form a rest for the rear seats and formed with shallow sides of immense strength, due to a special construction which turns the floor-sides into box-girders. So low is the floor (but not the ground clearance) that the seats feel unusually high, and the driver has the sensation of sitting in a pre-war Grand Prix car. In point of fact, although the seats are 13½ ins. high in relation to the floor they are lower than on most cars. An average seat height above the ground, taken on a dozen comparable cars, is 26½ ins., while the Citroen figure is 24½ ins.

Across the back of the floor runs a massive tubular cross-member with, in the middle, a bracket which holds the fixed ends of the two transverse torsion bars, running one to each side of the vehicle, and free to twist at the outer extremities. Swinging from each free end is a trailing arm on the end of which is mounted the wheel, moving in a vertical arc and controlled by a direct-action Newton hydraulic shock absorber (which, incidentally, can be easily got at for refilling). The assembly is located transversely by a Panhard rod and the trailing arms are flexible so as to accommodate the necessary distortion when the vertical motion of the wheels is unequal.

The tubular cross-member is not merely bolted into the body sides but is carried in very long longitudinal castings of immense rigidity.

The back axle is in the form of an X-section beam which does not, of course, revolve, but maintains the desired vertical position of the wheels when cornering.

The body is built of steel pressings, welded to the floor where it is reinforced to take it. The sum total of body and floor, trussed and cross braced, is a steel cage of box-girder-like construction, of remarkable strength.

The welding process is continued ahead of the front screen, where the scuttle divides into the form of two open "jaws" to take the engine unit. Four long tubes in pairs run right forward through these jaw-like bulkheads to take the cradle of the front assembly.

This pressed-steel cradle carries the entire front assembly, and is in the form of a flattened U. Two torsion bars run through the bottom of the cradle longitudinally, one on each side, free to twist in bonded rubber bushes. The anchored ends run right back to a stout cross-bar between the jaws of the body and are serrated for adjustment.

Front Suspension

A single girder-type link is forked on to the front or free end of the torsion bar on each side, while above, just below the shoulders of the cradle, a pair of wishbones swing in unison with the link below, moving on lubricated and adjustable bronze bushes of great length. It is the good geometry of this suspension layout that successfully compromises between the opposed factors of constant wheel track and parallel motion. Front tyres do not show unusual wear rates.

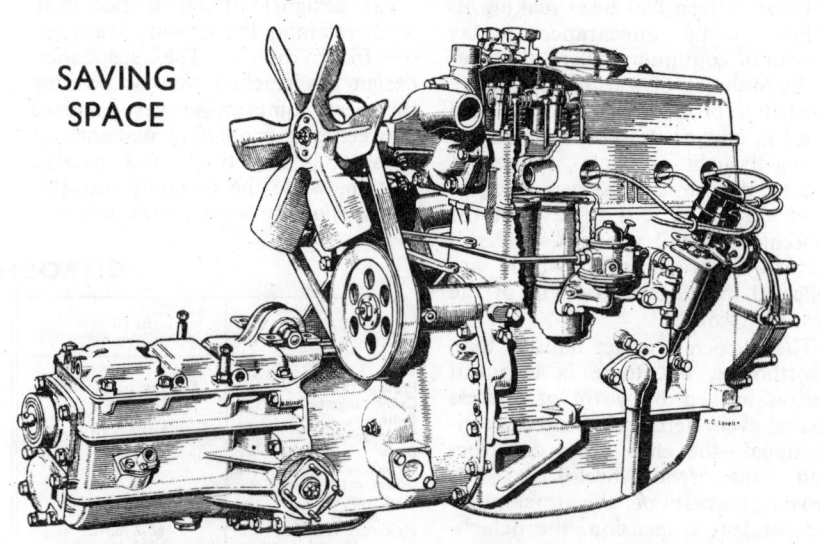

SAVING SPACE

The three-gear gearbox on the Citroen is mounted ahead of the front wheel centre as clearly shown in this drawing, the drive being taken over the crown wheel and pinion by an extension shaft shown below, left. The above drawing also indicates the principal features of the 1,911 c.c. engine which produces 56 b.h.p. Of particular interest is the way in which wet cylinder liners are inserted into the water space of the main crankcase cylinder casting.

The crown-wheel, bevel pinion and differential are compact and in unit with the gearbox, the bevel pinion being integral with the gearbox layshaft. Short transmission shafts run at right angles to the front wheels.

The swing arms top and bottom terminate in ball joints, swivelling on a large-diameter hemispherical casting which cups the double-universal joints of the short drive shafts and forms a sturdy knuckle joint. The universals, running on Hardy Spicer needle bearings, are spigoted together for axial rigidity. To the hemispheres the brake back-plates and stub axles, with the steering arms, are attached.

The suspension is again controlled by direct-action Newtons, mounted trans-

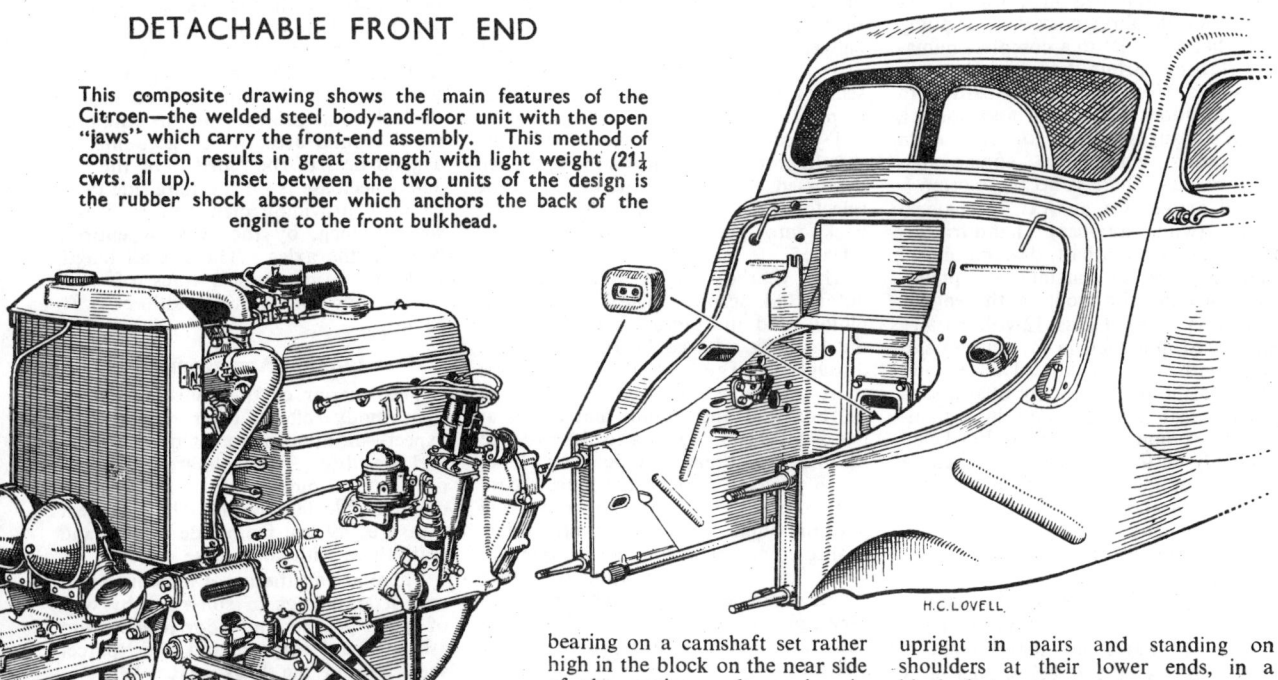

DETACHABLE FRONT END

This composite drawing shows the main features of the Citroen—the welded steel body-and-floor unit with the open "jaws" which carry the front-end assembly. This method of construction results in great strength with light weight (21¼ cwts. all up). Inset between the two units of the design is the rubber shock absorber which anchors the back of the engine to the front bulkhead.

versely at an angle of about 45 degrees.

Steering is by means of rack and pinion. The enclosed rack is fitted horizontally across the car, on a tube enclosed within a slotted sleeve with central ball joints locating the twin track rods. When the steering wheel is turned, the rack travels left or right and controls the track rods and wheels so that a very light, positive and self-centring steering action is obtained.

The gearbox protrudes through the middle of the U-shaped cradle, and is slung on a cross-member by a large rubber-lined mounting forming the forward engine anchorage, while the radiator is carried overhead.

The engine-gearbox unit is held in four mountings. There is the flexible mounting just mentioned, a large rubber shock absorber at the back which fits into a port in the dash, and on each side volute springs which sit on brackets formed in the steel bulkheads of the body-jaws.

The Power Limit

The power unit is of robust design, made for longevity and a medium power output at medium speeds which, in combination with the light weight of the car, produces a most lively performance. It is a four-cylinder engine with overhead valves, a capacity of 1,911 c.c. and bore and stroke of 78 mm. and 100 mm., delivering just under 56 b.h.p. at only 4,250 r.p.m with a compression ratio of 6.2 to 1.

The in-line overhead valves, set at 10 degrees from the vertical in the head, are opened by inclined pushrods, themselves raised by plunger-type tappets bearing on a camshaft set rather high in the block on the near side of the engine and running in three bearings. At the rear is the roller-chain camshaft drive. At the front end of the shaft an extension carries the big pulley which belt-drives the fan and dynamo; farther back the shaft drives the petrol pump and, at the rear, the distributor.

The crankshaft is carried in three bearings and is turned by light steel stamped connecting rods with white metal big-ends and bronze-bushed gudgeons. The aluminium, split-skirt pistons have concave crowns and carry four rings, the third of which is a grooved oil-ring and the lowest a slotted scraper.

Distinctive Cylinder Layout

The cylinders are a distinctive feature. They are merely barrels placed upright in pairs and standing on shoulders at their lower ends, in a block that amounts to a cast-iron box full of water, so that the coolant reaches almost the whole depth of the barrels. At the top they stand above the face of the block proud to about two thousandths of an inch. In this way, when the gasket is in place and the head is bolted down, the barrels are held immovable. To prevent any tendency to rotation, they are positioned in pairs, side by side, with the contacting tops flatted to make firm contact.

Thanks to this method, a better wear-resisting cast iron can be used for the barrels than could be used in block-casting in the more normal way.

Pressure lubrication is used throughout, including the rocker gear, where grooves cut in the rockers send oil to the inverted sockets which take the ball-ends of the pushrods and to the face-contact with the valve stems.

FLAT FLOOR and inter-axle seating are features of the Citroen. Simple and effective interior heating is arranged by a pipe picking up warmed air from behind the radiator and delivering it into the front compartment. Front seats are 22 ins. wide and 18 ins. deep. The rear seat is 49 ins. in width.

The 35 FA 1E Solex downdraught carburetter, mounted high on the off side, has a diaphragm accelerator pump and what amounts to a complete miniature starting carburetter. This latter is controlled from the facia board with three positions—start, run and normal, which changes over to the main carburetter when the engine has warmed up. There is no thermostat.

Fuel arrives via an A.C. mechanical pump (with auxiliary hand-trigger) driven by the camshaft, drawing petrol from the 9-gallon rear tank and pumping it around the front of the engine.

Ignition is by Lucas 12-volt battery (ensconced in a shelf in the scuttle) and coil, with automatic advance and retard to the distributor by dual means —centrifugal governor and inlet-manifold depression. The coil is bracketed accessibly on the near side of the scuttle a matter of inches away from the distributor. On the other side of the scuttle, equally to hand, are the fuses, cut-out, constant voltage regulator and brake fluid reservoir.

Cooling is by means of thermo-syphon and pump. The radiator holds 14 pints of water and the draught is from a six-bladed fan, belt-driven on a long spindle which also drives the big water pump.

Transmission

The transmission is extremely compact. The drive passes through a 9-in. single dry-plate clutch with flexible centre, to a three-speed gearbox (synchromesh on top and middle gears) with ratios of 4.3, 7.3, 13.1 and (reverse) 17.5 to 1. The two gear selector rods in the top of the box are locked into position by a special shaft which lies between them and which, spring loaded, drives two balls into corresponding sockets in the rods to hold them firmly in position. By a neat arrangement of a T-piece at the back end of this rod, the clutch pedal withdraws the rod when depressed, thus freeing the balls and the selector rods until allowed to spring back into place when the pedal is released.

The gearbox is in unit with the differential, crown wheel and bevel pinion, and the pinion is integral with the layshaft. Large bearings are used throughout. The transmission shafts are short and robust, with universals at the gearbox end and double universals at the driving end. Owing to the distance of the "umbrella" type gear lever which projects through the facia board, two rods are required to move the selectors, and by an ingenious interconnected compensating device, one rod shifts first and reverse while the other changes middle and top. The change is very light in consequence and behaves like any conventional gear-change system.

Lockheed hydraulic brakes are used all round, with 10-in. drums, plus an independent hand brake working cables to the rear wheels.

The wheels are very Continental, pressed-steel spoke type by Michelin with handsome "dish-covers," and are again unusually light but strong, carrying very broad-based "165 by 400" tyres.

The body style of the Citroen has not altered with the years. It is a smart, modern, saloon measuring only 5 ft. in height, with a clear 7 ins. underneath. The general contour is well formed and, with the front-wheel-drive layout, presents an uncommonly small frontal area, which accounts for a good deal of the "urge" of the car. The wheelbase is on the long side (9 ft. 6½ ins.), providing plenty of passenger space, which, by the way, is entirely between the axles. The overall length is 14 ft. 1 in., with a track of 4 ft. 4¾ ins. and an overall width of 5 ft. 5¼ ins.

Interior Arrangement

The interior of the four-door, four-light body offers what one would expect—an atmosphere of quiet quality and long life. The seats are in leather, deep and comfortable, and provide really ample leg room. The rear seat measures 49 ins. from side to side and seats three with some ease.

There are all the usual fitments, from walnut trimmings to sun-visor by way of glove pockets and door pockets and ashtrays and carpets, and wide parcel shelf behind the back seat. What is more important, the visibility is excellent and the feeling of controllability from the driving seat breeds immediate confidence.

As the rear seat finishes in front of the back wheel arches there is obviously plenty of luggage space in the boot. The spare wheel is slotted into the lid and covered with a metal sheath, while the wheel fastening is on the inside, so that, once locked, the wheel is thief-defying. To carry extra luggage or a trunk, the lid can be left open, in which case a flap lets down to bridge the space below and extend the platform. Under this bridge in the hinge-space of the lid go the tools.

A normal screw-type jack is supplied, fitting beneath special jacking pads on the rear wheel swing arms so that the wheels rise and not the body only, and there are similar pads for the front end.

Only one model of this sturdy car is in production at this time—the Light Fifteen saloon illustrated on these pages, priced at £573 3s. 11d. (of which £125 3s. 11d. goes in Purchase Tax). Two colour schemes are offered —black body and beige upholstery, or beige body and interior. Other colour schemes will be available later, including red and metallic grey.

The direct acting rack and pinion steering of the Citroen with (inset) the engagement of the steering column pinion with the rack enclosed within its tube. The twin track rods are ball jointed to the rack through a slotted sleeve.

A 1:40 scale section showing the principal dimensions. The luggage boot has a 21-inch platform and is 41 inches wide.

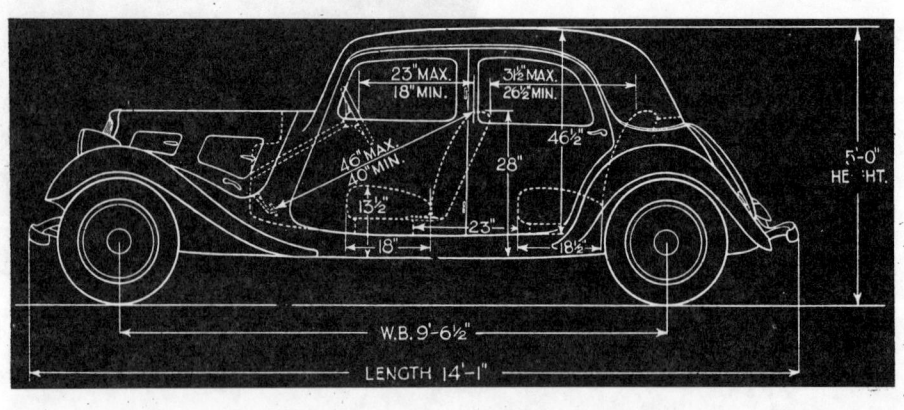

THE CITROEN LIGHT FIFTEEN
An Unusual Car with Chassis-less Construction and Front-wheel Drive

THE Citroen Light Fifteen saloon, a de luxe model of which was recently submitted to "The Motor" road test, is a car unusual both in technical specification and in the qualities which impress a driver or passenger. It is also a car which is not truly appreciated until many hundreds of miles have been covered, when the real reasons for unconventionality become evident and the tester realises why men who drive big annual mileages frequently run successive Citroens over a long period of years.

The whole character of the car is linked with such outstanding features as the use of front-wheel drive and chassis-less construction, features which have by now been incorporated in more than 200,000 British and French built examples of the marque.

A strong hint at what to expect is obtained when first entering the car, stepping right down into it through doors of normal height, and sitting on seats which are pleasantly upright yet, in fact, exceptionally low set. The correspondingly low, flat floor, obtained despite adequate ground clearance, is the prize for eliminating the propeller shaft and using the body as the frame of the car.

First Impressions

A driver coming as a stranger to the Citroen has at first to accustom himself to several new ideas. The upright driving position, giving a view of the two side lamps, imparts confidence, but initially, the pedals and gear change seem peculiarly sited. After some miles have been covered, however, a driver realizes that pedals of piano type, moving downwards rather than forwards, are extremely comfortable, and can be operated without the heel ever leaving the floor. Likewise, the facia-board gear lever, with vertical gate, proves perfectly convenient and facilitates entry from either side.

The gear change is unusual, a unique solution having been adopted to solve the problem of obtaining really light remote control—a problem which is not always ideally solved when steering-column gear levers are adopted. On the Citroen, the gear lever is normally locked immobile in whatever position it occupies, but is freed for the engagement or disengagement of gears when the clutch pedal is depressed. With this system a gear change is obtained which is pleasant and entirely simple, but the clutch pedal must always be depressed fully. The synchromesh mechanism operates well on the engagement of the top and middle ratios, provided the lever is not moved too rapidly, although with a little skill it is possible to make quite rapid up or down changes.

The suspension on the Citroen is by torsion bars, with damping by telescopic-pattern hydraulic shock absorbers. The front wheels, which transmit the engine power, are independently sprung, and at the rear a special form of radius-arm-located axle beam serves as an anti-roll stabilizer.

By present-day standards, the riding is moderately firm and well damped, excessive speed over a hump-backed bridge failing to evoke any bounce. At moderate speeds, with the car lightly loaded, there is a fair amount of up-and-down movement of the rear of the car, but at high speeds or with four people aboard the ride levels out most commendably. In any case, however, bumps and potholes of frightening proportions can be tackled at any speed, and singularly fail to influence the car.

Exceptional Cornering

The speed at which corners can be taken, on either wet or dry roads, is altogether exceptional. Above-average vigour of handling, especially on fast, open curves, produces neither roll, sway nor tyre howl. If corners are entered at speeds of which no sports car would be ashamed, there is some audible protest from the tyres, and the car feels to drift outwards slightly, but there is no sign of any disconcerting tendencies and it is a determined driver indeed who manages to skid the car at all. With front-wheel drive, the steadiest fast cornering is obtained with the throttle open, but corners taken quite furiously with the car coasting or the brakes applied also felt perfectly safe.

The steering is reasonably high geared, almost entirely free of lost motion, but rather variable in feel. At high speeds it is beautifully precise and quite light, but it feels rather more sluggish at traffic speeds. The self-centring action is obtained mainly from the engine, so that on the overrun it is slight, but it can be restored by a burst of throttle and is very marked when accelerating in a low gear. There is some vibration to be felt through the rigid two-spoke steering wheel, especially if a hill is climbed at around 30 m.p.h. in top gear, but the car is one which leaves the driver unfatigued after a big mileage.

The engine and transmission, it must be said, did not, at least on the test car, come up to the highest modern standards of silence and smoothness. Induction and exhaust noise are seldom audible inside the car, which cruises most pleasantly about 50 m.p.h. Towards the higher speeds, however, which are readily reached and sustained, there is a fair amount of gear noise.

In keeping with the road-holding qualities, however, is the performance, which is most usefully high. The car accelerates well from rest, with use of the three-speed gearbox. On the open road, too, where top gear is seldom abandoned, there is really excellent power instantly available for acceleration or hill-climbing, and the car seems tireless even if driven at almost its maximum speed.

The smoothness of running at low speeds, in town or elsewhere, is a little

SAFE CORNERING.—The ability of the Citroen to take corners at high speed is one of its best characteristics. Cornering at a more modest pace, the model tested shows off its lines in a pleasing setting.

SQUAT FROM THE SIDE.—The construction of the Citroen provides a low, flat floor, yet gives adequate ground clearance. Here, the long wheelbase emphasizes the car's flattened appearance.

disappointing. Lost motion in the rather flexible throttle linkage appeared to contribute to a certain amount of snatch experienced when crawling through heavy traffic, and the sort of small throttle opening required for a sustained 35 m.p.h. sometimes produced a slight hesitancy or surge.

A certain lack of refinement at low speeds must perhaps be regarded as the price paid for obtaining, at a cash outlay which must be regarded as extremely moderate, the ability to make long journeys at average speeds far above the normal without the car suffering rapid deterioration or the passengers feeling any strain. The Citroen is quite evidently a car which, although offering the spaciousness required by family motorists, will make its strongest appeal to those who do a lot of motoring and do not want to waste more time than is essential.

Economy Pace

The rather unusual results obtained in our fuel consumption tests substantiate this impression that the car is most at home when travelling briskly. It may be noted from the data page that, whereas many cars of comparable size give their best economy at a speed as low as 20 m.p.h., the economy pace for the Citroen is around 35 m.p.h. The overall consumption figure recorded during our tests, approximately 23 m.p.g., is creditable in view of the conditions, which included very fast runs and the negotiation of tracks rarely used by motor traffic.

Front wheel drive being sometimes criticized as giving limited adhesion on steep hills, we deliberately drove the Citroen well off the fairway. In rain, a winding, cobble-stone hill was easily climbed by the fully laden car, with a re-start on the steepest corner. A series of hills with rough and loose surfaces of large stones produced some wheelspin, accompanied by mild steering wheel "fight," but the only gradient which stopped the car was one used for testing military vehicles, a hill which would stop almost any other family car.

In keeping with the character of the car is the general standard of equipment, presentable and very practical but devoid of frills. The head lamps, raised to an unfashionable, but very sensible, height above the road, give an excellent driving light, the red tell-tale glasses of independent side lamps are both within the driver's field of vision, and interior and facia lights are provided.

For cold weather, there is a simple form of heater, a duct extending from the radiator to a controllable vent above the feet of the front seat passenger. For warmer weather, the de luxe saloon tested had winding door windows, sliding roof, scuttle ventilator, and a windscreen which opens for a short distance. During the mild weather in which our test was made, a certain amount of engine warmth was noticed in the front of the car, but it was found that the sliding roof could be left open almost continuously and did not cause unwanted draughts.

The front seat is in two individually adjustable halves, which can be set together for the carriage of a third passenger, and the rear seat is also able to accommodate three people. In view of the suitability of the car for fast driving, it is unfortunate that no arm-rests or handholds are provided in the rear compartment.

The rear luggage locker is of moderate capacity, and a fold-down panel extends the floor when loads which prevent the lid from closing are to be carried. The spare wheel, mounted on the locker door, is secured from within the locked compartment. Inside the car there is a facia cubby hole, sloped sensibly to keep things in place, and pockets are provided on the front doors.

No cold weather was experienced during our tests, but after nights spent in the open the Citroen invariably started at the first touch of the button. Thereafter, the idling with the Solex starter carburetter control pulled out was certain, and the car could be driven straight away.

From a somewhat dubious first impression, strongly flavoured with doubt as to whether the unusual features of the Citroen were justified, we rapidly came to a much more appreciative frame of mind. After upwards of 600 miles with the car, over highways, byways and city streets, we parted with it reluctantly, having acquired much of that enthusiasm for the marque which regular users so often reveal.

DETAIL SHOTS.—(Left) The rear luggage locker, of moderate capacity, carries the spare wheel, secured from inside, on its lid. (Right) An interior view, showing the facia layout and controls.

The Motor Road Test No. 4/48

Make: Citroen. **Type:** Light Fifteen. **Makers:** Citroen Cars Ltd., Trading Estate, Slough, Bucks.

Dimensions and Seating

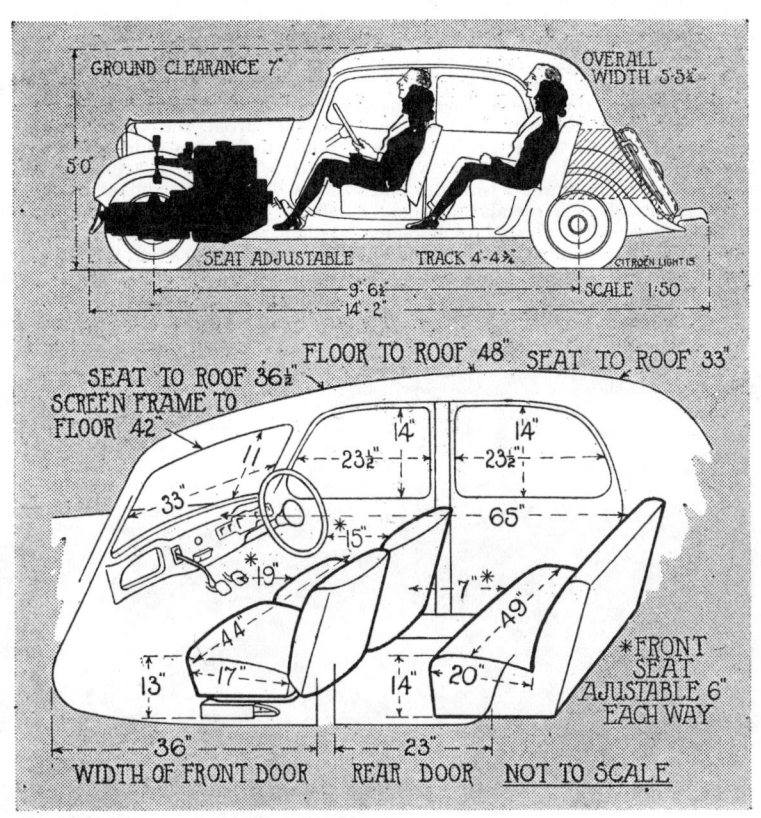

(Diagram showing side and interior views of the Citroen Light 15 with dimensions: Ground Clearance 7", Overall Width 5'5¼", Seat Adjustable, Track 4'4¾", 9'6½", 14'2", Scale 1:50, Seat to Roof 36½", Floor to Roof 48", Seat to Roof 33", Screen Frame to Floor 42", various interior measurements, Width of Front Door 36", Rear Door 23", Front Seat Adjustable 6" Each Way, Not to Scale)

In Brief

Price: £575. Plus purchase tax £160 9s. 6d. = £735 9s. 6d.
Capacity .. .. .. 1,911 c.c.
Road weight unladen .. 21 cwt.
Front/rear weight distribution 55/45
Laden weight as tested .. 23½ cwt.
Fuel consumption .. 22.9 m.p.g.
Maximum speed .. 73.3 m.p.h.
Maximum speed on 1 in 20 gradient .. .. .. 62 m.p.h.
Maximum top-gear gradient 1 in 9½
Acceleration 10-30 on top .. 10.6 secs.
0-50 through gears 14.5 secs.
Gearing, 17.4 m.p.h. in top at 1,000 r.p.m.
66.2 m.p.h. at 2,500 feet per minute piston speed.

Specification

Engine
Cylinders 4
Bore .. .. .. 78 mm.
Stroke .. .. .. 100 mm.
Cubic capacity .. 1911 c.c.
Piston area .. .. 29.6 sq. in.
Valves .. .. Pushrod o.h.v.
Compression ratio .. 6.25
Max. b.h.p. 55.7
at.. .. .. .. 4,250 r.p.m.
B.h.p. per sq. in. piston area 1.88
Piston speed at max.b.h.p. 2,790 ft./min.
Carburetter .. .. Down-draught Solex 35 F.A.I.E.
Ignition .. .. Lucas Coil
Sparking plugs .. Champion J8B
Fuel pump .. .. A.C. mechanical

Transmission
Clutch .. .. Single dry plate
Top gear 4.3
2nd gear 7.3
1st gear 13.1
Final drive .. .. 9/31 Spiral bevel

Chassis
Brakes .. .. Lockheed hydraulic
Brake drum diameter . Front 12 ins., Rear 10 ins.
Friction lining area .. 97.5 sq. ins.
Tyres .. .. Michelin 165 x 400
Steering gear .. Rack and pinion

Performance Factors (at laden weight as tested)
Piston area, sq. ins. per ton .. .. .. 25.2 sq. ins.
Brake lining area, sq. ins. per ton .. .. .. 86.5 sq. ins.
Litres per ton-mile .. 2,810

Fully described in "The Motor," March 20th, 1946

Test Conditions

Warm, light breeze, dry concrete surface, Pool petrol.

Test Data

ACCELERATION TIMES on Two Upper Ratios

	Top	2nd
10-30 m.p.h.	10.6 secs.	5.9 secs.
20-40 m.p.h.	10.1 secs.	6.0 secs.
30-50 m.p.h.	11.0 secs.	8.0 secs.
40-60 m.p.h.	15.6 secs.	—

ACCELERATION TIMES Through Gears
0-30 m.p.h. .. 6.0 secs.
0-40 m.p.h. .. 9.5 secs.
0-50 m.p.h. .. 14.5 secs.
0-60 m.p.h. .. 23.4 secs.
Standing ¼-mile .. 22.8 secs.

MAXIMUM SPEEDS
Flying Quarter-mile
Mean of four opposite runs .. 73.3 m.p.h.
Best time equals .. .. .. 77.6 m.p.h.
Speed in Gears
Max. speed in 2nd gear.. .. 52 m.p.h.

BRAKES AT 30 m.p.h.
0.23 g. (=130 ft. stopping distance) with 25 lb. pedal pressure.
0.44 g. (=68½ ft. stopping distance) with 50 lb. pedal pressure.
0.65 g. (=46 ft. stopping distance) with 75 lb. pedal pressure.
0.80 g. (=37½ ft. stopping distance) with 95 lb. pedal pressure.

FUEL CONSUMPTION
Overall consumption for 343 miles, 15 gallons, equals 22.9 m.p.g.
26.5 m.p.g. at constant 20 m.p.h.
29.5 m.p.g. at constant 30 m.p.h.
29.5 m.p.g. at constant 40 m.p.h.
26.5 m.p.g. at constant 50 m.p.h.
23.5 m.p.h. at constant 60 m.p.h.

HILL CLIMBING
Max. top-gear speed on 1 in 20 .. 62 m.p.h.
Max. top-gear speed on 1 in 15 .. 57 m.p.h.
Max. top-gear speed on 1 in 10 .. 44 m.p.h.
Max. gradient climbable on top gear, 1 in 9½ (Tapley 235 lb. per ton).
Max. gradient climbable on 2nd gear, 1 in 5¾ (Tapley 390 lb. per ton).

STEERING
Left- and right-hand lock .. .. 40 ft.
2½ turns of steering wheel, lock to lock.

Maintenance

Fuel tank: 9 gallons. **Sump:** 8 pints, S.A.E. 40. **Gearbox and Differential:** 3 pints, medium gear-oil. **Radiator:** 14 pints. **Chassis lubrication:** By grease-gun every 500 miles to 17 points. **Ignition timing:** 8° B.T.D.C. **Spark-plug gap:** 0.025 in. **Contact-breaker gap:** 0.012 to 0.015 in. **Valve timing:** Inlet opens 3° B.T.D.C., closes 45° A.B.D.C., exhaust opens 45° B.B.D.C., closes 11° A.T.D.C. **Tappets (hot):** Inlet 0.006 in., exhaust 0.008 in. **Front-wheel toe-out:** 1/16 in. to 3/32 in. at wheel rims. **Camber angle:** 1° 30'. **Castor angle:** 2° 45'. **Tyre pressures:** Front 22 lb.; Rear 24 lb. **Brake fluid:** Lockheed. **Battery:** 12 volts 57 amp.-hours.

NEW CARS DESCRIBED

The Citroen Light

Production of Well-known An[glo-French Car]
Radical Changes in Desi[gn]

MANY will be interested to learn that production of the front-wheel-drive Citroen has recommenced at the Slough works of Citroen Cars, Ltd. Although this car is, of course, basically of French origin, it is an Anglo-French production, as it is largely built at Slough, certain major components being imported from France, but much of the material, minor components and fittings being produced in this country. Under present conditions the proportion of British material is even greater than it was pre-war.

It has for long been the Citroen policy not to introduce seasonal changes of design, but to incorporate modifications as and when experience has shown them to be desirable. At present only the Light Fifteen is being produced in this country, and it is unchanged in general specification from that of the 1939-40 model. Nevertheless, it is an essentially up-to-date design, as when it was first conceived it possessed special features which were considerably ahead of the times. Amongst these special features are the integral construction of chassis and body, giving strength and lightness, independent front suspension and torsion bar springing—both of which are now becoming more generally used—and front-wheel drive.

The basis of the body construction is the floor unit, which has the side edges formed into strong box girders. The front of this also forms the lower part of the scuttle, and is reinforced by four tubes which project forward to receive the front suspension unit. To this floor the steel pressings which form the body structure proper are welded. Thus the whole becomes a section of a hollow box girder which, therefore, combines great strength and rigidity with lightness.

The front suspension unit consists of a pressed steel cradle which carries the superimposed wishbone links supporting the king pin and stub axle assemblies with the wheels and the universally jointed driving shafts. This unit is received by the tubes projecting from the scuttle assembly, there being two tubes at each side, and it is bolted in position, so that, when necessary, it can be easily removed and replaced. The torsion bars are attached to the lower wishbones and project rearwards, their ends being serrated to engage with serrated anchorages formed in a cross-member which is bolted into the body structure. The method of anchoring the ends of the torsion bars provides a ready means of adjustment.

Hydraulic dampers check the action of the torsion bars.

Flexibly mounted in the normal position is the power unit, which has the clutch and gear box projecting in front of it. Cylinder jackets and crankcase are formed of a single iron casting which is very rigid, and the cylinder barrels are detachable and are of a special wear-resisting and corrosion-proof cast iron. Thus, although cylinder wear is reduced to a minimum, when it does occur the barrels are easily withdrawn and renewed. They are held in place by the detachable cylinder head, carrying overhead valves operated by push-rods and rockers from the camshaft, set rather high on the near side. The camshaft is carried in three bearings and is driven by a double roller chain from the crankshaft, which is counterweighted, balanced both statically and dynamically, and supported in three bearings. Light alloy split-skirt pistons carry four rings; two are used for oil control.

Lubrication is by a gear-type pump to all bearings and also to the timing chain and the rocker arms, an oil tell-tale light being fitted on the instrument board. The cooling water is circulated by a pump, belt driven from the camshaft and carrying a six-bladed fan on its spindle. A simple system of car heating is provided; close behind the near side of the radiator there is fitted a pressed steel funnel, through which hot air passes to an inlet in the dash.

A Solex downdraught carburettor is supplied by a mechanically operated fuel pump from the 9-gallon tank, and it is fitted with a combined air filter and silencer of large size. The distributor is driven from the upper end of the oil pump drive and is provided with automatic advance mechanism. Lucas 12-volt electrical equipment is fitted, the dynamo being driven by the fan belt.

In unit with the engine is a dry single-plate clutch with a flexible centre to give smooth engagement, and a three-speed gear box and spiral bevel final drive, the latter being positioned between the gear box and the clutch. The rear axle is a simple beam of cruciform section carrying the wheels and it is fitted with diagonal radius rods. The rear suspension is likewise by adjustable torsion bars, which, however, are set parallel to the

Appearance of the four-door four-light saloon is unchanged. Note the absence of running boards.

Instruments are grouped in front of the driver. The gear lever projects from the facia board.

SPECIFICATION

Engine.—15.08 h.p., 4 cylinders, 78×100 mm. (1,911 c.c.).

Transmission.—Dry single-plate clutch, and three-speed gear box. Synchromesh on top and second. Ratios: Top 4.3; second 7.3; first 13.1 to 1. Front wheel drive.

Steering.—Rack and pinion. Turning circle 40ft.

Brakes.—Lockheed hydraulic.

Suspension.—Torsion bar. Independent at the front.

Carburettor.—Solex downdraught.

Tank Capacity.—9 gallons.

Electrical Equipment.—12-volt coil and distributor.

Tyres.—165×400 mm. Michelin super low pressure. Broad-base wheels.

Main dimensions.—Wheelbase: 9ft. 6½in. Track: 4ft. 4¾in. Ground clearance: 7in. Weight: 20½cwt.

Fifteen

...ch Car Recommened : No ...Improvements in Detail

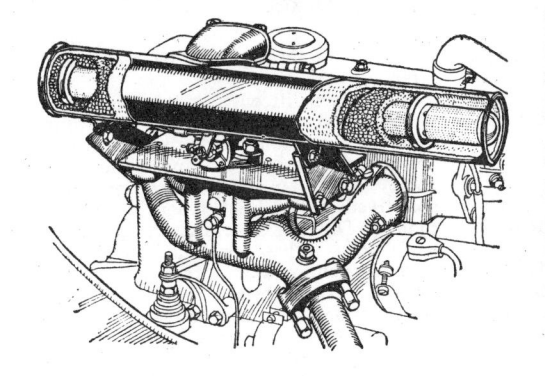

A large air filter and silencer is fitted to the carburettor.

W de doors give easy access to front and rear seats. The floor is totally unobstructed, a matter of great convenience.

axle beam, controlled by hydraulic dampers.

Brakes are Lockheed hydraulic with the hand brake acting independently on the rear wheels through the usual cable operation. The wheels are the pressed steel Michelin broad-base type and carry 165 × 400 mm. low-pressure tyres. The steering is by rack and pinion with spiral gear teeth. The rack operates direct on the push and pull rods, which also form the track rod. The arrangement is simple, light and gives very accurate steering unaffected by movement of the road wheels.

Features of the four-door, four-light, all-steel body are its roominess, its unobstructed floor and the easy access to both front and rear seats. Moreover, the occupants sit within the wheelbase, a fact which contributes to the comfortable riding of the car. The front seats are separate and both adjustable, and the rear seat comfortably accommodates three passengers.

Upholstery is of leather in normal fluted style, and the instrument board and window fillets are of polished walnut. The instruments are neatly grouped in a rectangular panel in front of the driver and include speedometer, ammeter, petrol gauge and clock, whilst as the gear lever projects through the centre of the facia board the floor of the front compartment is also unobstructed. The screen opens under the action of a centrally placed control, useful pockets are formed on the door trim and a large parcels shelf is provided at the back of the rear seat. The equipment also includes sun visor, interior light, ashtrays, rear blind and a glove recess in the facia board.

Ample luggage accommodation is provided by a locker in the rear panel with a folding extension flap for use when the lid of the locker is in the open position. In appearance the car is unchanged, and it is distinctive by the absence of running boards, by its sloping radiator grille and low bonnet line with large ventilator panels in the sides of the bonnet. The spare wheel is sunk in the rear panel with a cover which provides a theft-proof fixing. Chromium-plated full-length bumpers are fitted at front and rear, and the specification includes twin horns as well as dual screenwipers and Trafficators.

The price of the saloon is £448, plus purchase tax £125 3s. 11d., a total of £573 3s. 11d.

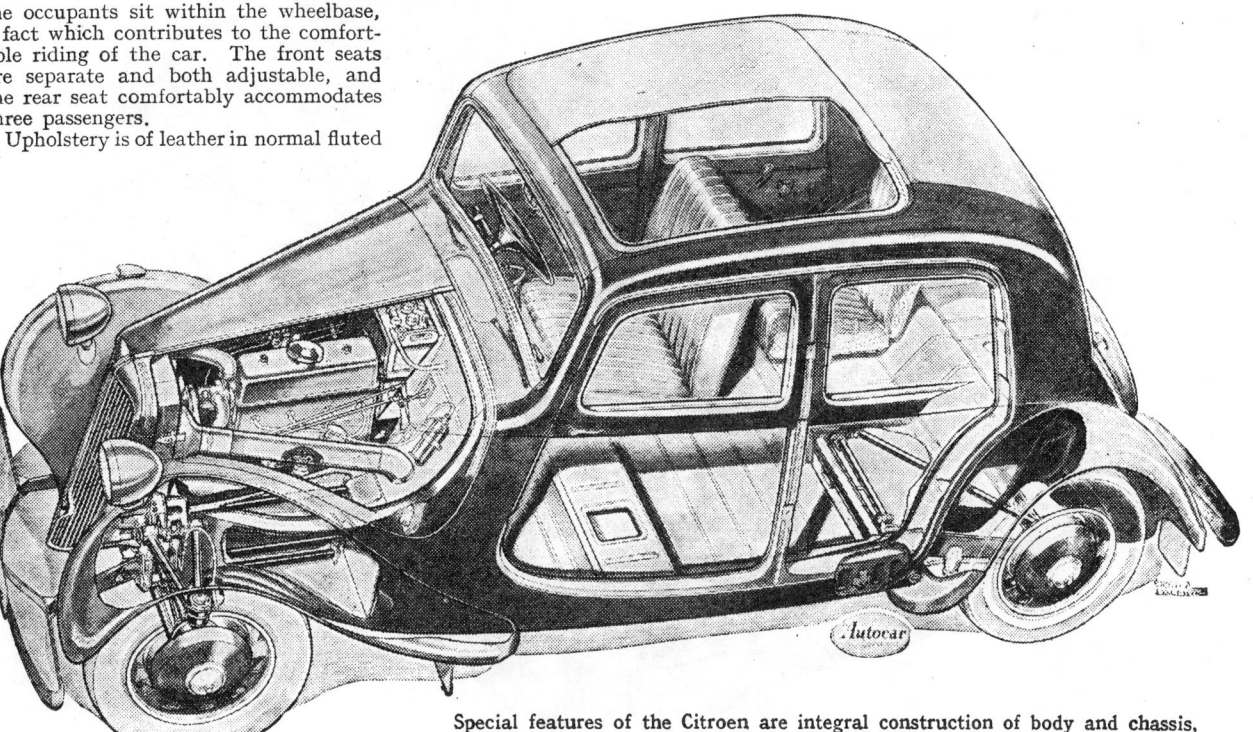

Special features of the Citroen are integral construction of body and chassis, independent front suspension, front wheel drive and torsion bar springing.

NEW CARS DESCRIBED

CITROEN SIX for BRITAIN

Successful Larger Model to Supplement Light Fifteen : New Small Car Expected Soon on French Market

THE Citroen Six, which would have appeared on the British market in 1940 but for the war, is to be included in the 1949 production programme of the factory at Slough. This car follows the general layout of the well-known four-cylinder models, with integral body and chassis structure, front-wheel drive, and independent front suspension by torsion bars, but it is appreciably larger, having six-seater bodywork and a 2.8-litre six-cylinder engine giving a maximum of 76 b.h.p.

The car was originally described in *The Autocar* of August 18, 1939, and the good impression it made in the course of a short road test has since been confirmed by its growing reputation for stability and effortless performance. Various small modifications have been incorporated in the latest model, which will be seen at the London Show. The direction of rotation of the engine has been reversed, so that it is now the same as on other cars.

Compact Power Plant

The transmission layout differs in detail from that of the four-cylinder model, as the engine is carried well forward, with its first cylinder over the differential casing in order to cut the length of power unit and transmission assembly to a minimum. The clutch shaft passes over the top of the differential and the drive passes down through spur gears to a three-speed gear box with synchromesh between second and top gears. On the four-cylinder models the engine lies right behind the differential and the drive passes straight across to the gear box without the intermediate step down.

The gear box of the Six is provided with its own eccentric-vane oil pump which lubricates all bearings under pressure, and there is the usual Citroen arrangement whereby the selector locks are connected to the clutch pedal so as to ensure light operation for the facia-mounted gear lever.

On a pedestal above the transmission casing is a belt-driven dynamo partly shrouded within the spinner of a massive six-bladed fan which is carried on the armature shaft. Alongside is the water pump, driven by a separate belt, and delivering cool water from the base of the radiator to a manifold connecting with the passages round the hardened cylinder barrels.

Both inlet and exhaust manifolds are on the off side of the engine, the inlet system being cast in aluminium with a substantial hot-spot, and fed from a Solex twin-choke downdraught carburettor. The sump oil level is shown by a float-operated indicator. In accordance with usual Citroen practice the car will be trimmed and finished in England and will have British 12-volt constant-voltage-control electrical equipment and British accessories.

Interior trim will be in leather, with bench-type seats having folding central arm-rests at both front and rear. A sliding roof will be available. The bumpers are a new design of massive appearance, with stout over-riders, and they incorporate frames for the number plates. The price is to be announced later.

The Light Fifteen four-cylinder model is continued without change. The four-cylinder Citroen is undoubtedly one of the classic designs of automobile history. It was far ahead of its time when conceived by André Citroen, a Frenchman of Dutch extraction, in 1932, and since 1934 it has continued in production without major alteration. Total output has reached a figure of somewhere about a quarter of a million despite the hiatus imposed by the war.

The British version of the Light Fifteen is trimmed in leather and has a split-bench type of front seat with the two halves separately adjustable. Both cars, besides being sold in Britain, will be exported to numerous foreign markets which have shown a liking for the anglicized version of this famous French *marque*.

It is expected that an entirely new small Citroen will be shown for the first time at the Paris Salon early in October, but is unlikely to be available on the British market for some time. The Citroen designers were said to be working on a smaller model than the present range even before the war, and reports suggest that mechanically it will have much in common with the cars now in production.

A cutaway view of the Citroen Six. The cylindrical housings on the drive shafts contain rubberized couplings which cushion the drive.

No. 1355 CITROEN LIGHT FIFTEEN SALOON

The Autocar ROAD TESTS

DATA FOR THE DRIVER

CITROEN LIGHT FIFTEEN

PRICE, with four-door four-light saloon body, £575, plus £160 9s 6d British purchase tax. Total (in Great Britain), £735 9s 6d.
RATING: 15.08 h.p., 4 cylinders, overhead valves, 78 × 100 mm, 1,911 c.c.
TAX (in Great Britain), £10.
BRAKE HORSE-POWER: 55.9 at 4,250 r.p.m. **COMPRESSION RATIO**: 6.2 to 1.
WEIGHT, without passengers: 21 cwt 3 qr 21 lb. **LB. per C.C.**: 1.29.
TYRE SIZE: 165 × 400 broadbase on bolt-on perforated steel wheels.
LIGHTING SET: 12-volt. Automatic voltage control.
TANK CAPACITY: 10 gallons; approx. fuel consumption range, 23-28 m.p.g.
TURNING CIRCLE: 41ft (L and R). **MINIMUM GROUND CLEARANCE**: 7in.
MAIN DIMENSIONS: Wheelbase, 9ft 6½in. Track, 4ft 4¾in (front and rear).
 Overall length, 14ft 2in; width, 5ft 5in; height, 5ft 0in.

ACCELERATION

Overall gear ratios	From steady m.p.h. of		
	10 to 30	20 to 40	30 to 50
4.30 to 1	12.7 sec.	11.9 sec.	12.6 sec.
7.30 to 1	6.7 sec.	7.1 sec.	9.6 sec.
13.10 to 1	—	—	—

From rest through gears to :—
 30 m.p.h. .. .. .. 7.0 sec.
 50 m.p.h. .. .. .. 15.6 sec.
 60 m.p.h. .. .. .. 20.6 sec.

Steering wheel movement from lock to lock: 2½ turns.

Speedometer correction by Electric Speedometer :—

Car Speedometer	Electric Speedometer	Car Speedometer	Electric Speedometer
10	= 12	50	= 47
20	= 20.5	60	= 56
30	= 29.5	70	= 65.5
40	= 37.5		

Speeds attainable on gears (by Electric Speedometer)
	M.p.h. (normal and max.)
1st	20—26
2nd	40—50
Top	76

WEATHER: Dry, mild; light wind.
Acceleration figures are the means of several runs in opposite directions.

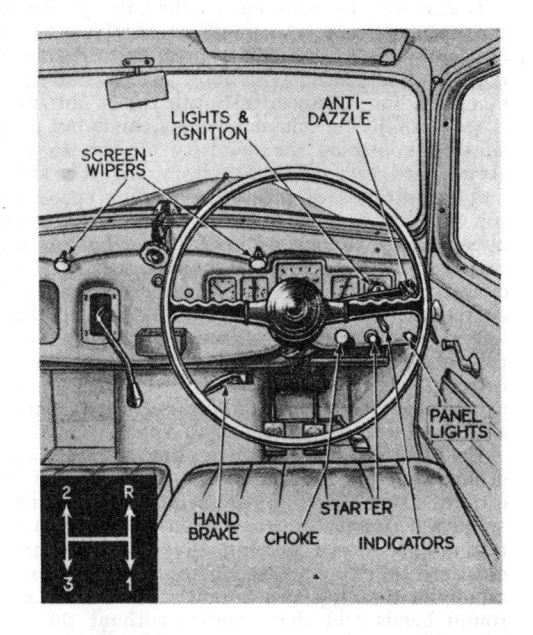

ANYONE who is interested in car design cannot fail to be struck on approaching the Citroen afresh after a lapse of years by thoughts concerning the extent to which features embodied in this car have been followed elsewhere. Fourteen years ago Citroen in France adopted front-wheel drive and a design which as a whole was then almost revolutionary, and certainly ahead of its time for a car destined to be made in large numbers. Features which the Citroen then displayed as originalities, or comparatively so, were torsion bar suspension of all four wheels, independently in front, combined construction of body and chassis, rack and pinion steering, detachable cylinder barrels, and flat floors in both compartments.

As is well known, the Citroen as seen in Great Britain and exported to practically every British territory, large and small, is assembled in this country, and is finished and equipped on British lines in a totally different way from the purely French version. The current edition is a car in no way fundamentally different from that which was being built before the war; as will be realized, when it was first introduced the design was advanced, and it is one that has stood the test of time, gaining the front-wheel drive model a good name all over the world among those who have had the opportunity of knowing this car intimately.

An even more extensive test than is usual, such as has now been made by *The Autocar*, reveals a car of highly individual type, which today one is even more inclined than before the war to describe as unique, in spite of the fact that a number of its design features, excepting the front-wheel drive, are nowadays to be found in principle in other current cars. This Light Fifteen, which is at present the only model being produced in England, has an overhead valve four-cylinder engine of nearly 2-litre capacity. The total weight is low and therefore the power-to-weight ratio is decidedly favourable, and the car can run on a quite high top gear ratio, with the result that engine speed is still moderate at the higher road speeds, and wear and tear are reduced.

This car's true character might not have been revealed

"THE AUTOCAR" ROAD TESTS ... continued

had it not been possible on this occasion to take it over a long-distance journey under present traffic-free conditions in England. Pre-eminent is its apparent ability to travel all day at between 50 and 60 m.p.h. when that sort of motoring is required. Here, clearly, is an inbuilt characteristic derived from motoring conditions imposed by the long straight roads of France, where not so much a very high maximum speed is required as the ability to keep up 60 and thus put, say, 500 miles into a twelve-hour run. At the end of a long fast journey the highest possible respect had been formed for the Citroen's rugged qualities, its ability to stand up to sustained wide throttle openings, its reasonable petrol consumption even when driven in this style, and its almost nil oil consumption, a recording in the region of 5,000 m.p.g. being noted in this direction. Furthermore, after the severe treatment in question the engine note had not in any way altered, the tick-over was still quiet, the valve gear had not developed clatter, the water level had not altered, and no oil leaks were evident.

Suitability for Rough Work

It can well be believed, as the Citroen firm claim, that the car particularly attracts as regular users those who have to make long, fast journeys, and those in other countries who are faced with long stretches of bad road surfaces, as, for instance, in South Africa, for its suspension is remarkable. Mechanically this car is not so quiet or so silkily smooth as the majority of cars to which one is accustomed today, nor are its strongest points flexibility and smoothness of pick-up from low speed on the high top gear ratio. On the other side of the picture, as has already been indicated, it is, however, outstanding for a car of moderate size and running costs. It is a French outlook on motoring—and basically the car is French, of course, although assembled and finished in this country—not to be bothered about some mechanical noise provided that a car is fast and roadworthy. In these respects the Citroen is typical, and it appeals strongly as a thoroughly practical, sturdy, go-anywhere and withal inherently likeable form of transport.

Its suspension is of outstanding merit in insulating the occupants from shock, almost irrespective of surface conditions, and in addition it is taut and provides complete stability. This factor, in conjunction with firm and accurate fairly high geared steering, gives the car the feel of the old-time sports car without the riding discomfort associated with that type of machine. The Citroen can be swung round bends and sharp curves without the slightest hint of heeling over, whilst the tyres feel as if they are glued to the road.

The driver sits well up to a large single-spoked rigid wheel, more upright, owing to the shape of the back rest,

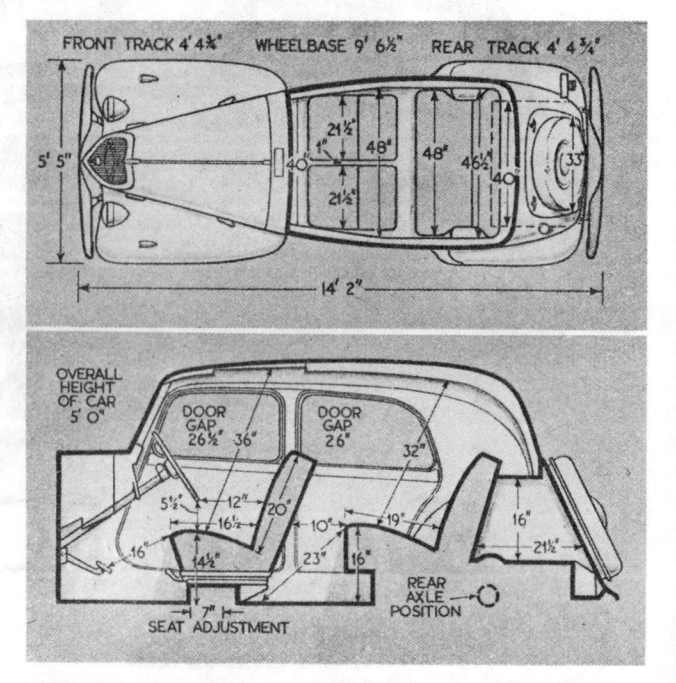

Measurements are taken with the driving seat at the central position of fore and aft adjustment. These body diagrams are to scale.

than is usual today, in a position where he feels that he has complete command over the car in all circumstances. The view outwards is excellent, with both wings visible and good vision over a non-obstructive bonnet. Within a few minutes of acquaintance the driver feels that he can place the car exactly where he wants it and put it through the narrowest of gaps with certainty.

Thus with a maximum speed in reserve exceeding a genuine 75 m.p.h. on level ground, and a readiness to get quickly up to 65 or so and to cruise at 65 to 70 m.p.h., one has a car which can achieve some rather startling average speed performances, especially under present conditions in England. Averages of over 40 m.p.h. from point to point are the Citroen's normal mode of progression, and interesting performances recorded in this connection, even when making allowances for a somewhat optimistic mileometer, were 92 miles covered at night in two hours' total time inclusive of an eight-minute stop, and a recording hardly credible for this country—27 miles put into thirty minutes. When a top speed on the level of 76 m.p.h. was recorded by electric speedometer the car's speedometer showed a reading of just over 80, but on a slight down grade a speedometer reading of 86 was seen, and it still did not seem that the car was being overdriven.

Emphasis which has been laid upon this car's being seen to best advantage on a straightaway journey should not be construed as meaning that it is intractable or unpleasant for normal "potter" motoring or in town traffic. Its top gear range is good down to quite low speeds, and it is seldom brought off top of the three-speed box by a main road gradient. In town, indeed, it is handier than

Wide door openings and flat floors at both front and rear, sunk below the level of the door sills, are features of the comfortable and roomy four-light body. The seats give firm support right up to the shoulders and the separate front seats almost meet, thus providing occasional three-seater accommodation in front, the gear lever and hand brake not forming obstructions.

most modern cars owing to the excellent visibility and the precise control that one has over it, making the driver perfectly happy when at close quarters with other vehicles. Second gear will pull it away in most circumstances where top does not suffice, and the synchromesh between second and top gears is effective provided that the clutch pedal is fully depressed; first gear is not easily engaged quietly when the car is in motion.

The precise advantages gained from front-wheel drive, and any corresponding disadvantages, are to some extent technically controversial. What can be said from this extended renewed experience of the foremost exponent of this principle, in terms of number of cars built during the past fourteen years, is that a driver without knowledge of the construction might well remain unaware from the Citroen's road behaviour that it did not follow the normal practice of rear-wheel transmission. During the test bends were taken at all varieties of speeds; deliberate attempts were made to introduce instability by letting the engine overrun in a bend instead of taking the car round on the pull, and wet roads were experienced for considerable distances, but at no time was there any unusual feeling of insecurity. If the throttle foot is lifted abruptly in a bend some slight reaction on the steering is noticeable, but not to a degree that presents any worry or problem.

Control Layout.

A point of unconventionality is the pedals, which involve a downward pressure instead of the usual approximately horizontal movement. As regards the clutch pedal in particular, a projection into the driving compartment is apt to be touched by the toe of the left shoe, and a driver strange to the car is not altogether happy at first with this movement, but it can certainly be said that no problem is presented in this connection after a day or two spent with the car. The gear lever, projecting towards the left of the instrument board, is more convenient than it may appear at first sight and has more obviously apparent movements, in an open gate, than is general today. The hand-brake lever could well be within closer reach.

The main brakes, hydraulically operated, although not of special pattern as regards shoe design, are exceptionally effective, outstandingly smooth in application, need only light pedal pressure, and bring the speed down with velvet-like but really powerful certainty. They remained efficient, too, after hard driving. The steering is on the heavy side for low-speed turning, but really scores for its precision and lack of transmitted road shocks at speed; it has some castor action.

The seats prove very comfortable, although they are not specially soft, and the upholstery is in good quality leather. A most convenient and practical grouping of horn, dip and traffic signal switches is made on an arm on the right of the steering column. Absence of obstructing controls makes it easy for the driver to get in and out by either front door, and also permits the occasional seating of three in front, for the separate front seat cushions and back rests all but touch. A central arm rest is not provided at the rear, but even with unusually rapid cornering methods a back-seat passenger travelling alone is not thrown sideways.

Equipment includes a sliding roof, a sun vizor for the driver, ash trays, a useful instrument board cubby hole and a rear-window blind. The view given by the driving mirror could be better. An excellent provision is a simple but effective form of interior heater. Warm air is led in from immediately behind the main radiator through a duct with its outlet, which can be shut off immediately, in front of the front passenger. No trace of fumes was noticed from this system.

An almost instantaneous start was obtained at all times from cold and the engine quickly warmed up to its work. The main head lamp beam is excellent for fast driving, and a good beam satisfactory to oncoming drivers as well as to the Citroen driver is given in the dipped position.

Quite useful enclosed luggage space is provided, or for extra accommodation the lid can be left open, a metal plate then dropping to fill the space between the lid and the floor of the boot, and thus act as a platform. The spare wheel is in the lid of the luggage locker and the tools are in a separate compartment reached by opening this lid.

From the front the front-wheel-drive Citroen does not display to the world that its transmission pulls the car instead of pushing it in the conventional manner. In rear view, too, the car is trim. Indeed, from any angle it looks what it is, a purposeful, functional design, carrying more than the bare necessities of equipment, but without frills and pointless decorations.

1949 CARS
Front-wheel-drive CITROENS—

BIG BROTHER.—Distinguished from the Light Fifteen by scaled-up dimensions and different bumpers, the new Six Cylinder combines spaciousness with speed

—in Two Sizes

A Six-cylinder Saloon Now Being Produced in England Similar in Layout to the Light Fifteen which Continues Unchanged

IMITATION, the sincerest form of flattery, has been enjoyed by Citroen Cars, Ltd., ever since they announced their first front-wheel-driven model in 1934, steadily increasing numbers of manufacturers having followed their example in the use of monocoque construction, independent suspension by torsion bars, and renewable cylinder liners. For 1949 the well-proven four-cylinder Light Fifteen model is being continued without change, but a larger six-cylinder car is joining it on the production lines at Slough.

Both cars, although basically of French design, are built in England, for the British market and for export to many parts of the world. Body pressings and many mechanical parts are imported from the parent factory, but a very large proportion of British parts are incorporated in the Slough-built models.

The Light Fifteen model, having undergone evolutionary development over a long period of years without drastic design changes, is so well known as to need little description. Structurally, it is based on a welded steel body, the box section members of which form a light yet rigid and strong frame. Elimination of the orthodox chassis and the absence of a propeller shaft enable a very low floor level to be attained, so that moderate height does not imply restricted headroom.

The complete engine and transmission system form a single unit ahead of the windscreen, readily removable for major service operations. Power is taken from the front end of the crankshaft, through a single dry-plate clutch to the upper shaft of the gearbox, thence through a pair of gears appropriate to whichever of the three forward ratios and one reverse ratio is selected to the

CITROEN DATA

Model	Light Fifteen	Six Cylinder
Engine Dimensions :		
Cylinders	4	6
Bore	78 mm.	78 mm.
Stroke	100 mm.	100 mm.
Cubic capacity	1,911 c.c.	2,866 c.c.
Piston area	29.63 sq. ins.	44.44 sq. ins.
Valves	Pushrod o.h.v.	Pushrod o.h.v.
Compression ratio	6.2 to 1	6.2 to 1
Engine Performance :		
Max. b.h.p.	55.7	76
at	4,250 r.p.m.	3,800 r.p.m.
Max. b.m.e.p.	114 lb./sq. in.	118 lb./sq. in.
at	2,000 r.p.m.	2,000 r.p.m.
B.h.p. per sq. in. piston area	1.81	1.71
Peak piston speed, ft. per min.	2,790	2,620
Engine Details :		
Carburetter	Solex downdraught	Solex twin-choke downdraught
Ignition	Coil	Coil
Plugs : Make and type	Champion J8B	Champion J8B
Fuel pump	AC mechanical	AC mechanical
Fuel capacity	10 gallons	15 gallons
Oil capacity	8 pints	12 pints
Cooling system	Pump and fan	Pump and fan
Water capacity	14 pints	21 pints
Electrical system	12-volt	12-volt
Battery capacity	57 amp./hrs.	57 amp./hrs.
Transmission :		
Clutch	Single dry plate	Twin dry plate
Gear ratios :		
Top	4.3	3.875
2nd	7.3	5.62
1st	13.1	13.24
Rev.	17.5	15.87
Prop. shaft	Nil	Nil
Final drive	9/31 spiral bevel	8/31 spiral bevel
Chassis Details :		
Brakes	Lockheed hydraulic	Lockheed hydraulic
Brake drum diameter	Front, 12 ins. Rear, 10 ins.	Front, 12 ins. Rear, 12 ins.
Friction lining area	88 sq. ins.	143 sq. ins.
Suspension :		
Front	Torsion bars and wishbones	Torsion bars and wishbones
Rear	Torsion bars and axle	Torsion bars and axle
Shock absorbers	Hydraulic (telescopic)	Hydraulic (telescopic)
Wheel type	Steel disc	Steel disc
Tyre size	165/400 Broadbase	185/400 Broadbase
Steering gear	Citroen rack and pinion	Citroen rack and pinion
Steering wheel	3-spoke	3-spoke
Dimensions :		
Wheelbase	9 ft. 6½ ins.	10 ft. 1½ ins.
Track :		
Front	4 ft. 6 ins.	4 ft. 10½ ins.
Rear	4 ft. 5¼ ins.	4 ft. 10½ ins.
Overall length	14 ft. 2 ins.	15 ft. 11 ins.
Overall width	5 ft. 5¼ ins.	5 ft. 10 ins.
Overall height	5 ft.	5 ft. 1 in.
Ground clearance	7 ins.	7 ins.
Turning circle	43 ft.	45 ft. 6 ins.
Dry weight	21 cwt.	26 cwt.

Performance Data :		
Piston area, sq. ins. per ton	28.2	34.1
Brake lining area, sq. ins. per ton	84	110
Top gear m.p.h. per 1,000 r.p.m.	17	20
Top gear m.p.h. at 2,500 ft./min. piston speed	66	76
Litres per ton-mile, dry	3,140	3,315

lower gearbox shaft, at the rear of which the final drive bevel pinion is formed. Universally jointed shafts extend outwards from the bevel gear and differential unit to the independently sprung front wheels, smooth transmission being ensured by the pairing of Hookes joints to form constant-velocity universals.

Front suspension is by a pair of long transverse wishbones on each side of the car, the lower wishbones being splined to fore-and-aft torsion bar springs; damping is by telescopic shock absorbers. At the rear, transverse torsion bars located below the back seat are linked by trailing radius arms to the extremities of an axle beam, this being a forging of X section designed to be rigid in bending yet provide a desirable small degree of flexibility in torsion. Lateral location

FRONT END REFINEMENTS.—The six-cylinder model incorporates suspension wishbones of built-up construction, and each wheel is driven through a rubber-in-torsion cushion unit as shown on the left.

CONCENTRIC.—Gearbox length is reduced by arranging for one gear cluster to slide on the outside of the synchromesh clutch assembly, as shown in the drawing, the compact transmission thus attained being indicated by the photograph lower left.

of the body relative to the rear axle is provided positively by means of a transverse stabilizer rod.

The new six-cylinder model bears a very strong family resemblance to the Light Fifteen, in both appearance and mechanical specification, although it is of larger overall dimensions.

Components in Common

Many components are common to both the engines, which employ overhead-valved cylinders of identical bore and stroke so that the six-cylinder model has precisely 50 per cent. more displacement and piston area than the four. There are, however, considerable changes in the transmission layout, which have saved valuable space and avoided any great increase in the length of the power unit.

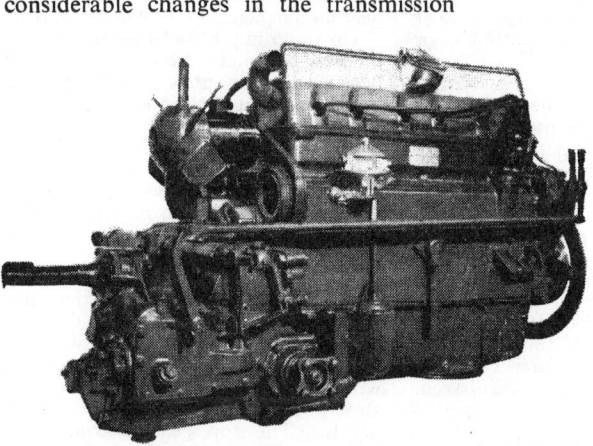

The "six" transmission incorporates, firstly, a twin dry-plate clutch at the nose of the engine crankshaft, from which a pair of gears transmit power to an intermediate shaft at a lower level. The pairs of gears providing the three forward ratios take the drive from this shaft onto a shaft below it which carries the bevel pinion at its rearward end.

The modest diameter of the twin-plate clutch and the use of an intermediate shaft enable the clutch to be located almost directly above the bevel gears, so that the nose of the engine is farther forward relative to the front wheels than is that of the more compact four-cylinder type. Unwanted frontal overhang is avoided by the ingenious length-economizing layout of the gearbox, the first- and reverse-gear pinion sliding on the splined outside of the second-third gear synchro-dog clutch assembly, which, in turn, slides on the gearbox lower shaft

A number of details contribute to the attainment of complete smoothness of running on this model. A torsional vibration damper is mounted at the rear of the crankshaft, rubber-cushion drive units are incorporated in the drive shafts to the front wheels, and a twin-choke Solex carburetter ensures even mixture distribution.

The torsion-bar suspension system is on similar lines to that of the Light Fifteen, as is the self-adjusting rack-and-pinion steering gear. Braking power has been increased, however, to suit extra speed and weight, by an increase in rear brake drum diameter, and by the use of two leading-shoe Lockheed hydraulic front brakes. Broad-base tyres are fitted to both models.

An established favourite, well known for sturdiness, good road holding and lively road performance, the Light Fifteen will continue to form the major part of the Citroen output, but the Six Cylinder which is to be produced in smaller numbers should appeal strongly to those who, accepting inevitable higher cost and fuel consumption, seek greater speed, refinement and space.

The price of the Light Fifteen remains unchanged, and no price is yet announced for the Six.

The... CITROEN FIFTEEN

A car of character, brisk performance, and offering ample accommodation

THE Citroen Light Fifteen Saloon is unusual in technical specifications and in its road qualities. Its unconventional features do not become fully appreciated before the coverage of many miles on the road. Then the reasons behind the general design become apparent and one can appreciate why so many motorists remain faithful to this marque.

The unique character of the car is linked with front-wheel drive, chassisless construction, and torsion bar suspension, all features introduced some years before the war.

It is a tribute to its designer that although this model is substantially the same as when introduced it looks right up to date, of modern individualistic appearance, but resisting the "new look." It has character all its own.

PERFORMANCE

Acceleration in m.p.h.	Top	Second
20-40	10.2 secs.	6.1 secs.
35-50	11.1 secs.	8.1 secs.
Through Gears		
0-30 m.p.h.	6.1 secs.	
0-40 m.p.h.	9.6 secs.	
0-60 m.p.h.	23.6 secs.	

The first entry of the driver into the car displays unusual features. Through wide doors one steps down and sits on seats upright, but set low. The absence of the propeller shaft for rear drive plus the employment of the body as a frame provides a flat floor front and rear.

Practical Features

The driver has unusual visibility. Seated upright, which is the correct position, he enjoys full sight of both wings and side lamps, and of the rather high mounted headlamps. Initially, pedals of the piano type appear both unusual and strange. But depressed downwards they are comfortable and can be operated without moving the heel from the floor.

The gear change, too, is unusual, with the lever in a wide gate centrally disposed on the dash, but conveniently to hand. Light remote control is obtained by the locking of the lever in whichever position it occupies. Full depression of the clutch pedal frees the lever for engagement or disengagement of the gears. This method gives a gear change which is light and simple, but comparatively slow. In top and second the synchromesh mechanism operates easily, but it is difficult to avoid gear scrape when engaging low gear. It is also important to remember that in the upward ranges gear engagement calls for full clutch depression.

Suspension is firm and well damped. With full confidence, and without passenger jolting, the driver can speed over hump-backed bridges, deep potholes, or the roughest of surfaces.

The front-wheel drive enables corners to be taken at sports car speeds without ill effect. The correct procedure is to keep the engine going hard. Vigorous handling round open curves provokes neither sway, roll, or tyre scream.

At speed the steering is precise and light, but on the sluggish side when traffic crawling. There is a certain amount of vibration transferred to the two-spoked steering wheel when pulling in the lower ratios on hills. But generally speaking the steering is of the fatigueless type, and gives no indication that the front wheels are both steered and driven.

Fron Wheel Drive Impressions

The front-wheel drive gives a new conception of motor travel, especially up hills. One is pleasantly conscious of being pulled by an invisible string. But the general impression gives the feeling that come what may the Citroen can tackle any road conditions powerfully and with certainty. Up hills the power is obvious. Main road gradients of 1 in 10 were climbed at 43 m.p.h. and a 1 in 20 at 60 m.p.h. Front-wheel drive is sometimes criticised on the score of limited adhesion on hills. Deliberately we chose some single figure gradients with loose surfaces. There was no suspicion of wheelspin or slip.

Up to 55 m.p.h. both engine and transmission are quiet. Above that speed there is a certain amount of gear whine which is not unpleasing to those mechanically minded.

Performance is of a high order. The car accelerates readily from rest with full use of the three-speed gearbox. The tenacious engine shows reluctance to come off top gear. There is a surge of power on tap for hills or on the level.

At traffic speeds there is a certain lack of refinement in the form of transmission snatch. But that is a minor detail compared with the ability of the car to maintain really high averages in comfort, and with the economy of some 29 m.p.g. at a steady 40 m.p.h., and 26 m.p.g. at 50 m.p.h. The top gear maximum is 75 m.p.h., with a useful 52 m.p.h. in second. There are 2½ turns from lock to lock.

The model tested was the de luxe saloon with sliding roof, opening screen, and air conditioning. The equipment was practical but without unnecessary elaboration. The high mounted headlamps gave a good beam for fast driving, and the visible red tops of the side lights were comforting.

An inexpensive six-seater—the 2.8-litre Citroen Six.

BRITISH CITROENS

TWO POPULAR MODELS WITHOUT MAJOR ALTERATION : MINOR IMPROVEMENTS TO THE LIGHT FIFTEEN

IF the announcement is once again made that the Citroen Light Fifteen and six-cylinder models are to be continued substantially unchanged for another year, this is certainly not likely to have any adverse effect on the brisk demand experienced for the two models assembled at the British factory at Slough. The Light Fifteen has now undergone no fundamental change in mechanical specification or external appearance since the design left the drawing board seventeen years ago, yet the car is still eagerly bought by discerning drivers in many different countries. This year it has proved capable of making best performance in the Alpine Trial and winning the most important team prizes in the Alpine Trial and the International Evian Alpine Rally. A different kind of tribute to its ubiquity, performance and nimble behaviour has recently been paid by the most expert Continental jewel thieves, whose devotion to the Light Fifteen has gained them the name of *"Traction-Avant Bandits."*

After seventeen years it is right up-to-date in specification, and in point of numbers sold it is still one of the most popular cars produced in Europe, surely an achievement without parallel in automobile history. Although the basic design has remained unchanged there has, of course, been a steady process of improvement and refinement and this has continued during the past year. The latest cars to be assembled, trimmed, equipped and finished in Britain have a new type of pedal for brakes and clutch, with a longer arm and higher pivot which gives a more conventional arc of movement and eliminates the need for pressing sharply downwards on the pedals, which was a feature of past Citroens. This has been combined with a rod instead of a cable to control the clutch, and in redesigning the operating levers it has been possible to give a smoother and more progressive action. The new brake pedal pivot has made it possible to move the Lockheed master cylinder to a more accessible position. The clutch itself now has nine springs instead of six and the crown wheel has recently been stiffened. The engine now has a new Vokes air cleaner and silencer which gives better accessibility for minor engine adjustments and appreciably reduces noise from the carburettor intake. It has a felt filter element which is easily removable for cleaning. The cars finished in Britain are, of course, equipped with a British 12-volt electrical system and the latest models have a cable rack screenwiper. A hardened drive dog is now used on the distributor.

The six-cylinder follows the same general layout as the Light Fifteen with an integral body-chassis structure, front-wheel drive, independent front suspension by torsion bars, and a dead trailing axle with torsion bar springing at the rear. The 2.8-litre engine delivers 76 b.h.p., giving a high performance, and the car is finding a ready sale in countries which require a car of American performance and body space, but which have to pay for their imports in sterling rather than dollars.

LIGHT FIFTEEN SPECIFICATION

Engine.—4 cylinders, 78×100 mm, 1,911 c.c., 55.7 h.p. at 4,250 r.p.m.

Transmission.—3-speed gear box. Front-wheel drive. Facia gear change lever. Gear ratios: 4.3, 7.3 and 13.1 to 1.

Suspension.—Front, independent wishbones and longitudinal torsion bars. Rear, trailing axle with transverse torsion bars.

Brakes.—Lockheed hydraulic.

Electrical Equipment.—Lucas 12-volt. 57 ampère-hour battery.

Main Dimensions.—Wheelbase, 9ft 6½in. Track, 4ft 6in. Overall length, 14ft 2in. Overall width, 5ft 5¼in. Height, 4ft 11¼in. Weight, 2,300 lb.

Price.—£570 plus British Purchase Tax £159 1s 8d. Total £729 1s 8d.

CITROEN SIX

Engine.—6 cylinders, 78×100 mm, 2,866 c.c., 76 b.h.p. at 3,800 r.p.m.

Transmission.—Gear ratios: 3.87, 5.62 and 13.24 to 1.

Main Dimensions.—Wheelbase, 10ft 1½in. Track, 4ft 10½in. Overall length, 15ft 11in. Overall width, 5ft 10in. Height, 5ft 1in. Weight, 2,912 lb.

Otherwise as Light Fifteen.

Price.—£850 plus British Purchase Tax £236 17s 3d. Total £1,086 17s 3d.

Latest modifications to the Citroen Light Fifteen as assembled in Britain include longer pedal arms with a higher pivot point giving a more normal pedal travel. The brake cylinder is moved to a more accessible position inside the main structure and the clutch is now operated by rod instead of cable. The new Vokes air cleaner and silencer is shown.

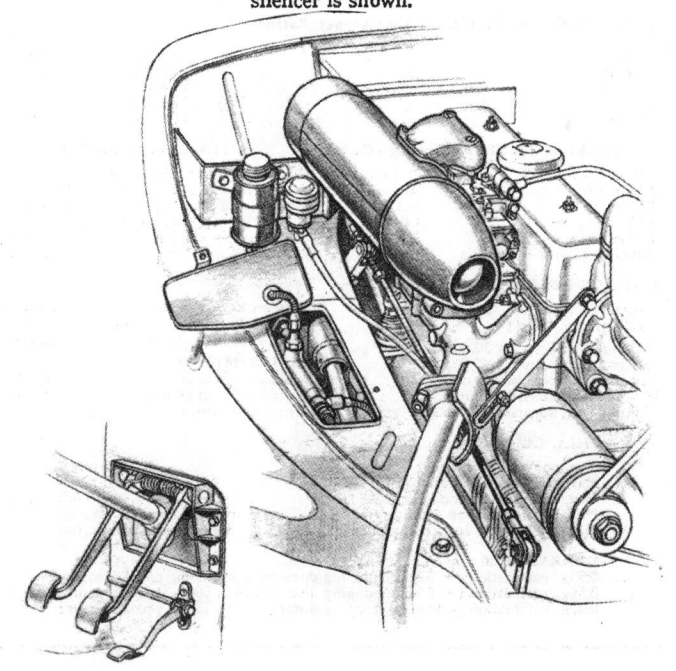

The Motor Road Test No. 21/49

Make: Citroen.   **Type:** Six-cylinder
Makers: Citroen Cars Ltd., Trading Estate, Slough, Bucks

Dimensions and Seating

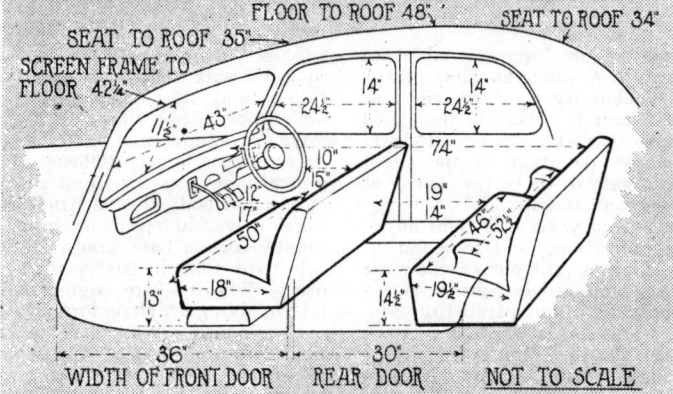

(Diagram: Side and plan views of Citroen Six-cylinder showing ground clearance 7", overall width 5'10", height 5'1", seat adjustable, track 4'10½", wheelbase 10'1½", overall length 15'11", scale 1:50. Interior dimensions: screen frame to floor 42¼", seat to roof 35" front / 34" rear, floor to roof 48", width of front door 36", rear door 30".)

In Brief

Price £850 plus purchase tax £236 17s. 3d. equals £1,086 17s. 3d.
Capacity .. .. 2,866 c.c.
Unladen kerb weight 26½ cwt.
Fuel consumption 19.1 m.p.g.
Maximum speed 81.8 m.p.h.
Maximum speed on 1 in 20 gradient .. 69 m.p.h.
Maximum top gear gradient 1 in 9.9
Acceleration 10-30 m.p.h. in top 9.5 secs.
0-50 m.p.h. through gears 12.6 secs.
Gearing 20 m.p.h. in top at 1,000 r.p.m., 76 m.p.h. at 2,500 ft. per minute piston speed.

Specification

Engine

Cylinders	6
Bore	78 mm.
Stroke	100 mm.
Cubic capacity	2,866 c.c.
Piston area	44.4 sq. ins.
Valves	Pushrod o.h.v.
Compression ratio	6.2/1
Max. power	76 b.h.p.
at	3,800 r.p.m.
Piston speed at max. b.h.p.	2,620 ft. per min.
Carburetter	Solex 2-choke downdraught
Ignition	Coil
Sparking plugs	Champion J8B
Fuel Pump	S.E.V. Mechanical

Transmission

Clutch	Twin dry plate
Top gear (s/m)	3.875
2nd gear (s/m)	5.62
1st gear	13.24
Propeller shaft	Nil (Front wheel drive)
Final drive	8/31 spiral bevel

Chassis

Brakes	Lockheed hydraulic
Brake drum diameter	12 ins.
Friction lining area	143 sq. ins.
Suspension: Front	Torsion bars and wishbones I.F.S
Rear	Torsion bars and axle
Shock absorbers	Newton telescopic
Tyres	Michelin Pilote, 185 × 400

Steering

Steering gear	Rack and pinion
Turning circle	45½ ft.
Turns of steering wheel, lock to lock	2¼

Performance factors (at laden weight as tested).
Piston area, sq. ins. per ton .. 29.6
Brake lining area, sq. ins. per ton .. 96
Specific displacement, litres per ton/mile 2,870

Fully described in "The Motor," September 29, 1948

Maintenance

Fuel tank: 15¼ gallons. **Sump:** 12½ pints, S.A.E. 30. **Gearbox and differential:** 5 pints, S.A.E. 90 E.P. gear oil. **Steering gear:** Heavy grease. **Radiator:** 21 pints. (2 drain taps). **Chassis lubrication:** By grease gun every 600 miles to 9 points. **Ignition timing:** 6° before t.d.c., fully retarded. **Spark plug gap:** 0.015-0.020 in. **Contact breaker gap:** 0.015 in. **Tappet clearances (hot):** Inlet 0.006 in. Exhaust 0.008 in. **Front-wheel toe-in:** TOE OUT 0-0.08 in. **Camber angle:** 0°30'-1°30'. **Castor angle:** 0°, ±15'. **Tyre pressures:** Front 20 lb. Rear 22 lb. **Brake fluid:** Lockheed. **Battery:** 12-volt, 57 amp.-hr.
Ref. B-F/29/49.

Test Conditions

Cool, dry weather with little wind: tarmac surface. Pool-grade petrol.

Test Data

ACCELERATION TIMES on Two Upper Ratios

	Top	2nd
10-30 m.p.h.	9.5 secs.	5.6 secs.
20-40 m.p.h.	9.5 secs.	5.9 secs.
30-50 m.p.h.	10.4 secs.	7.2 secs.
40-60 m.p.h.	12.3 secs.	
50-70 m.p.h.	17.5 secs.	

ACCELERATION TIMES Through Gears

0-30 m.p.h.	5.2 secs.
0-40 m.p.h.	8.3 secs.
0-50 m.p.h.	12.6 secs.
0-60 m.p.h.	19.4 secs.
0-70 m.p.h.	31.0 secs.
Standing quarter-mile	21.6 secs.

FUEL CONSUMPTION

Overall consumption for 343.9 miles, 18 gallons =19.1 m.p.g.
25.0 m.p.g. at constant 30 m.p.h.
23.5 m.p.g. at constant 40 m.p.h.
22.0 m.p.g. at constant 50 m.p.h.
20.5 m.p.g. at constant 60 m.p.h.
17.0 m.p.g. at constant 70 m.p.h.

HILL CLIMBING (at steady speeds)

Max. top gear speed on 1 in 20 .. 69 m.p.h.
Max. top gear speed on 1 in 15 .. 63 m.p.h.
Max. top gear speed on 1 in 10 .. 44 m.p.h.
Max. gradient on top gear .. 1 in 9.9 (Tapley 225 lb./ton)
Max. gradient on 2nd gear .. 1 in 5.6 (Tapley 395 lb./ton)

BRAKES at 30 m.p.h.

0.93g. retardation (= 32½ ft. stopping distance) with 75 lb. pedal pressure.
0.56g. retardation (= 54 ft. stopping distance) with 50 lb. pedal pressure.
0.27g. retardation (= 111 ft. stopping distance) with 25 lb. pedal pressure.

MAXIMUM SPEEDS

Flying Quarter-mile
Mean of four opposite runs .. 81.8 m.p.h.
Best time equals .. 83.3 m.p.h.
Speed in Gears
Max. speed in 2nd gear .. 55 m.p.h.
Max. speed in 1st gear .. 26 m.p.h.

WEIGHT

Unladen kerb weight .. 26½ cwt.
Front/rear weight distribution 60/40
Weight laden as tested .. 30 cwt.

INSTRUMENTS

Speedometer at 30 m.p.h. .. 6% fast
Speedometer at 60 m.p.h. .. 4% fast
Distance recorder .. 2% fast

THE CITROEN SIX

An Interesting Newcomer to the British Market, Especially Designed to Go Far and Fast

SINGLENESS of purpose is the predominant impression left on those members of our staff whose pleasure it recently was to drive the Citroen Six. It is manifestly intended, not as a more refined version of the same manufacturer's sturdy and popular "Light Fifteen," but as a car scaled up in all its dimensions to suit the especial requirements of the long-distance traveller.

In every dimension except height, the Citroen Six rates as a large car by European standards. It is wide enough to seat three people very comfortably on each of its bench seats when the central armrests are folded away, although the unusual gear lever droops far enough from the facia panel to touch a central passenger's knees. It is long enough for rear seat passengers to appreciate a foot-rail on the floor, in contrast to the modern idea of letting them put their toes under the front seat. And, although the seats are upright and the car's overall height very modest, the low, flat floor allows headroom for a tall passenger to be ample in front and adequate above the rear seat.

Stable Equilibrium

The whole appearance of the car is unusual, in that whereas fashion has tended to crowd bodies onto chassis of limited wheelbase and track, this design slings the body low between four wheels which are spread out to the very extreme corners of the car. It is an arrangement which elementary mechanics suggest should make for stability, and such indeed proves to be the case, for even violent manœuvres produce only the smallest amounts of pitch during braking or roll on corners. Few sports cars, even, are so little affected by awkward adverse road cambers such as are often and inevitably encountered on corners.

Riding qualities as well as stability are influenced by large wheelbase and track dimensions, and although many other modern designs have obviously been inspired by this particular layout of torsion-bar springing the results are a little out of the ordinary. In so far as such matters can be defined in words, the movement of the car is perhaps slightly quicker than that of other models equally flexibly sprung, more a matter of two small motions as front and rear wheels successively encounter a bump than of the whole car rising and falling bodily once—it is a definite but restrained, well-damped motion which should have, happily, little effect on those so unfortunate as to be subject to car sickness.

The six-cylinder engine which very well fills the car's long bonnet uses the same cylinder dimensions as the better known "four," so that there is 50 per cent. more swept volume, but the actual power increase is 36 per cent. only. The enlarged engine develops its required maximum power at less than 4,000 r.p.m., and correspondingly gains in flexibility at low speeds, so that the car will run through quite surprisingly slow town traffic in top gear. It is not a completely smooth engine, there being at times some vibration transmitted through the rigid steering wheel, but it pulls hard and without snatch over an extremely wide speed range, certainly extending from below 10 m.p.h. to over 80 m.p.h.

On the sort of cross-country journey which especially suits the car, there is a very strong temptation to forget about the gear lever completely. Outside town, there always seems to be ample power for overtaking other traffic in the highest of the three ratios, and much the same is true in passing through towns, provided a complete stop is not enforced by traffic: the engine pinks quite readily on Pool petrol at moderate speeds, but in town, rather than change down, one turns the ignition timing control (which adjoins the facia-panel gear lever) towards "retard" and stays in top gear.

It must be admitted that this practice is somewhat encouraged by the rather deliberate nature of the gear change. The gear lever is locked immobile until the clutch pedal is depressed fully, so that although the synchronized changes into 2nd and top gears are easy, it is impossible to hustle them by any running together of clutch and gear lever movements. First gear provides power enough to spin the front wheels on a

LOW SLUNG.—Generous wheelbase and track dimensions combine with low build to make the Citroen an exceptionally stable car.

FULL HOUSE.—Set high up, and almost completely filling the long bonnet, the engine is flanked by the gear control linkage, fuel pump and ignition distributor.

WELL SPACED.—The floor is dropped below the level of the doors, and the rear seat positioned to give generous knee and foot room.

dry road, second gives power to climb any ordinary hill and allows a useful 50 m.p.h. to be attained, and top carries the car steadily up hills of the 1-in-10 order of steepness.

Anybody accustomed to driving pre-1950 Citroen front-drive models finds on new cars such as this a very much more comfortable pedal layout, but the use of pedals hung from high up on the scuttle remains an unusual feature. With use they are found quite comfortable, however, although initially the left foot is apt to rub against the central bulkhead as the clutch is depressed.

Our initial usage of the Citroen was in London and around Sussex lanes, where it behaved very well, although the turning circle was rather large in diameter: it is an asset that the front wings stand up in full view on either side of the long bonnet, although with some settings of the adjustable driving seat the near-side parking lamp is hidden behind the driving mirror. Low-speed service, however, is not the car's proper sphere, as under these conditions the driver has to handle quite unusually heavy steering.

Our incursions into country ways did show up circumstances in which front wheel drive, often accused of being at a disadvantage on slippery surfaces, in fact proved an asset. Turning at a gateway in a narrow lane, keeping the driving wheels on hard ground, proved much easier than usual when it was a case of reversing into the gateway.

Open road was what really suited the Citroen, however, the desirable condition being not smoothness, straightness or width, but simply a clear enough view ahead to make fast cruising safe. Under such conditions, the car automatically covered the ground very quickly indeed, to an even greater extent than the over-80 m.p.h. maximum speed and good acceleration figures would suggest. There are, of course, many faster cars on the road, but in practice we encountered nothing outside restricted areas which was not quickly overhauled, passed and left.

In so far as a cruising speed can be specified for this car, we would mention a genuine 70 m.p.h. (a little more than 72 m.p.h. on the speedometer) on English roads, a speed which is quickly

Citroen Six - - Contd.

attained, instinctively maintained despite normal gradients and corners, and is not noisy, extravagant or consciously fast. At somewhat lower speeds some transmission noise is audible, something rather like the humming of roadside telegraph wires on a windy day: higher cruising speeds are perfectly acceptable to the car when conditions are suitable, but acceleration is much less rapid above 70 m.p.h.

Responsive at Speed

Used in this fashion, as it is very evidently intended to be used, the car ceases to be subject to any criticism on the ground of heavy controls. The steering takes it precisely where the driver wishes it to go, on the straight or on a curve, and needs no real effort. The hydraulic brakes are little noticed, but in fact slow the car promptly and surely whenever required in response to a comfortably moderate pedal pressure, although at high speeds some vibration of the pedal is felt. The way in which the car changes character, from being relatively heavy on the helm in town to being responsive to a light touch when road conditions allow it to "get up onto the step" inevitably

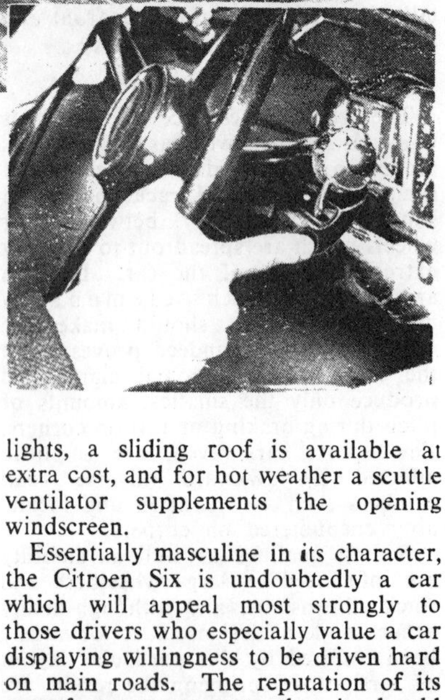

COMMAND CENTRE.—Mounting of the gear lever on the facia panel is convenient and gives an unobstructed floor. Neat also is the grouping of trafficator, head-lamp dipping and horn switches to the right of the steering wheel.

invites comparison with comparable sized sporting cars of what is fondly remembered as the "vintage" period.

The influence of front wheel drive on the car's character is not especially easy to distinguish, although it has certainly helped in the attainment of a very low floor level, and may contribute to heavy steering at low speeds, since a good deal of weight is carried on the front wheels. Acceleration away from a standstill on wet wood blocks is quite brisk, and only fierce use of the power on sharp turns induces much change in the feel of the steering: the old f.w.d. advice to accelerate round bends still seems to apply, however, as

if the car is coasted round a corner at the (very high) speed which just induces tyre howl a little opening of the throttle instantly silences this noise.

Foggy weather conditions relented frequently enough during our test to let us appreciate head lamps of very much above average quality—but closed in frequently enough for us to be frustrated by an opening windscreen which opened insufficiently far to provide direct forward vision: however, an unobstructed front compartment facilitates the driving of a car with right-hand steering from the left-hand seat! The same wintry conditions allowed us to appreciate first-touch-of-the-button cold starting, using only the half-way setting of the two-stage Solex mixture enrichment control—a performance which was even repeated when a long evening journey with a broken dynamo belt was the prelude to a cool night of outdoor parking.

The car is sensibly equipped, the facia cubby hole and four map pockets, for example, being supplemented by a pocket on one side of the scuttle and a useful parcel trough behind the rear seat cushion. The wide luggage locker is not very deep, but there is proper provision for leaving the door open when bulky loads are carried. There are a pair of usefully bright interior lights, a sliding roof is available at extra cost, and for hot weather a scuttle ventilator supplements the opening windscreen.

Essentially masculine in its character, the Citroen Six is undoubtedly a car which will appeal most strongly to those drivers who especially value a car displaying willingness to be driven hard on main roads. The reputation of its manufacturers suggests that it should be extremely durable, as well as effortless in feel, in the hands of fast drivers.

Current fashions have left the Citroen untouched outwardly, and looking what it is, a no-nonsense, "real" car. It is interesting to reflect on the number of its 15-year-old basic design features that have been followed in the meantime.

No. 1395.—2.8-LITRE CITROEN SIX SALOON

The Autocar ROAD TESTS

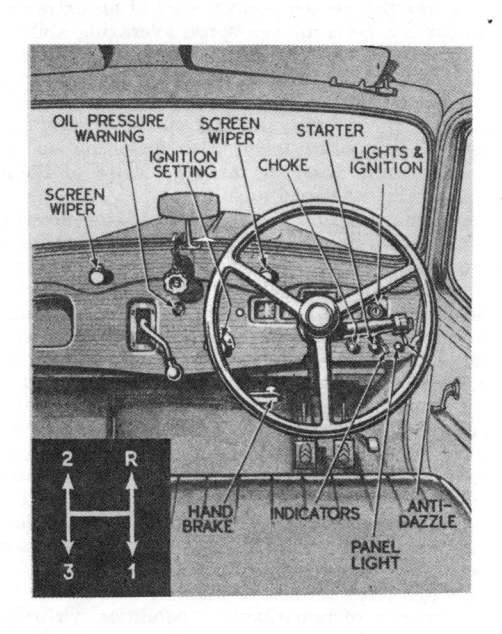

DATA FOR THE DRIVER
2.8-LITRE CITROEN SIX

PRICE, with four-door saloon body, £850, plus £236 17s 3d British purchase tax. Total (in Great Britain), £1,086 17s 3d.
RATING: 22.6 h.p., 6 cylinders, overhead valves, 78 × 100 mm, 2,867 c.c.
BRAKE HORSE-POWER: 76 at 3,700 r.p.m.   **COMPRESSION RATIO**: 6.4 to 1.
MAX. TORQUE: 137.8 lb/ft at 2,000 r.p.m. 20 m.p.h. per 1,000 r.p.m. on top gear.
WEIGHT: 27 cwt 2 qr 2 lb (3,082 lb).   **LB. per C.C.**: 1.07.   **B.H.P. per TON**: 55.24.
TYRE SIZE: 185 × 400 on bolt-on steel disc wheels.   **LIGHTING SET**: 12 volt.
TANK CAPACITY: 15 Imperial gallons: approximate fuel consumption range, 16-21 m.p.g. (17.7—13.5 litres per 100 km).
TURNING CIRCLE: 45ft 0in (L and R).   **MINIMUM GROUND CLEARANCE**: 7in.
MAIN DIMENSIONS: Wheelbase, 10ft 1½in. Track, 4ft 10½in (front); 4ft 9½in (rear). Overall length, 15ft 11in; width, 5ft 10in; height, 5ft 1in.

ACCELERATION

Overall gear ratios	From steady m.p.h of		
	10 to 30 sec.	20 to 40 sec.	30 to 50 sec.
3.875 to 1	9.9	10.4	11.1
5.62 to 1 ..	6.0	6.5	8.0
13.24 to 1 ..	—	—	—

From rest through gears to:—

	sec.		sec.
30 m.p.h.	6.8	60 m.p.h.	21.9
50 m.p.h.	14.4	70 m.p.h.	32.8

Steering wheel movement from lock to lock: 2¼ turns.

Speedometer correction by Electric Speedometer:—

Car Speedometer	Electric Speedometer m.p.h.	Car Speedometer	Electric Speedometer m.p.h.
10	8.5	50	46.5
20	18.5	60	57
30	28	70	67
40	37	80	78

Speeds on gears (by Electric Speedometer)	M.p.h. (normal and max)	K.p.h. (normal and max)
1st ..	19—24	30.6—38.6
2nd ..	50—60	80.5—96.6
Top ..	83	133.5

WEATHER: Dry, mild; wind light.
Acceleration figures are the means of several runs in opposite directions.

BRIEF road experience of the six-cylinder Citroen shortly before the war remained clearly in mind over ten years on account of the high performance which it so obviously possessed, coupled with the handling qualities conferred by design features which even then had shown themselves of great value in the four-cylinder front-wheel drive models. These can now be seen to have stood the test of time most successfully, and an inherent ruggedness makes the car seem specially able to withstand very hard work without suffering.

Strongly favourable impressions of a quite exceptional car, instantly created by recent renewal of acquaintance with the Six, have persisted throughout a test even more than usually comprehensive in mileage covered and variety of conditions. This present experience has materially increased the high respect in which this front-wheel-drive car was already held by members of The Autocar staff. It is rather remarkable, in passing, that this should be so, when it is remembered that the basic design is some fifteen years old.

An experienced driver who enjoys for its own sake the handling of a powerful and responsive car, and who fully appreciates real accuracy of control and real stability, finds it difficult not to over-enthuse about the Citroen Six. It is lively, eager, purposeful, feels taut and solid, allows no side sway whatsoever, and travels up to a maximum exceeding a genuine 80 m.p.h. with the same aplomb, which is shared by the driver and passengers, as it displays in the fifties. The inherent safety factor is exceptionally high and it is a car that gets its driver "out of a jam"— of his own or other people's making.

The six-cylinder engine is smooth without being silky, and the performance suffers little from the fitting of a three-speed gear box, as for long past on Citroens, for the intermediate ratio gives a genuine 60 maximum, and 50 m.p.h. comfortably, whilst the top-gear range is extremely good in spite of the ratio being decidedly high at less than 4 to 1. There is flexibility on this top gear down to below 10 m.p.h., from which the engine will pull away with real power. Slow traffic can be threaded without its being actually necessary to drop to second gear frequently, although this is a ratio which the driver is tempted to use for its slicing acceleration or for storming a 1 in 6-7 gradient (a regularly used hill with such a maxi-

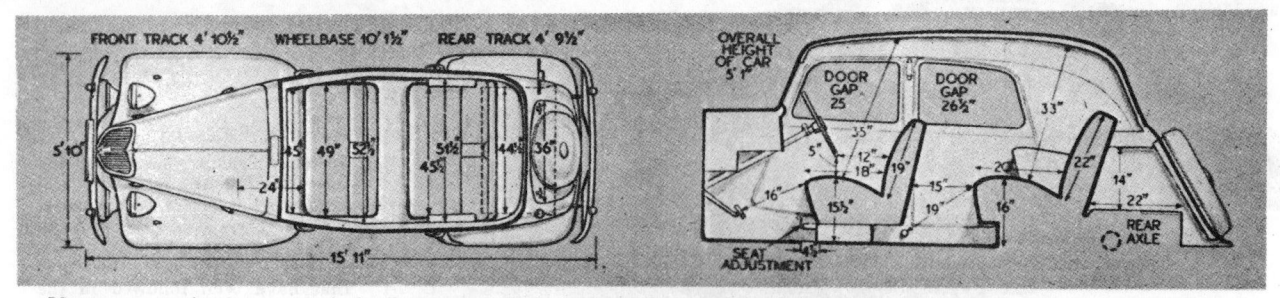

Measurements in these scale body diagrams are taken with the driving seat in the central position of fore and aft adjustment and with the seat cushions uncompressed.

mum was climbed at not less than 40 m.p.h. and the brakes had to be used hard for an acute corner at the summit).

Ordinary main road slopes are taken mostly accelerating or holding the 60 to 70 m.p.h. rate which it could obviously maintain all day along the unrestricted stretches of road found in its native country, or on another continent. This reference to nationality should be accompanied by the remark that, as with the four-cylinder for many years past, the Six is assembled and finished in England, and fitted with a proportion of British accessories.

Outstanding in the performance of an extremely impressive car has been its high-speed averaging abilities without special efforts to record unusual figures. Early in the test 50 miles were obtained in one hour over an English main-road route that strikes a balance between straight sections free from speed limit, the odd village or two, and the outskirts of a county town, also including on the day in question more than the average share of hold-ups from heavy traffic and road repairs.

After dark, came a still more striking illustration. On a journey which was partly cross-country, including a good many speed limits, a brief stop for change of drivers, another halt totalling fully ten minutes for replacement of a head-lamp bulb, and without definite intention to put up the highest average possible, it was discovered by the chance coincidence of round-figure mileometer readings that 100 miles had been covered in two hours' total time, not running time. It was almost startling that, although maximum speed had not been used, so fine a recording should have been made under those particular conditions. On other occasions plus-50 miles were put into an hour. And regularly it was found to be one of those rare cars that improve on one's best times for a familiar journey, long or short, the more difficult the route the more striking being the improvement.

Enough has been said to indicate that here is no ordinary car, but one of remarkable capabilities, virtually a sports car as regards the good points that are usually implied by this somewhat outdated term, yet built in quantity, and with a comparatively low power output taken from its comfortably sized overhead valve engine.

It was suggested earlier that from an experienced driver's point of view it was difficult not to over-praise this model —as a "man's car" this was meant to be understood. The debit side to such an assessment must be fairly stated, which is that it is not so quiet overall as is now usual and lacks one or two minor amenities that have come to be considered as normal equipment. In addition the steering is frankly heavy at low speed and for turning round, making it perhaps not a woman's car from the driving point of view, whilst the steering lock is at times noticeably restricted, although this is a matter to which adjustment can be quickly made as regards most occasions involving a sharp turn on familiar ground. In the eyes of a driver to whom the merits so plainly stressed will appeal, such features would probably be readily accepted, if indeed they were seen at all as disadvantages.

This steering, high-geared, is as accurate as could be wished and delightfully firm and safe feeling; bends can be taken on an exact course, helped by the lack of undesirable give in the suspension. There is some castor return action in the steering, which feels the drive torque a little, whilst slight wheel shake is apparent at times. From numerous experiments at varying speeds and on widely differing surfaces it is of little or no importance whether the engine is kept on the pull or allowed to go on to the overrun on bends, or, again, whether braking is indulged in while cornering, although the transfer to overrun is slightly harsh if the throttle is released suddenly. It is highly questionable whether a driver previously unfamiliar with Citroens would be aware that the front wheels were driven.

There happens also to be the experience of a considerable mileage over icy roads, when the driver found himself able virtually to disregard the conditions, and, in fact, only to be fully aware of the state of the surface when he got out of the car. Over frozen snow and ice no more concentra-

The folding central armrest in the one-piece front seat is seen, as also are the pedals of pendant type, the facia-mounted gear lever, behind which in the photograph is the ignition setting control that is a feature of the Six; the pull-and-push handbrake lever and the now unusual central control allowing limited opening of the windscreen for ventilation purposes. The roomy rear compartment has an entirely flat floor, as also there is in front. Leather upholstery, a folding central armrest, comfortable elbow rests and a foot rail go with lesser amenities such as useful pockets in the back of the front seat.

Except for added width, a higher bonnet line, and more massive bumpers with deep overriders, the Six has few exterior differences from the four-cylinder models at present better known in Britain.

Enclosed luggage space is not large by current standards, but the lid can be left open as seen. A metal extension piece folds down as a platform level with the floor of the compartment to make extra accommodation possible at this position. All tools and the starting handle are carried in a stout box stowed in a space below the luggage compartment.

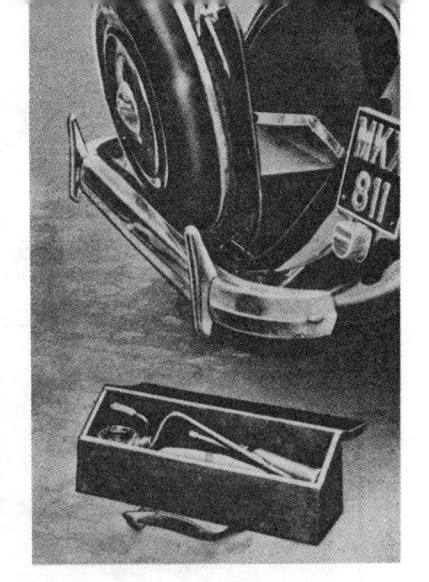

tion was called for than on wet roads with ordinary rear wheel drive, although in conditions of a standing start on snow or ice on an appreciable gradient, wheelspin can occur.

The suspension—independent in front by torsion bars, and at the rear non-independently by torsion bars with a dead axle—absorbs shock extremely well. There is slight, but very restricted up-and-down motion on some surfaces, but never pitching, and as regards both front and rear seats a really bad surface is ironed out, as it were, without the car seeming to receive unreasonable shocks. Unusual praise goes to the hydraulically operated Lockheed brakes, which mostly with only toe pressure do what is wanted, and very smoothly, and which with a firmer application perform prodigies of deceleration, again always squarely and smoothly.

As for long past on the Citroen, the gear lever operates in an open gate to the left of the facia board centre. It proves a remarkably light, convenient and positive control, and the synchromesh is thoroughly effective with full depression of the clutch pedal. The pendant-type pedals present slight unfamiliarity, although the movement is now more nearly in the usual forward direction, instead of purely downward as formerly, and they offer no practical disadvantage beyond the fact that the right leg lacks support at the part-throttle position.

The front seat is in one piece and, in spite of the gear lever position, could accommodate three people on occasion. Adjustment is easy and light by a centrally placed handle. It provides an upright seating position of the kind which usually appeals strongly to the type of driver already referred to in these impressions. In any case this is valuable for the confidence it promotes, when coupled, as in the Citroen, with the wheel at exactly the right angle, a comparatively short and not too high bonnet allowing a forward view on to the road close to the car, and vision of the left wing when required. The view all round is good, even though a blind spot is noticeable on occasion towards the left, formed by the low-mounted driving mirror—which gives a good view behind; the windscreen pillars are thick, and the top rail of the windscreen is rather low. It is easy to put one's head out of the driving-side window and one of the virtues of this car is the ability, so quickly felt, to put it anywhere required, at low speed as well as high.

At the centre of the instrument board is an ignition setting control. It may not be meant for continual use, but can be set towards retard for town work and a fairly advanced position for open-road running, thus checking the pinking which is quite pronounced on Pool petrol when accelerating hard with an advanced setting in use. The plain and simple instruments include an ammeter, but not an engine water thermometer, a little surprisingly, whilst an oil pressure gauge is replaced by a green warning light. The speedometer proved unusually accurate at the top end of the speed range.

Horn, traffic signal and dipper switches are very convenient in a group on an arm carried by the steering column. The pull-and-push hand-brake control is convenient and operates effectively on a steep gradient. The horn note is strong, but melodious, and the beam from the large externally mounted head lamps is satisfyingly powerful, especially in these days, although the dipped beam was set too close for maximum convenience. A simple form of interior heater is standard for the British market, and a sliding roof can be supplied if required at £8 18s 10d extra, including purchase tax. Luggage accommodation is restricted, although provision is made for running with the lid of the compartment open. The spare wheel, although carried externally in the lid, is secured against theft when the compartment is locked.

Manually operated choking gives starting from cold with a certainty that quickly inspires confidence on this important matter. This applied in spite of the car standing in the open overnight on several occasions at varying temperatures, the windscreen being frost covered one morning.

A longitudinally hinged bonnet of light construction, the inside of which is sprayed with sound-deadening material, gives a now almost forgotten accessibility of the straightforward overhead valve six-cylinder engine and of its auxiliaries. Convenience of the sparking plugs, ignition distributor, petrol pump and visible filter bowl, and of the large snap-catch oil filler is perfect. Oil level is indicated by a float-operated pointer, not seen here, and the radiator cap has a quick lever action. At the right, beneath the screen-wiper motor, is seen the spring-cap-plugged connection for a simple hot-air heater, not fitted to the car tested. The other side of the engine is not so tidy, but there is good accessibility of the electrical fuse box and of the hydraulic brake system reservoir.

The Motor Road Test No. 3/51

Make: Citroen.
Type: Light Fifteen Saloon.
Makers: Citroen Cars Ltd., Trading Estate, Slough, Bucks.

Dimensions and Seating

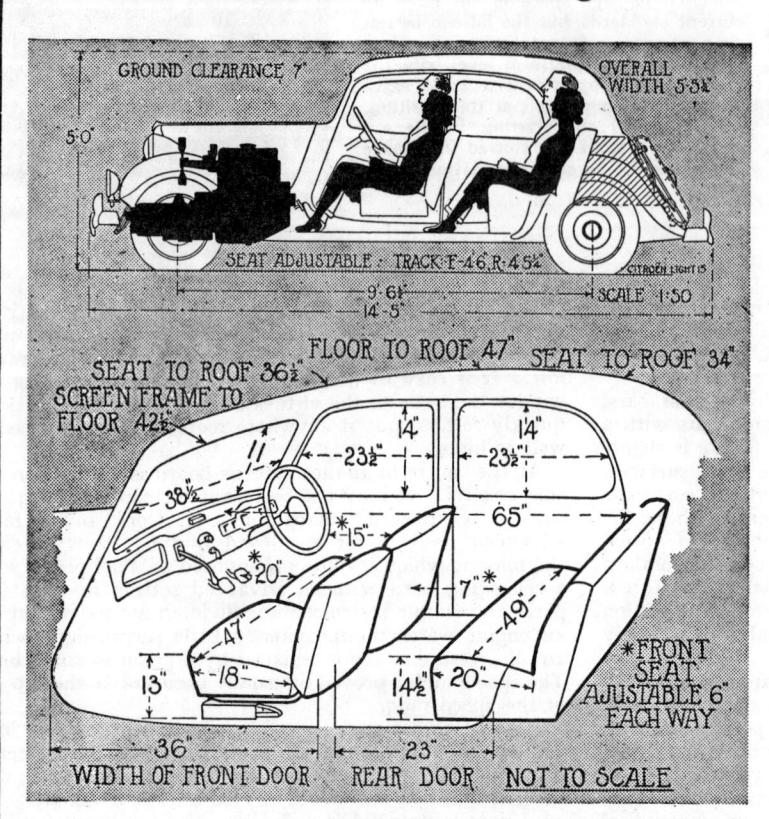

(Diagram: side and plan views of Citroen Light 15, scale 1:50. Ground clearance 7". Overall width 5'3½". Height 5'0". Seat adjustable. Track F 4'6", R 4'5½". Wheelbase 9'6½". Overall length 14'5". Seat to roof 36½", floor to roof 47", seat to roof 34". Screen frame to floor 42½". Interior dimensions 38½", 11", 23½", 14", 23½", 14", 65", 15", 20", 47", 7", 49", 13", 18", 14½", 20". Width of front door 36". Rear door 23". Front seat adjustable 6" each way. Not to scale.)

In Brief

Price (With sliding roof, etc.) £635 plus purchase tax £177 2s. 10d. equals £812 2s. 10d.
Capacity .. .. 1,911 c.c.
Unladen kerb weight .. 22 cwt.
Fuel consumption .. 25.2 m.p.g.
Maximum speed .. 72.6 m.p.h.
Max. speed on 1 in 20 gradient .. 54 m.p.h.
Maximum top gear gradient 1 in 12.1
Acceleration
 10-30 m.p.h. in top .. 11.1 secs.
 0-50 m.p.h. through gears 15.7 secs.
Gearing 17.4 m.p.h. in top at 1,000 r.p.m., 66.2 m.p.h. at 2,500 ft. per min. piston speed.

Test Conditions

Cold, wet weather with strong, gusty wind. Smooth tarmac surface. Pool grade petrol.

Test Data

ACCELERATION TIMES on Two Upper Ratios

	Top	2nd
10-30 m.p.h.	11.1 secs.	6.3 secs.
20-40 m.p.h.	11.0 secs.	6.9 secs.
30-50 m.p.h.	12.5 secs.	9.4 secs.
40-60 m.p.h.	20.0 secs.	—

ACCELERATION TIMES Through Gears

0-30 m.p.h. .. .. .. 6.3 secs.
0-40 m.p.h. .. .. .. 9.8 secs.
0-50 m.p.h. .. .. .. 15.7 secs.
0-60 m.p.h. .. .. .. 29.7 secs.
Standing Quarter-Mile .. 23.2 secs.

FUEL CONSUMPTION

35.0 m.p.g. at constant 20 m.p.h.
33.5 m.p.g. at constant 30 m.p.h.
31.0 m.p.g. at constant 40 m.p.h.
27.5 m.p.g. at constant 50 m.p.h.
23.5 m.p.g. at constant 60 m.p.h.
Overall consumption for 504 miles, 20 gallons = 25.2 m.p.g.

HILL CLIMBING (at steady speeds)

Max. top gear speed on 1 in 20 .. 54 m.p.h.
Max. top gear speed on 1 in 15 .. 48 m.p.h.
Max. gradient on top gear .. 1 in 12.1 (Tapley 185 lb./ton)
Max. gradient on 2nd gear .. 1 in 6.2 (Tapley 355 lb./ton)

BRAKES at 30 m.p.h.

0.82g retardation (=36½ ft. stopping distance) with 90 lb. pedal pressure
0.60g retardation (=50 ft. stopping distance) with 50 lb. pedal pressure
0.35g retardation (=86 ft. stopping distance) with 25 lb. pedal pressure

MAXIMUM SPEEDS

Flying Quarter Mile
Mean of four opposite runs .. 72.6 m.p.h.
Best time equals .. 78.9 m.p.h.

Speed in Gears
Max. speed in 2nd gear .. 51 m.p.h.
Max. speed in 1st gear .. 30 m.p.h.

WEIGHT

Unladen kerb weight .. 22 cwt.
Front/rear weight distribution .. 56/44
Weight laden as tested .. 25¼ cwt.

INSTRUMENTS

Speedometer at 30 m.p.h. .. 8% fast
Speedometer at 60 m.p.h. .. 12% fast
Distance recorder .. 6% fast

Specification

Engine
Cylinders 4
Bore .. .. 78 mm.
Stroke .. .. 100 mm.
Cubic Capacity .. .. 1,911 c.c.
Piston area .. .. 29.6 sq. in.
Valves .. .. Pushrod o.h.v.
Compression ratio 6.5/1
Max. power .. .. 55.7 b.h.p.
 at .. .. 4,250 r.p.m.
Piston speed at max. b.h.p. 2,790 ft. per min.
Carburetter .. Solex 32PBI downdraught
Ignition .. .. Coil
Sparking plugs .. .. Champion H.10
Fuel pump .. .. AC Mechanical

Transmission
Clutch .. .. Single dry plate
Top gear (s/m) 4.3
2nd gear (s/m) 7.3
1st gear 13.1
Propeller shaft .. .. Nil (f.w.d.)
Drive shafts .. 2 needle roller universals in each
Final drive .. .. 9/31 spiral bevel

Chassis
Brakes .. .. Lockheed hydraulic
Brake drum diameter
 front .. .. 12 ins.
 rear .. .. 10 ins.
Friction lining area .. 80½ sq. ins.
Suspension
 front .. Torsion bar and wishbones I.F.S.
 rear .. Torsion bars and axle beam
Shock absorbers .. Newton telescopic
Tyres .. .. Michelin, 165 × 400

Steering
Steering gear .. .. Rack and pinion
Turning circle .. .. 43 feet
Turns of steering wheel, lock to lock .. 2¼

Performance factors (at laden weight as tested)
Piston area, sq. in. per ton 23.3
Brake lining area, sq. in. per ton .. 63
Specific displacement, litres per ton mile 2,590
Fully described in "The Motor," September 29th, 1948

Maintenance

Fuel tank: 10¼ gallons. **Sump:** 7 pints, S.A.E. 20 or 30. **Gearbox and Differential:** 3 pints, medium gear oil. **Steering gear:** Shell Retinax A. **Radiator:** 14 pints. (2 drain taps). **Chassis lubrication:** By grease gun every 1,000 miles to 17 points. **Ignition timing:** 12° b.t.d.c. **Spark plug gap:** 0.025 in. **Contact breaker gap:** 0.012 in. **Valve timing:** I.O., 3° b.t.d.c. I.C., 45° a.b.d.c.; E.O., 45° b.b.d.c.; E.C., 11° a.t.d.c. **Tappet clearances:** (Hot) Inlet 0.006 in., Exhaust 0.008 in. **Front wheel toe out:** 1/16–3/32 in. at rims. **Camber angle:** 1° ± 30'. **Castor angle:** 1° 30' ± 15'. **Tyre pressures:** 20 lb. all round. **Brake fluid:** Lockheed. **Battery:** 12 volt, 57 amp-hour. **Headlamp bulbs:** 36/36 watt, 12 volt.

Ref. B-F/20/51

The Citroen Light Fifteen

A Front-Wheel-drive Car of Renowned Stability Shows Improved Fuel Economy in its Latest Form

"TEN years ahead of its time" was the claim made for the front-wheel-drive Citroen when it was introduced in 1934, a claim which has since proved to be almost too modest.

Since that time, a steady process of detail improvement has maintained the popularity of cars of the same basic layout, and a 1951-series model from the British factory at Slough recently reminded us once more of the unique roadworthiness which has made the Citroen famous.

Powered by a 2-litre engine, and seating four passengers normally plus one or two more should need arise, the Light Fifteen, can be compared dimensionally with models from all the principal British car factories. It is interesting to note, however, that taking the average figures for the 2¼-litre models produced by the "big six" British manufacturers, as a basis for comparison, the Citroen is 13 inches longer in wheelbase, 1 inch wider in track, and 3¼ inches lower in overall height.

Long, low and wide, the car is obviously designed first and foremost for stability, and amongst quantity production models it may be said to have pioneered the idea that flexible springs are not incompatible with good cornering qualities. Whilst even softer springing systems are now in use, the Citroen remains a very comfortable car indeed, and its stability on wet roads is unsurpassed.

Adherence to a basic design throughout a long period of years, with changes which whilst substantial are in no sense fundamental, is a policy with both advantages and disadvantages. On the credit side of the balance, there is a background of experience which allows cars to be built with the certainty that very large mileages can be covered without major service work being required. On the debit side, it must be admitted that it has not apparently been found possible to obtain the highest modern standards of mechanical silence and smoothness.

Like most cars of unusual character, the Citroen does not reveal its full charm to a stranger on first acquaintance; it does however make friends slowly but surely, gradually revealing virtues which, in the eyes of many long-distance motorists, more than counterbalance its imperfections. Covering upwards of 1,250 miles in the course of our test, including more than one day in which a mileage of around 300 was incidental to other activities, we parted reluctantly with this latest example of the marque.

Entry to the four-door steel saloon body is unusual, in so far as it is necessary to step over door sills which are approximately 7 inches above floor level, and also because the floor inside the car is both very low (10 inches above ground level) and completely flat. Low floor level, made possible by the use of front-wheel drive and so the virtually complete elimination of under-floor mechanism, means that the car's modest overall height is attained without sacrifice of ground clearance, seating height or headroom dimensions.

All the seats are provided with upright backrests, a fact the driver particularly should appreciate, and the individual front seats move or lock exceptionally easily on their slides. The angles and cushioning of the seats are both deserving of high marks for providing comfort throughout long journeys, but a car with fast cornering qualities should give passengers more lateral support, either by central armrests or by shaped seat backrests, than is at present provided.

Control arrangements have been subtly improved since we last sampled this model three years ago, the pedals in particular having been re-arranged to provide very much lighter and more natural operation. The steering, geared at fractionally more than two turns from extreme to extreme of a none-too-compact lock, is typically rack-and-pinion in its absolute precision. The brakes are equal to all demands normally made on them, either for powerful or for gentle retardation according to road surface conditions. The gearchange alone fails to be altogether pleasing in feel, although it must in fairness be said

ECONOMY AND DURABILITY—Recent modifications to the accessible 2-litre engine include an easily cleaned air filter and new pump-type Solex carburetter.

LOCKER DETAILS—The luggage compartment door carries the spare wheel, but does not open so far that its weight is embarrassing. Tools are carried in a fibre case fitted into a locker-floor recess.

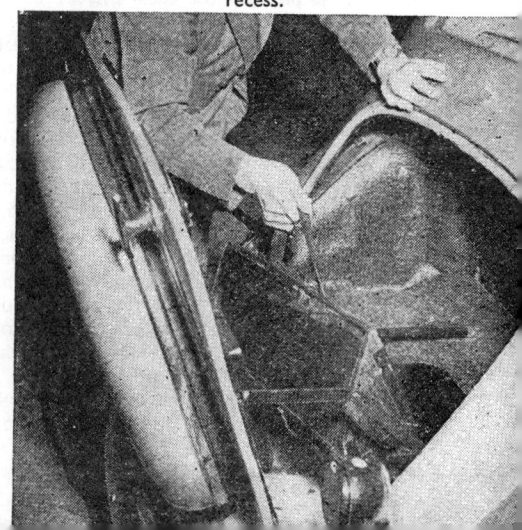

EASIER PEDALLING—Re-design of the control pedals makes the latest Citroen Light Fifteen a much pleasanter car to drive than its predecessors.

TRIPLE ACTING—This finger-tip control is pressed to sound the horn, turned to switch on parking or headlamps, and moved forwards for headlamp dipping.

The Citroen Light Fifteen - - - Contd.

that when handled in a normally unhurried manner it is light and adequately assisted by synchromesh mechanism.

The steering is heavy at times, the self-centering action being forceful during acceleration from low speeds, but it allows an experienced driver to "feel" variations in road slipperiness—icy weather conditions prevailed during part of our test. If this steering heaviness is inevitable with broad-base tyres running at low pressure, then many will consider it a small price to pay for exceptional road holding thereby attained. It is difficult to divide credit for good road holding between Citroen car and Michelin tyres, and perhaps unfair to attempt any division since each has been developed to suit the other and the two companies concerned are now linked together financially, but there can be few cars in which the normal cornering speed in wet weather is so high.

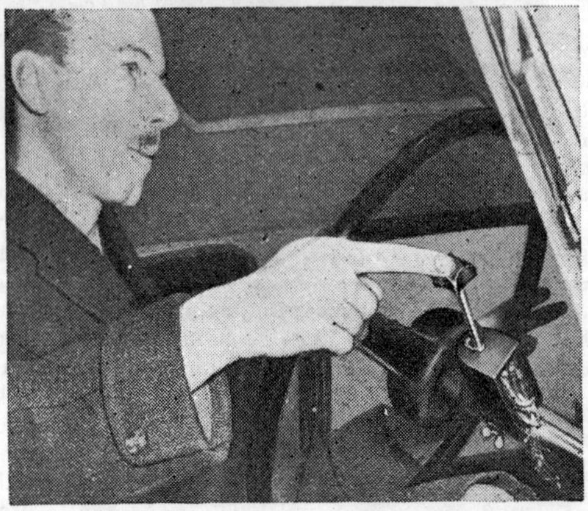

Roll on corners may occur with this car, in a strictly geometrical sense, but it is so slight that neither driver nor passengers are ever really conscious of it. The same is true of the nose-dipping during hard braking which afflicts many softly-sprung cars of more compact dimensions.

Riding characteristics depart appreciably from the fashionable "boulevard ride," inter-axle distribution of most of the car's weight resulting in a rather sharper motion, particularly during leisurely negotiation of rough roads, than is felt in cars of greater k^2/ab ratio. Overall, however, riding characteristics are such as to please most occupants of either front or rear seats, the sort of large-amplitude low-frequency motion which some people find particularly discomforting being happily absent, and there being no cause to avoid rough or unmade roads.

In terms of measured figures the Citreon no longer ranks as a particularly fast car, its maximum speed of 72.6 m.p.h. and 0-50 m.p.h. acceleration time of 15.7 seconds being good rather than outstanding. On the road, however, it remains one of those cars which overtakes many and is overtaken by few—during the 500-odd miles included in our fuel consumption measurement, only two cars overtook the test model outside of speed-limit areas.

Habitually good average speeds result partly from roadworthiness which minimises both fatigue and delays for corners or rough surfaces. It also results from power unit characteristics which make most of the car's maximum performance easily available, there being no call for frequent gear changes such as are needed with some cars of continental design.

The car cannot be described as silent or mechanically smooth by 1951 standards, but nevertheless does not give any indication that it objects to the sort of hard driving to which French drivers habitually subject cars from the parent factory. The engine does not make itself heard appreciably in top gear, but there is enough wind noise to make conversation tiring if 60 or 70 m.p.h. cruising speeds are adopted. Also, there is vibration transmitted through the rigid-spoked steering wheel, sufficient on the test model to encourage the use of gloves by a sensitive driver. The really pleasant cruising speed is 50 m.p.h. true speed, or nearer to 60 m.p.h. by a rather boastful speedometer, there being a huge reserve of power at this smooth and quiet gait.

At the opposite end of the scale, the engine with its volute-spring mountings idles with exceptional smoothness. The engine starts easily from cold; after the car had stood in the open through a night which covered the windows with ice, the first touch of the starter produced response and the second touch set the engine running steadily in readiness for driving away. Some use of the choke was advisable during the first mile, but rich mixture was not needed for subsequent re-starts even after many hours of outdoor parking.

Our usage of the car extended well off the beaten track, and despite its low build it was evident that there was good ground clearance in what are normally the most vulnerable places. For climbing steep and slippery hills, front wheel drive is at something of a disadvantage, the weight distribution being least favourable when the car is fully loaded, but it should always be possible to get up a freakish grade in reverse-gear. This is not, however, a procedure to be adopted unnecessarily, as it reveals unpleasant reversed-castor action in the steering.

Any normal steep hill (we sampled many such in Dorset by-ways) is of course as easily surmounted with front as with rear wheel drive. For stopping or re-starting, the handbrake is amply powerful to hold the car on any gradient, but it cannot be described as accessible and with a twist-to-release ratchet it appears liable to get knocked off accidentally.

Welcome Equipment

Some unusual and welcome items of equipment are incorporated in the design, amongst them an opening windscreen and a sliding roof (a cheaper model without the latter is also produced). Dual sun visors can be used to shade either windscreen or side windows, and there are both map pockets and a facia panel cubby hole. The front door window winder handles, unfortunately, seem especially placed to catch the cuffs of a driver's overcoat, and the rear windows do not drop far down into the doors.

A very simple fresh-air heating system is provided, in the form of a duct taking air from behind part of the radiator into the car. At times this system delivers very useful warmth, but with no fan to assist airflow the heat output drops away as the car is slowed or stopped by traffic.

Most buyers of Citroens are motorists who have already owned earlier cars of the type, a fact which indicates that the cars attractions are in no sense superficial. To such, it can safely be said that the 1951 model is very considerably better than the Citroens they have already liked, not merely an even more pleasant car to drive, but also one which has been made fully 10 per cent. more economical of fuel under conditions of ordinary usage.

STEP DOWN—The completely flat floor of th roomy rear compartment is dropped down betwee the side members of the chassis-less steel saloo body.

SEE THEM ON STAND 146 AT THE MOTOR SHOW EARLS COURT

In the long run... the real test of any car... THE FRONT WHEEL DRIVE CITROEN is at its very best! With its powerful o.h.v. engine it is faster off the mark, maintains a higher average speed than most, and gets you there faster and with greater safety! *Independent front suspension* and *torsion bar springing* smooths out and holds any road in wonderful fashion. The immensely strong *one-piece chassis-body* is so roomy and luxurious that travel fatigue is ruled out! The "Light Fifteen" and the "Six Cylinder" Saloons have stood the test of time and gained an enviable reputation amongst motorists all over the world... be sure to see them on Stand 146 at the Earl's Court Motor Show.

PRICES
"LIGHT FIFTEEN" Saloon from £1,067 1 2
"SIX CYLINDER" Saloon from £1,525 18 11
(Including Purchase Tax)

You can't beat **CITROEN** *in the long run!*

CITROEN CARS LTD., SLOUGH, BUCKS.

Telephone: 23811   Telegrams: Citroworks, Slough

An Owner's Impressions of the
CITROEN 11 CV "NORMALE"

THE Citroën 11 CV "Normale" is seldom seen in this country as the factory at Slough concentrates on the Light Fifteen (11 CV Legere) and the six-cylinder models. However, in France and Belgium the "Normale" is well known; it possesses the engine of the Light Fifteen with the interior dimensions of the Six. There is a small sacrifice of performance owing to the heavier weight, but this is more than compensated for by the additional room available. To the casual observer it is distinguishable from the Light Fifteen by the longer covering over the starting handle slot, and at the same time cannot be confused with the Six owing to its shorter bonnet, and smaller wheels of 165 by 400, as opposed to 185 by 400.

Although it was early January when I took delivery, I did not fill up with antifreeze as I was heading for Morocco, where it would have been superfluous. The run from Brussels to Paris served to acquaint me completely with the car, and I was immediately impressed by the excellent view from the driving seat, taking in the off-side lamp and, with very little effort, the wing as well.

Front-wheel drive causes a certain steering heaviness at slow speeds, and it was not till the car was run-in that I was really able to appreciate its advantages to the full. I would venture to say there are few, if any, family cars of the price of the Citroën (in France about £600) which can compare with it for road-holding and security in wet weather. The torsion bar suspension is a trifle hard at slow speeds and when the car is empty, but with a full load and at a steady sixty, on the type of undulating roads found in France, it is really first-class. The flat floor, front and back, enables six people to be carried in comfort, and the boot, supplemented by the usual roof-rack—practically universal on Continental cars—permits ample baggage to be carried for a long holiday. I have always found that the flat roof-rack offers greater security than the curved type, and with a canvas cover and elastic "claw" straps it solves the luggage problem most satisfactorily. Vastly better than overloading the rear springs by piling suitcases on the open lid of the boot, all too common a sight in England.

Having commenced with Mobiloil in the sump, I have persistently stuck to it, having a complete oil change every 2,000 miles. I believe it is a mistake to mix oils, especially those containing detergent additives.

The three-speed gearbox has one drawback, it is virtually impossible to change from second to first without the most distressing noises, and with the car almost stationary. The engine, however, is remarkably flexible and in second it is possible to crawl along with no sign of labouring or "snatch."

Although I had 1,500 miles ahead of me I was resigned to a slow trip as I feel no car is "run-in" for continuous speed at less than 2,000 miles; however, the roads encountered in Spain were so bad that speeding was "out" anyway. The weather was cold and clear until we arrived at the foot of the Guadarrama range between Burgos and Madrid. Here the snow was thick, and a heavy blizzard raged so that we could rarely see ten yards beyond the bonnet. I kept my eye on the tail light of a new Cadillac, which conveniently carved a way through the white wall ahead. To my consternation it suddenly stopped, and remained "stopped." We got out to see what was wrong and at once the biting wind lashed snow into our faces, and cut mercilessly through our coats. I purposely left the engine running, as in that temperature water freezes surprisingly quickly! The Cadillac was jammed in heavy snow, its rear wheels spinning in helpless futility as its hydraulic transmission desperately tried to cope. The road was completely blocked, and as the Spanish don't appear to possess snow-ploughs it probably remained so for many days. Fortunately for the Cadillac a large truck appeared and dragged it out tail first. There was nothing for it but to return and seek another road, and after twenty miles at 7 m.p.h. over the most appalling surface we finally arrived in Sergovia. We still had to cross the Guadarramas and the locals cheerfully informed us that the road was covered in snow. However, we decided to "bash on" and fortunately found that although it was lying thickly, there were no drifts in this area. I remembered people who had told me that front-wheel drive was "absolutely 'u.s.' old boy" when the tractive conditions were bad; "now," I thought to myself, "we shall see." Needless to say, such fears are unfounded. The car climbed like a bird, passing many with "orthodox" transmissions, and showed no trace of wheel spin at any time, even under the most icy conditions. On the descent the steering was steady as a rock, and any "sliding" tendency could be corrected at once. So dumbfounding the pessimists we arrived in Madrid in time for an excellent supper. Thanks a lot, André.

At later stages of the journey, when the roads were persistently shocking, I found the steering rather high-geared, and too many road shocks found their way up the column, so that after six hours' driving I had an ache in both wrists. But I must stress that the conditions are not likely to be found in the course of normal motoring.

After several thousand miles through French and Spanish Morocco, experiencing every known kind of surface, from excellent tarmac to appalling ruts and pot holes, I made the return trip.

We left Gibraltar at midday on a Thursday and arrived in Paris for dinner on Sunday. This time can be improved on, no doubt, although I am sorry for the suspension of the car that does it.

After 10,000 miles I have had but one adjustment to make, necessitating a garage; some lateral play had developed in the upper wishbone of the near-side front wheel. This no doubt was accentuated when I was forced to take a curb-stone at fifty to avoid a "clot" on a donkey whose steering gear had apparently failed. Otherwise, apart from regular oil changes and pressure greasing every 2,000 miles I have had no expenses whatever.

In my car, and I presume on the English model as well, the squab of the rear seat can be lifted from its two supporting hooks; and it is possible to make a comfortable bed for two by placing the rear seat on two wood blocks on the floor, and after reversing the squab so that it fits, putting it beside the seat. With the boot open there is ample room to stretch full length. I have frequently slept in my car in this manner, and find it very comfortable. The luggage can be shifted to the front seats for the night.

Petrol consumption going fast is around 21 m.p.g., and oil negligible so far. The headlamps are excellent, and powerful enough up to 60 m.p.h., which is the comfortable cruising speed of the car. This may not appear high by present-day standards, but as it can be maintained indefinitely over any roads, even in bad weather, one scores over those cars which are basically faster but get an attack of the "dithers" the moment it starts to rain. For those who prefer it, and have the cash, it is possible to buy a four-speed gearbox by Reda in Paris for £75, but except for competition driving, it is hardly worth the expense. Tyres should be changed round approximately every 3,000 miles to ensure even wear, as the front wheels have a heavy task coping with steering and driving together. It is advisable to stick to the broad-base type of tyre, although not necessarily to any one make. For hard work the Michelin X is completely satisfactory.

I found the engine made no complaints about fuel (compression ratio 6 to 1) except in Spain when somebody poured in paraffin instead of petrol and charged me just the same!

Starting has always been sure, but always on the third attempt for some reason, and remarkably little use of the choke is needed, even on the coldest days.

After 7,000 or 8,000 miles many Citroëns develop their "noise," a slight "clack, clack" when the engine is idling. This is not serious and is caused through lack of grease, or the right type of grease, in the water pump. I unscrewed the nipple and forced Mobilgrease No. 6 into the "works" by hand, as it is so thick as to choke the average gun. The "noise" has now disappeared.

The operation of the pedals with a downward movement (since modified in the English model) appears rather odd at first, but one soon gets used to it, and de-clutching with the heel becomes as easy as with the ball of the foot. The ground clearance is too low for "Colonial" use over bad tracks, but is quite adequate for European use. The hydraulic brakes are sure and powerful but could be lighter in operation.

So in conclusion I would say that for those seeking a thoroughly "practical" car, bereft of frills but possessing all that is necessary for fast motoring with plenty of space available for friends or luggage, the Citroën 11 CV "Normale" (15 h.p.) holds many attractions.—I. A. S. C.

NEW CARS DESCRIBED

CITROENS CONTINUED

The popular four-cylinder Citroen, known in France as the Onze Legère, and in Britain as the Light Fifteen, receives its first major appearance change in 18 years with the addition of a luggage locker at the rear.

New Luggage Lockers the Main Change on Long-established Models

DESPITE reports that have been published in the French Press regarding new Citroen cars with flat six engines and hydraulic suspension systems, the Citroen company, biggest of the private enterprise French car manufacturers, has announced that the cars at this year's Paris Salon will be the well-known current models with detail modifications. The most obvious change is the addition of a new luggage locker to the four-cylinder two-litre cars and the big three-litre Six. The locker has been created by hinging a deep steel pressing to the existing rear panel, extending the floor of the luggage space and moving the bumper rearwards. It not only allows the spare wheel and a fair amount of luggage to be carried under cover, but it also makes a useful contribution towards modernizing the lines, which have now remained practically unchanged for 18 years, a remarkable record of manufacturing continuity.

Several other detail improvements are announced by the Paris factory. The interior has been given a more pleasing appearance by painting the frames around the windscreen, side windows and rear window to harmonize with the upholstery. The instrument panel is given a two-tone finish to match, and the control layout has been improved by a symmetrical grouping of switches and a more easily visible arrangement of speedometer, fuel gauge and ammeter. Instrument illumination is now controlled by a rheostat.

The windscreen wiper has at last been moved from its old position above the windscreen. The motor is now concealed behind the instrument panel, and the wiper arms are pivoted on the scuttle. This does not interfere with the ability to open the windscreen, a feature retained on Citroens that is rapidly disappearing from other cars.

A particularly simple air heating system is used on Citroens, incorporating a duct that picks up fresh, warm air from just behind the radiator. The entry opening for the warm air is down by the clutch pedal (in the left-hand-drive French version), thus to vary the flow it had previously been necessary to get out of the car. The new arrangement allows the flow of warm air to be controlled from the driving seat. Parking lamps, visible from front and rear, are now placed over the centre door pillars, and four separate flashing indicator lamps are provided, two on the front wings and two on the rear quarter panels.

Seating improvements include a higher back rest for the rear seat. The standard upholstery is still in cloth, but the parts most likely to wear are now protected by synthetic leather grained material. Two vizors are provided in the front; they can be used as a protection against light coming through the windscreen or from the side.

Production of the 2CV, the remarkable little two-cylinder front-wheel-drive economy car, is still increasing and has now reached a figure of approximately 500 per week. No modifications to this model have been announced for 1953.

The Citroens assembled and marketed in Britain have, of course, a more luxurious interior than the standard French models, including full leather upholstery throughout and a polished wood facia. Few of the modifications to the French models are therefore applicable, but a decision on whether the new luggage locker will be incorporated is expected shortly.

A new view of the 2CV, Citroen's small economy car with flat twin, air-cooled engine of 375 c.c. This view shows clearly the four seats on tubular frames, the flat floor and the fabric covers which roll back to open the roof and the luggage and spare wheel compartment. Right: Chief innovation on the larger French Citroens for 1953 is a luggage locker which also houses the spare wheel. It is illuminated when the lid is open at night. Both four- and six-cylinder models will have it.

ROAD TEST

A TRULY EXCELLENT MOTOR CAR

"Motor Sport" Forms a High Opinion of the Citroen Six

WHEN the Citroen Six came to us for road test it came for a considerable period, as if in compensation for not having had a Press car from the Slough factory since before the war—of course, we have borrowed various Citroens in between, for to deny oneself completely the pleasure of driving these cars would indeed be monastic. But the lengthy trial of the six-cylinder version meant that it could be experienced under a wide variety of conditions, road, load, weather and circumstance, including taking it down to cover the Land's End Trial.

As a result of this extended test, our enthusiasm runs high. We can recall no other car as roomy as this which has proved so enjoyable to drive in sports-car style. Certainly, after a four-figure mileage it qualified as that exclusive class of car which we return with real reluctance.

Taking over this Citroen in London's competitive rush-hour traffic we were reminded of Kent Karslake's tribute to a very different form of automobile, the Hispano-Suiza, of which he once wrote that it has the rare knack of feeling far less bulky than measurements prove it to be. So with the Citroen, for although one is conscious that it is a *wide* car (actually 5 ft. 10 in. wide and nearly 16 ft. long), so accurate is the control, so excellent the visibility from behind the wheel, that one straight away on first acquaintance pushes through the traffic at a speed which appears quick and is actually faster than it seems. Gain the clearer roads and this liking of the Citroen for deceptive speed increases. Very soon the car makes firm friends with its driver and is cruising at a genuine 65-70 m.p.h., more if circumstances call for it, if not in entire silence, certainly with seven league boots—comfortable, sure and very safe.

It might be said, in an entirely complimentary fashion, of this French conceived, British assembled and equipped car, that it is the last of the vintage vehicles. By this we mean that it achieves prodigies of performance for its modest output of 27 b.h.p.-per-litre, is high-geared (20 m.p.h. per 1,000 r.p.m.), possesses many very practical features of detail layout, and in handling is distinctly a man's motor car.

The performance factor can be represented for the moment as a maximum in excess of 80 m.p.h., 60 m.p.h. in second gear of the three-speed box and acceleration of the 0-50 in 12½ sec. order—this on 2.8-litres of very roomy five/six-seater saloon, with a wheelbase of just over 10 ft., weighing 1.65 tons.

The handling qualities constitute, perhaps, the greatest charm of this refreshingly unusual car. The front-wheel-drive can be dismissed as almost unnoticeable except for a happy knack that using more throttle on a corner proves an asset instead of promoting a disaster and for violent castor return on the steering out of tight corners. The torsion-bar suspension is somewhat softer than expected, allowing some up and down motion, transmitted as separate effects of front and rear wheels following irregularities over rough roads, bad surfaces being ironed out admirably in consequence. Yet, have no illusions, this tendency to softness promotes no handling headaches. The roll-factor when cornering fast is low and at all times, until rear-end breakaway occurs, the Citroen Six understeers. Moreover, the rack and pinion steering is so absolutely free from lost-motion that the car can corner with extreme precision, driving along winding lanes thus becoming a 60 m.p.h. affair, when in most cars you would feel you were really pressing along at 45 m.p.h. The only way in which the Citroen reminds eager drivers that, after all, it is a family saloon and not a sports-car is when the rear wheels slide under *extreme* cornering methods. As, however, the steering is so high-geared—another " vintage " attribute—asking only two turns lock-to-lock, this is of no particular moment to anyone accustomed to sliding his corners. Incidentally, this high-gearing is further emphasised by the 18 in. dia. steering wheel.

It has to be admitted that the steering is heavy, due to the aforementioned strong castor action—very heavy when manoeuvring and still quite heavy when cornering at moderate speeds. But as speed is increased above 20 m.p.h. the steering alters surprisingly, becoming light and smooth in action. The turning circle is rather large (45 ft.). The size and width of rim of the two-spoke steering wheel lends strength to our " vintage " comparison, making the Citroen more a man's than a lady's car. There is a little return motion at times, becoming excessive under light application of the brakes, but no column vibration.

For our part, the excellence of the cornering and road-holding—the latter extends to well-damped suspension, producing arrow-landings after hump-back bridges—allied to the comfort of the riding characteristics and roomy interior, effectively offset the muscular effort required to handle the car briskly.

Coming to the practical features of detail layout, there is, for instance, the tin box in the luggage locker enclosing a second petrol filter, to ensure clean fuel lines, the presence of tiny ventilator doors at the fronts of the bonnet-sides to admit cool air, the simple but effective quick-action radiator filler, the accessible under-bonnet battery, the simple heater turned off by putting a bung in the delivery pipe, the snap-shut oil filler on the valve cover, the extension tube for the starting handle, separate tool bag, etc. The engine, of the same size as that of the famous Light Fifteen but with two additional " pots," pushes out its 76 b.h.p. at a leisurely 3,800 r.p.m.; and 2,500 f.p.m. piston speed = 76 m.p.h. It is finished in agricultural fashion, red paint, but no polish on the valve cover, but it and its fascinating appurtenances fill the under-bonnet space in satisfactory fashion. The asbestos shields for carburetter and battery, screw-down greaser for the water-pump, float-type oil level indicator, cowled fan, general under-bonnet accessibility and easy action of the bonnet fasteners did not escape us.

There is one aspect of the Citroen which is neither " vintage " nor modern practice and perhaps the one weak point in a brilliantly-conceived car. We refer to the dashboard gear-lever for the three-speed gearbox. This lever can be moved only if the clutch pedal is fully depressed. From the aspect of preventing the gears from jumping out of mesh this is a good feature, but a bulge in the bulkhead tends to impede one's clutch foot, so that the pedal is not always depressed to the floor and a bungled change results. This is accentuated because the pedal is set further to the off side than is usual. As a mediator, the lever's travel is not unduly inconvenient; there is synchro-mesh; but the change is not the equal of even a normal steering-column control. The keen driver sometimes regrets the lack of four speeds, too, although these objections are mediated by the mile-a-minute maximum in second gear and the engine's ability to pull away fairly happily from about 10 m.p.h. in top, in spite of the high gearing. Here one is reminded of the hand ignition control in the centre of the facia, which encourages lazy drivers to slog along with a minimum of gear-changing, killing " pinking " by its ready aid.

In any other car we might make much of this somewhat tricky gear-change, but the other splendid qualities of the Citroen allow us to dismiss it with the observation that the car is at its best devouring the open—but not necessarily straight!—road, although rather more exacting as a town carriage.

It remains to enlarge on a few matters of detail. We can give full marks to so many—the comfort of the generously-upholstered leather seats, the leg-room, front and back, with foot-rail for the rear-seat passengers, the real wood instrument panel, the S.G.D.G. extension to the right of the steering wheel carrying controls for Lucas lamps, dipper (for both side and headlamp positions) and twin Lucas horns,

[Motor Sport Copyright
FRONT-DRIVE.—*The Citroen Six has a dignified, slightly aggressive appearance, handles splendidly and is usefully spacious.*

etc. Alas, the designer departs from the vintage tradition by omitting a water thermometer and substituting a warning light for an oil gauge. The window winders work well, the doors shut nicely. Visibility, we have said, is excellent, the wheel low-set, both front wings visible, although the large rear-view mirror on the facia sill was obstructive (since replaced, however, by a smaller mirror). The instrument lighting is excellent and separate from the side-lamp circuit. The luggage locker is not very large, but the lid, carrying the covered spare wheel, falls back to carry extra luggage on a small platform and in the car's country of origin a roof-rack would offer the complete solution. There is a lined cubby-hole, lidless but of adequate depth, and elastic-top pockets of the most sensible sort in scuttle doors and front-seat backs. The horn note is delightfully "Continental." There are self-cancelling indicators. The hanging pedals are comfortable to operate and all work lightly, clutch action is good, the Lockheed brakes entirely adequate, silent, and free from vices.

The screen opens for ventilation, but not for fog (rare in France ?). A sliding roof is available as an extra.

The engine is smooth without being "silky," scarcely runs-on after the hardest collar-work, starts instantly (with Bugatti-like ring of starter on open flywheel) and gives 16/17 m.p.g. under hard-driving conditions (22, "pottering"). From 55 m.p.h. onwards it makes a "wind in the wires" sound, but otherwise the car is outstandingly quiet, although the speedometer-drive made an irritating noise (its needle also swung badly, making the logging of acceleration

---

THE CITROEN SIX SALOON

Engine: Six-cylinder, 78 by 100 mm. (2,866 c.c.). Push-rod o.h.v.; 6.5 to 1 compression-ratio. 76 b.h.p. at 3,800 r.p.m.
Gear ratios: 1st, 13.24 to 1; 2nd, 5.62 to 1; top, 3.875 to 1.
Tyres: 185 by 400 Michelin Pilote on steel disc wheels.
Weight: 26½ cwt. (less occupants and with one gallon of petrol).
Steering ratio: 2 turns, lock to lock.
Fuel capacity: 15 gallons. Range approximately 300 miles.
Wheelbase: 10 ft. 1½ in.
Track: 4 ft. 10¾ in.
Overall dimensions: 15 ft. 11 in. by 5 ft. 10 in. (wide) by 5 ft. 1 in. (high).
Price: £980 (£1,525 18s. 11d. with p.t.).

PERFORMANCE DATA

Speeds in gears:
    1st .. 30 m.p.h.   2nd .. 60 m.p.h.
              Top .. 84 m.p.h.

Acceleration (speedometer corrected):
    0—30 m.p.h. in 5 sec.   0—50 m.p.h. in 12.3 sec.
    0—40 m.p.h. in 9 sec.   0—60 m.p.h. in 16.6 sec.
            s.s. ¼-mile in 22 sec.

Makers: Citroen Cars, Ltd., Trading Estate, Slough, Bucks.

---

figures difficult; it was also about 5 m.p.h. fast throughout). The front wheels will spin momentarily on a dry road in bottom gear when accelerating fiercely. The headlamps are good but not exceptional; the anti-dazzle excellent.

As apt as stepping down on to the low floor is stepping out from the wide doors, unimpeded by running boards, and the external appearance is as imposing as it is distinctive...

We hope we have written sufficient to emphasise that the Citroen Six is a car quite out of the ordinary run of cars, very fast in point of average speed, very comfortable, fascinatingly individualistic, essentially practical, with a spacious interior, yet possessing handling qualities to delight enthusiasts and render this an outstandingly safe vehicle in their hands. At the basic price of £980 it represents excellent value, for to its more obvious good qualities must be added the safety of Citroen steel construction and the durability of their wet-liner engine and, indeed, of the car as a whole.—W. B.

By way of summing up we append the experiences of Cecil Clutton, well-known as a car connoisseur, with one of these Citroens over an appreciable mileage in private ownership. He writes :—

As soon as it was announced, I decided that the six-cylinder Citroen was what I wanted for every-day motoring, but I did not acquire one until 1950. JOP 623 was just out of quarantine and came to me with 10,000 miles on the clock. I have driven it for a further 20,000 after which I see no reason to change my original opinion and there is no car for which I would willingly exchange it, regardless of money. There are many more exciting cars, but none I would prefer for daily use embracing business and pleasure motoring. For me, such a car must be capable of cruising between 70-75 m.p.h. without fuss and of accelerating from 0-60 m.p.h. in not more than 20 seconds. It must be tough, durable and easily serviced. It must corner better than most and have absolutely positive steering.

It is staggering what people will put up with in the way of spongy steering. After two years of fiasco it took a Stirling Moss to point out that the B.R.M.'s steering was no better than the average American saloon, and to suggest a course of rack-and-pinion.

All my requirements are met by the Citroen. It does not corner as well as some cars, including the Light Fifteen because, with 60 per cent. of its weight at the front, it is not so well balanced. This endows it with understeer, but not to any embarrassing extent since, coupled with the superbly positive steering, it can change direction with great rapidity. If, when cornering at the limit, the foot is taken off the accelerator, this understeer turns into oversteer; but this is only embarrassing during limit cornering, when exactly the same applies to rear-wheel-driven cars. Owing to the heavy front-end the Six becomes rather soggy at limit cornering and cannot be "diced" like the Fifteen, but on wet roads the front wheels break away more readily and fun can be had. Either on wet or dry roads it can out-corner all but the very best.

On wet or icy roads the engine can be used to drag the car out of an emergency, but if several emergencies succeed each other in rapid succession a day of retribution eventually intervenes. Under such conditions it is best to declutch, when the f.w.b. effect of deceleration is no longer present.

Although the steering is heavier than some, I would willingly put up with much heavier steering than the "Six's" for the sake of such positive accuracy.

The "Six" is not a perfect car—there is, of course, no such thing. It is a pest having to lubricate the front universals and king-pins every 600 miles. Front wheel judder under braking can be alarming, although never dangerous. It can be mitigated by balancing but never cured. The early Fifteens, which had much lighter wheels than those now fitted, had no such trouble, so the weight of the now standard wheels is probably the cause of the trouble. The engine runs very cool and at freezing air-temperature it cannot be got over 40 deg. C. I consider that a muff is therefore essential, as also is an oil-gauge and water thermometer.

The sump oil-level indicator is an incurable liar, but may easily be converted to dip-stick.

Petrol consumption has been disappointing and I cannot better 16-17 m.p.g. with my methods of driving.

The change from second to bottom, involving a drop from 5.6 to 13.2 to 1, is not easy, and between 20-30 m.p.h. there is no really effective ratio, though second will pull away from walking pace.

With the single-carburetter breathing fades out noticeably above 3,500 r.p.m. (50 and 70 m.p.h.), and I believe two carburetters effect a great improvement, which would be particularly valuable in second.

The Citroen Six will certainly go down in history as one of the really great cars, in the best "30/98" tradition.

THIS MONTH'S CONTINENTAL RACES

F.I. = Formula I cars, i.e., up to 1½-litres s/c., up to 4½ litres non-s/c.
F.II. = Formula II cars, i.e., cars up to 500 c.c. s/c., up to 2 litres non-s/c.
C. = Formula Libre, i.e., any type of racing car.
S. = Sports cars.
T. = Touring cars.

4th.	Mille Miglia (*S.T.*), Italy.
18th.	Swiss G.P. (*F.II*), Berne.
22nd.	Luxembourg G.P. (*F.III*).
25th.	Eifel Meeting (*F.I, II, III, S.*), Nurburg Ring.

BID FOR VICTORY?

It was announced in the *News Chronicle* of April 10th that : "Two B.R.M.s are to take part in this year's Ulster Tourist Trophy race."

CONNAUGHT PRICES

We are informed by Connaught Engineering that the price of the L3/SR Connaught is now £1,945 18s. 11d. with p.t. (basic, £1,250) and that of the chassis £959 1s. 8d. (basic, £750).

NEW CARS DESCRIBED

The Big Fifteen retains the familiar Citroen characteristics despite the addition of 7in. to the chassis length and the new luggage locker at the rear—the first major change in appearance for 18 years.

BRITISH CITROEN'S BIG FIFTEEN

Family Model New to Post-war British Market: Six-cylinder Price Reduced

THE range of cars presented by the British Citroen factory at Slough for 1953 comprises the well-known four-cylinder Light Fifteen and the 2.8-litre six-cylinder car, plus a large four-door family saloon known as the Big Fifteen, which is new to the post-war British market. This model, which is already popular with family men in France, has the same four-cylinder 2-litre engine and the same front-wheel drive layout as on the popular Light Fifteen, but it has a much larger body capable of seating six people on two wide and well upholstered bench-type seats. The wheelbase is 7in longer than that of the Light Fifteen, and the track is 4½in wider, giving a car with a carrying capacity similar to that of the six-cylinder model but with lower operating costs.

Status Quo

No significant mechanical changes have been made to these popular and well-known front-wheel-drive cars, which, with their independent torsion bar front suspension, overhead valve engines, removable cylinder liners and light, rigid unit structures, have now been in production virtually unchanged in essentials for 18 years. All models will, however, have a new hinged luggage locker, as illustrated in *The Autocar* description of the French parent factory's programme on August 22.

The British models, which are luxuriously upholstered in real leather, are now available with bench-type front seats on all models; this type is optional on the Light Fifteen. The facia gear lever is admirably suited to this utilization of space for seating. Folding centre arm-rests are provided at front and rear.

Detail changes include a new exhaust silencer with a revised method of suspension, which renders it more efficient, a new pattern steering wheel and, on the four-cylinder engines, the new Lucas D.M.2 distributor. The direction indicators, which, on the cars assembled in Britain, consist of semaphore arms recessed in the body centre pillars, are now operated by a pneumatic time switch, and there are combined stop and tail lamps on each rear wing.

A Zenith petrol filter is incorporated in the fuel tank, and the warm air supply from the simple interior heater, which collects warm air from behind the radiator, is now regulated by a control on the facia. The cars are finished in black or in metallic-chrome cellulose.

Despite the improvements incorporated, the basic price of the six-cylinder saloon is now £940 plus tax, showing a reduction of £40 compared with the 1952 model.

Continuation of the basic design of the Citroen is a tribute to the "rightness" of its individuality and unorthodoxy.

SPECIFICATIONS

LIGHT FIFTEEN

Engine.—4 cyl, 78 × 100 mm (1,911 c.c.), 55.7 b.h.p. at 4,250 r.p.m.
Transmission.—3-speed gear box. Front-wheel drive. Facia gear lever. Overall ratios: 4.3, 7.3 and 13.1 to 1.
Suspension.—Front, independent, wishbones and longitudinal torsion bars. Rear, trailing axle with transverse torsion bars.
Brakes.—Lockheed hydraulic.
Main Dimensions.—Wheelbase, 9ft 6½in. Track, 4ft 6in. Overall length, 14ft 2in. Overall width, 5ft 5¼in. Height, 5ft. Weight, 2,380 lb.
Price.—£685 plus British purchase tax £382 1s 2d. Total, £1,067 1s 2d. Sliding roof £10 extra plus p.t.

BIG FIFTEEN

Engine, Transmission and Suspension.—As Light Fifteen.
Dimensions.—As Six, except overall length 15ft 6½in. Weight dry, 2,548 lb.
Price.—£740 plus British purchase tax £412 12s 3d. Total, £1,152 12s 3d.

SIX

Engine.—6 cyl, 78 × 100 mm (2,866 c.c.), 76 b.h.p. at 3,800 r.p.m.
Transmission.—Overall ratios: 3.87, 5.62 and 13.24 to 1.
Main Dimensions.—Wheelbase, 10ft 1½in. Track, 4ft 10½in. Overall length, 15ft 11in. Overall width, 5ft 10in. Height, 5ft 1in. Weight, 2,940 lb.
Otherwise as Light Fifteen.
Price.—£940 plus British purchase tax £523 14s 6d. Total, £1,463 14s 6d.

Right: With folding central arm-rest, the wide bench-type front seat has room for three abreast, and the large doors make entry and exit easy despite the low build of the car.

Left: New to the range of cars presented by the Citroen British factory at Slough is the hinged luggage locker, the capacity of which is clearly shown. A light is fitted inside the lid.

The well-known four-light body style with detachable wings and separate, external head lamps, features desired by many, is still retained. The familiar "wheel at each corner" characteristic is particularly noticeable at the rear.

CITROEN LIGHT FIFTEEN

ALTHOUGH no major changes have been made to the model since it was last tested by *The Autocar* in 1948, enthusiasts for the *marque* will no doubt be interested to see that the Citroen Light Fifteen has again been put through its paces. It is difficult to think of any other car that has changed as little as the front wheel drive Citroen over as long a period, and yet after more than seventeen years is in many ways not outdated. A number of detail changes in design has been made through the years, although the basic structure of the car remains the same.

For example, the three-piece track rod and slave lever type of steering used earlier has been replaced by a rack and pinion unit. Engine and transmission modifications include the change over to downdraught carburation, before the war, and also larger bearings in the final drive unit. Briefly, this car was from the start fundamentally right in design and construction to suit Continental motoring. It must be remembered that in its country of origin this is a quantity-produced popular model; on the other hand in England—where, of course, it is also assembled, with a considerable content of British components, for this and overseas markets—it is regarded as a car for the enthusiast. This is perhaps because of its marked difference in character from an average family saloon.

The Citroen is essentially a machine for motoring in, using the phrase in a specialized sense; it is robust and rugged and perhaps a little rough by some standards, but its somewhat vintage character is not spoiled by any desire to remind its occupants of either a gin palace or a stately home. It has an exceedingly high common-sense factor. It is happiest on the open road with long distances to cover in as short a time as possible, but this does not mean that it is unsuited

---

PERFORMANCE

CITROEN LIGHT FIFTEEN

ACCELERATION: from constant speeds.
Speed, Gear Ratios and time in sec.

M.P.H.	4.3 to 1	7.3 to 1	13.1 to 1
10—30	10.6	6.1	—
20—40	10.4	6.4	—
30—50	11.0	8.3	—
40—60	13.5	—	—

From rest through gears to:

M.P.H.	sec
30	5.7
50	14.1
60	22.1

Standing quarter mile, 22.1 sec.

SPEED ON GEARS:

	M.P.H. (normal and max.)	K.P.H. (normal and max.)
Top (mean)	75.5	121
(best)	76.0	122
2nd	40—50	64—80
1st	20—26	32—42

TRACTIVE RESISTANCE: 25 lb per ton at 10 M.P.H.

SPEEDOMETER CORRECTION: M.P.H.

Car speedometer	10	20	30	40	50	60	70	80	85
True speed	10	19	27	35	44	53	61	71	76

TRACTIVE EFFORT:

	Pull (lb per ton)	Equivalent Gradient
Top	166	1 in 13.5
Second	333	1 in 6.6

BRAKES:

Efficiency	Pedal Pressure (lb)
85.3 per cent	86
53 per cent	50

FUEL CONSUMPTION:
24.4 m.p.g. overall for 366 miles. 11.5 litres per 100 km.
Approximate normal range 22–25 m.p.g. (12.8–11.3 litres per 100 km).
Fuel: British Pool.

WEATHER: Dry surface. Wind light.
Air temperature 44 degrees F.
Acceleration figures are the means of several runs in opposite directions.
Tractive effort and resistance obtained by Tapley meter.
Model described in *The Autocar* of February 22, 1946.

DATA

PRICE (basic), with saloon body, **£685**.
British purchase tax, £382 1s 2d.
Total (in Great Britain), £1,067 1s 2d.
Extras: Sliding roof £10.
Fog lamp £4 7s 6d.

ENGINE: Capacity: 1,911 c.c. (116.6 cu in).
Number of cylinders: 4.
Bore and stroke: 78 × 100 mm (3.07 × 3.93in).
Valve gear: o.h.v. push rods and rockers.
Compression ratio: 6.5 to 1.
B.H.P.: 55.7 at 4,250 r.p.m. (43.3 B.H.P. per ton laden).
Torque: 90.4 lb ft at 2,200 r.p.m.
M.P.H. per 1,000 r.p.m. on top gear, 17.4.

WEIGHT (with 5 galls fuel), 22 cwt (2,464 lb).
Weight distribution (per cent), 54.8 F; 45.2 R.
Laden as tested: 25¾ cwt (2,878 lb).
Lb per c.c. (laden): 1.50.

TYRES: 165—400mm broadbase.
Pressures (lb per sq in): 18 F; 20 R.

TANK CAPACITY: 11 Imperial gallons.
Oil sump, 8 pints.
Cooling system, 14 pints.

TURNING CIRCLE: 45ft 3in (L and R).
Steering wheel turns (lock to lock): 2¼.

DIMENSIONS: Wheelbase 9ft 6½in.
Track: 4ft 4¾in (F); 4ft 4¼in (R).
Length (overall): 14ft 0in.
Height: 4ft 11½in.
Width: 5ft 5¼in.
Ground clearance: 7in.
Frontal area: 22.3 sq ft (approx).

ELECTRICAL SYSTEM: 12-volt 57 ampère-hour battery.
Head lights: Double dip, 36-36watt.

SUSPENSION: Front, independent, with wishbones and torsion bars.
Rear, trailing arms and torsion bars.

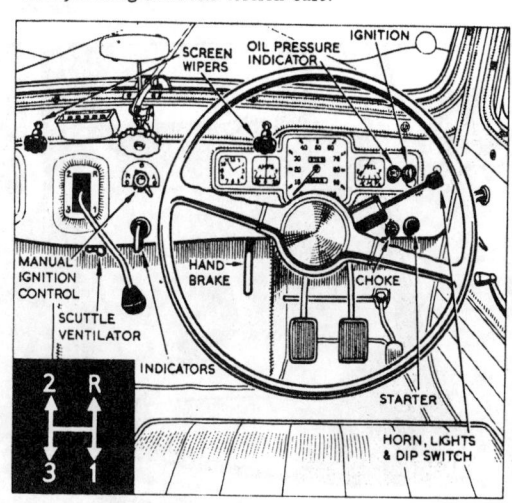

The Citroen chevron motif is attached to the rear of the radiator grille, and a small detachable plated pressing covers the hole in the bottom of the grille that accommodates the starting handle. The fog lamp is an extra fitting.

The body style is businesslike, with the minimum of external plated fittings. Light alloy splash plates are fitted to the leading edge of the rear wings to give protection against stones. Mud flaps are also attached to the underside of the front wings.

to traffic conditions. In spite of a maximum speed that is not above average for a modern 2-litre car, the Light Fifteen will put up very good averages on all types of journey; on a known run nearer fifty than forty miles is possible under suitable conditions. The four-cylinder engine is quite flexible on top gear, but it should not be allowed to slog. Main road hills of the normal variety with a gradient of 1 in 10 can be climbed on top gear, while second will cope with most of the steeper ones; a 1 in 5 hill was climbed with the car in a well-laden condition without changing down to first gear in the three-speed box.

Perhaps the most impressive thing is the way the Citroen can be cornered at speed with a complete feeling of security surpassed by few cars' behaviour. This is because it has a very wide track and long wheelbase relative to the size of the body; or, in other words, it has "a wheel at each corner." So fast can it be taken round even appreciable bends that in the hands of enterprising drivers some form of hand rail or a partition would be a useful addition to position the front passengers on left-hand bends. The unsprung weight is reduced to a minimum by reason of the torsion bar suspension system and also by the arrangement of the final drive whereby the crown wheel and differential unit are part of the sprung mass. Roll stiffness, too, is increased by the trailing arm system for the rear suspension, and the interaxle seating and general low build result in a low centre of gravity.

Steering by means of a rack and pinion unit is very positive and there is strong confirmation that the steering wheel is connected to the road wheels. With 2¼ turns from lock to lock the steering is geared for fairly direct operation rather than lightness, yet it is not unduly heavy, especially bearing in mind that the steered wheels are also transmitting power. The steering is a little heavier when the car is on drive than when it is on overrun. There is quite good self-centring action. Stability and a general feeling of confidence are further created by the understeer characteristics of the car as a whole. A slight amount of road shock is transmitted back through the steering wheel, and at certain speeds some vibration is also felt.

Gear Change and Clutch

The gear change lever, mounted on the facia panel, is in a somewhat unusual position, but not an inconvenient one. It is, in fact, in the same position for both left- and right-hand drive cars, which is unusual for gear change mechanisms that do not use a central lever mounted directly on the gear box. To prevent jumping out of gear, and to enable the selector spring loads to be reduced to a minimum, thereby permitting a light gear change, a form of positive interlock is fitted. This consists of a clutch-operated lock that prevents the gear box selector rods from being moved except when the clutch is depressed, and some care must be taken to depress the clutch pedal fully when changing gear; it is comparatively easy to beat the synchromesh. Also the clutch pedal must be pressed well down to the limit of its travel to ensure silent engagement of first gear when at rest.

A combination of well-damped torsion bars and large-section low pressure tyres ensures comfortable riding for all occupants over a very wide range of surfaces, including pavé and rough "colonial" sections, and there is no body

Because of the low floor level and integral construction there is a definite step down into the passenger compartment. The separate front seats are adjustable for leg length. Pockets are provided in the front doors. The rear seat is deep and well positioned within the wheelbase. The bottoms of the doors are trimmed with carpet, and wood cappings are fitted just below the window level; ashtrays are provided on both rear doors.

The spare wheel, carried on the outside of the luggage locker, is protected by a cover released from inside the lid. Tools are carried in a detachable box that fits in the bottom of the luggage compartment. To increase carrying capacity with the locker lid in the open position a hinged flap attached to the inside of the lid can be lowered to form a platform.

Measurements in these ⅛in to 1ft scale body diagrams are taken with the driving seat in the central position of fore and aft adjustment and with the seat cushions uncompressed.

sway or pitching under normal conditions. The hydraulically operated brakes have leading and trailing shoes at both front and rear. They are very powerful and operate in twelve-inch drums at the front, while ten-inch drums are used at the rear. No fade was experienced during the severe conditions imposed during performance testing. Also, the brake pedal pressure required for maximum efficiency is quite low. Under normal driving conditions the brakes leave less than usual to be desired and the operation has a light yet solid feel, whilst the actual application of the brakes is smooth yet reassuringly powerful. The hand brake, too, is quite powerful but the control is not as convenient to reach as could be desired.

Control

The driving position, helped by a low, flat floor, is very good, the seat being well positioned and fairly high in relation to the floor. This gives the driver the feeling that he is sitting on a seat and not on the floor carpet. The front seats are separately adjustable for leg length, and the cushions give good support; they all but meet to form virtually a one-piece seat in effect. No adjustment is provided on the steering column, but the wheel is very well placed and set at a comfortable angle. A large-diameter, solid type of wheel is used and its two horizontal spokes are ridged along the top edge. All the pedals are pivoted from the top. They are well placed and the throttle and brake pedals are conveniently placed to permit heel and toe gear changing. There is also ample room for the driver's left foot when it is off the clutch pedal.

Practical utility rather than idle luxury applies to most things about the Citroen, and this is so with the minor controls and fittings. An exceptionally neat and sturdy combination switch that looks like a small edition of a steering column gear lever is fitted to the right of the column, below the wheel. This operates the horn when it is pressed in, switches on the side lights when it is twisted 90 degrees, brings the head lights into operation when it is twisted another 90 degrees, and dips the lights when it is flicked down away from the steering wheel. Also, when the side lights alone are switched on it is possible to switch to dipped head lights by the convenient method of flicking the lever down.

A manually operated ignition control is fitted on the facia, and below it is a time-delay direction indicator switch. It would be better if this took a longer time to switch off, as unless operating it is delayed until the last moment the indicator will switch itself off before the turn concerned has been reached. The windscreen wipers are very well placed and wipe a very large proportion of the total glass area; the blades are parked off the screen and can be operated by hand should the motor fail.

Fresh air, both hot and cold, is supplied to the body interior. There are a simple ram type heater taking warm air from behind the radiator block, a scuttle ventilator, and an opening windscreen, and, as an optional extra, a sunshine roof can be fitted.

Perhaps because of the integral type of body shell construction the noise level is a little high, aggravated probably by the fact that top gear is obtained through a train of gears and is not a direct drive. Wind noise is low and the interior is well insulated from draughts. The horn has a Continental note of the type that is more effective than appears from inside the car. Double dip lighting is used and gives a good beam and spread of light. Starting from cold was usually instantaneous.

Considered as a whole the Light Fifteen is a very practical car for the driver who wants to cover ground effectively. It has a very solid and stable feel and corners as though it were on rails.

A shallow air cleaner with forward-facing intake fits neatly just below the bonnet. Control rods for the gear change mechanism run diagonally in front of the ignition leads. A simple form of fresh air heater has a duct which enters the body just below the ignition coil.

1953 CARS

A NEW OUTLINE now characterizes the Citroëns which have an enlarged luggage boot, here seen on the Big Fifteen model, an addition to the range for 1953. On the right is seen the considerable luggage space in the boot; the spare wheel may be tipped back for clear access to the baggage.

Three CITROËN Models

The "Big Fifteen" Rejoins Improved "Light Fifteen" and "Six Cylinder" Models in the British-built Range

MAIN features of the 1953 range of Slough-built Citroën cars are the provision of much increased luggage locker capacity on the "Light Fifteen" and "Six Cylinder" models, and the introduction of a new "Big Fifteen" model which combines "four-cylinder" economy with "six-cylinder" roominess.

1953 models will be very clearly distinguishable from earlier Citroëns, by a changed silhouette which goes with the increased luggage capacity already mentioned. The main body shell remains as hitherto, but the tail treatment now conforms to the type descriptively known in America as the "notch back," a boot projecting outwards below the waistline. The changes made at the back of the body have incidentally increased the overall length by two inches.

A lift-up lid of very modest weight gives access to the luggage locker, and is supported by a simple automatic strut with a press-to-release catch: provision is made for illuminating the locker interior at night from the number-plate lamp. As may be seen from the illustrations, the luggage capacity of this locker is very generous, amounting to 11½ cubic feet on the "Light 15" and rather more than this on the other, wider models.

Ingenious Stowage

Particularly ingenious is the provision made for securing the spare wheel. This is not attached to the car in the usual way, but simply rests in a tray where it leans against rubber buffers: when the locker lid is raised, the spare wheel can be tilted backwards to give clear access to the luggage space, but when it is tilted forward again and the locker closed, a spring-loaded pad on the lid inner surface presses on the tyre and holds the spare wheel quite secure.

Other, less conspicuous, improvements have also been put into production for 1953. For example, the car interior heater (which adds only £1 to the basic price of the car!) now has an on-off air valve controlled from the facia panel: the exhaust silencer and its mounting have been re-designed; a new type of Lucas

CITROEN "BIG FIFTEEN"

Engine Dimensions	
Cylinders	4
Bore	78 mm.
Stroke	100 mm.
Cubic capacity	1,911 c.c.
Piston area	29.6 sq. ins.
Valves	Pushrod o.h.v.
Compression ratio	6.5 : 1
Engine Performance	
Max. power	55.7 b.h.p.
at	4,250 r.p.m.
Max. b.m.e.p.	117 lb./sq. in.
at	2,200 r.p.m.
B.h.p. per sq. in. piston area	1.81
Peak piston speed ft. per min.	2,790
Engine Details	
Carburetter	Solex 32PBI d/draught
Ignition	12-volt coil
Plugs (make)	Champion H10
Fuel pump	AC mechanical
Fuel capacity	15 gallons
Oil capacity	8 pints
Cooling system	Pump and fan
Water capacity	14 pints
Electrical system	12-volt
Battery capacity	57 amp./hr.
Transmission	
Clutch	Single dry plate
Gear ratios: Top	4.3
2nd	7.3
1st	13.1
rev.	17.5
Prop. shaft	Nil (front wheel drive)
Final drive	9/31 spiral bevel
Chassis Details	
Brakes	Lockheed hydraulic
Brake drum diameter	Front 12 in., rear 10 in.
Friction lining area	80½ sq. in.
Suspension: Front	Torsion bar and wishbone I.F.S.
Rear	Torsion bars and axle
Shock absorbers	Telescopic
Wheel type	Steel disc
Tyre size	Michelin, 165 x 400
Steering gear	Rack and pinion
Steering wheel	2-spoke (optional spring-spoke)
Dimensions	
Wheelbase	10 ft. 1½ in.
Track: Front and Rear	4 ft. 10¼ in.
Overall length	15 ft. 6½ in.
Overall width	5 ft. 10 in.
Overall height	5 ft. 1 in.
Ground clearance	7 in.
Turning circle	45½ ft.
Dry weight	22¾ cwt.
Performance Data	
Piston area, sq. in. per ton	26.1
Brake lining area, sq. in. per ton	71
Top gear m.p.h. per 1,000 r.p.m.	17.4
Top gear m.p.h. at 2,500 ft./min. piston speed	66.3
Litres per ton-mile, dry	2,900

INCORPORATING the new luggage boot, and certain detail refinements, the Light Fifteen (*below*) and the Six (*right*) are being continued without basic change.

ignition distributor has been adopted, retaining manual adjustment of ignition timing controlled also by speed and throttle opening; a filter has been incorporated in the fuel system; the time switch for self-cancelling trafficators is now a silent pneumatic unit; metallichrome paintwork is being standardised for coloured cars.

A Popular Combination

The "Light Fifteen" is a very old favourite, the product of long years of evolution, and for four years past there has also been the "Six", with 50 per cent. greater engine capacity, and offering greater roominess consequent upon a longer wheelbase and wider track. Now a model which was popular before the war re-appears in modern guise, the "Big Fifteen" which combines the body dimensions of the 3-litre car with the more economical 2-litre power unit. It is safe to prophesy that this car will prove very popular in a year when economy is likely to be a common demand —it is a genuine six-seater with exceptionally generous rear-seat leg room, but being only 4¾ ins. wider than the "Light Fifteen" and heavier by an amount equivalent to the weight of one adult passenger, it should reveal much of the economical liveliness for which this car is well known.

In mechanical specification, the three cars have a great deal in common, and whilst certain features are relatively unusual in Britain it is fair to recall that these cars have been amongst Europe's best-selling models for more than 15 years. Steel bodies serve also as chassis frames, the suspension is by torsion bars and telescopic-pattern hydraulic dampers at front and rear, pushrod-o.h.v. engines are built in unit with three-speed gearboxes, and front-wheel drive is used to eliminate all mechanism from beneath floors which are exceptionally low and flat.

On road test by "The Motor," the two models which continue in improved form showed the following key performance figures:—

	"Light Fifteen"	"Six"
0-50 m.p.h.	15.7 secs.	12.6 secs.
30-50 m.p.h. in top gear	9.4 secs.	7.2 secs.
Maximum speed	72.6 m.p.h.	81.8 m.p.h.
Overall fuel consumption	25.2 m.p.g.	19.1 m.p.g.

GENUINE SIX-SEATER accommodation is offered by the body which the Big Fifteen and the Six Cylinder cars share. As may be seen here, rear-seat leg room is unusually generous.

SPECIFICATIONS

CITROEN LIGHT FIFTEEN

ENGINE.—Dimensions: Cylinders, 4; bore, 78 mm.; stroke, 100 mm.; cubic capacity, 1,911 c.c.; piston area, 29.6 sq. in.; valves, pushrod o.h.v.; compression ratio, 6.5 : 1. Performance: Max. power, 55.7 b.h.p. at 4,250 r.p.m.; b.h.p. per sq. in. area, 1.81. Details: Carburetter, Solex 32 PBI downdraught; ignition, 12-volt coil; sparking plugs, Champion H.10; fuel capacity, 11 gallons.

TRANSMISSION.—Clutch, single dry plate; overall gear ratios: top, 4.3; 2nd, 7.3; 1st, 13.1; rev., 17.5; propeller shaft, nil (front wheel drive); final drive, spiral bevel.

CHASSIS DETAILS.—Brakes, Lockheed hydraulic; friction lining area, 80.5 sq. in.; front suspension, torsion bar and wishbone I.F.S.; rear suspension, torsion bars and axle; shock absorbers, telescopic; tyre size, 165 by 400.

DIMENSIONS.—Wheelbase, 9 ft. 6½ in.; front track, 4 ft. 6 in.; rear track, 4 ft. 5¼ in.; overall length, 14 ft. 5 in.; overall width, 5 ft. 5¼ in.; overall height, 5 ft.; ground clearance, 7 in.; turning circle, 43 ft. 5 in.; dry weight, 21¼ cwt.

PERFORMANCE DATA.—Top gear m.p.h. per 1,000 r.p.m., 17.4; top gear m.p.h. at 2,500 ft./min. piston speed, 66.3; litres per ton-mile, dry, 3,100.

CITROEN 6-CYLINDER

ENGINE.—Dimensions: Cylinders, 6; bore, 78 mm.; stroke, 100 mm.; cubic capacity, 2,866 c.c.; piston area, 44.4 sq. in.; valves, pushrod o.h.v.; compression ratio, 6.5 : 1. Performance: Max. power, 76 b.h.p. at 3,800 r.p.m.; b.h.p. per sq. in. piston area, 1.71. Details: Carburetter, Solex 30PAAI dual downdraught; ignition, 12-volt coil; sparking plugs, Champion H10; fuel capacity, 15 gallons.

TRANSMISSION.—Clutch, twin dry plate; overall gear ratios: top, 3.87; 2nd, 5.62; 1st, 13.25; rev., 15.87; propeller shaft, nil (front wheel drive); final drive, spiral bevel.

CHASSIS DETAILS.—Brakes, Lockheed hydraulic; friction lining area, 143 sq. in.; front suspension, torsion bar and wishbone I.F.S.; rear suspension, torsion bars and axle; shock absorbers, telescopic; tyre size, 185 by 400.

DIMENSIONS.—Wheelbase, 10 ft. 1½ in.; front and rear track, 4 ft. 10½ in.; overall length, 15 ft. 11 in.; overall width, 5 ft. 10 in.; overall height, 5 ft. 1 in.; ground clearance, 7 in.; turning circle, 45¼ ft.; dry weight, 26¼ cwt.

PERFORMANCE DATA.—Top gear m.p.h. per 1,000 r.p.m., 20; top gear m.p.h. at 2,500 ft./min. piston speed, 76; litres per ton-mile, dry, 3,290.

The Motor Road Test No. 17/52

Make: Citroen **Type:** Big Fifteen

Makers: Citroen Cars Ltd., Trading Estate, Slough, Bucks.

Dimensions and Seating

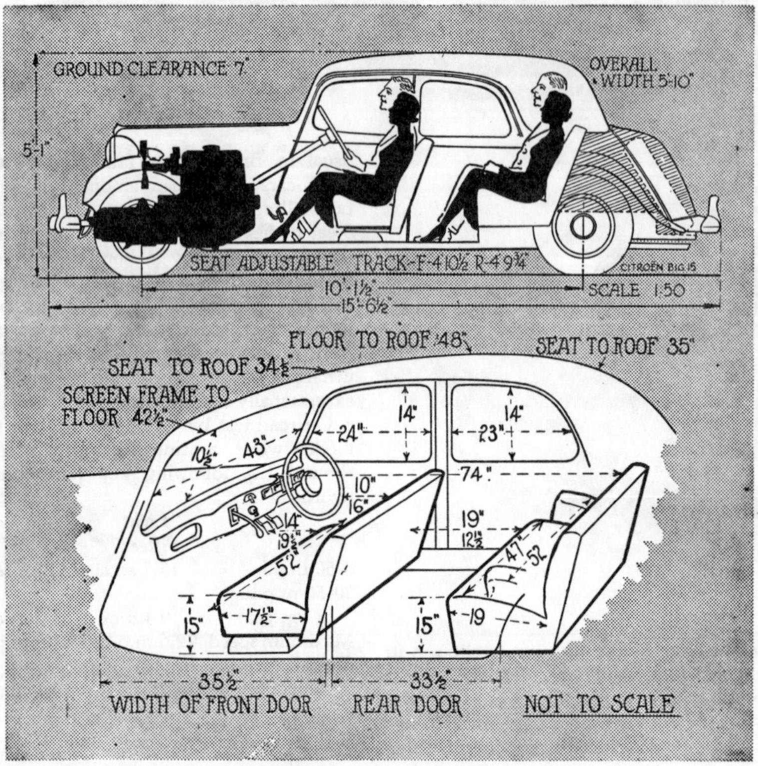

GROUND CLEARANCE 7"
OVERALL WIDTH 5'-10"
5'-1"
SEAT ADJUSTABLE TRACK-F-4'10½ R-4'9¾
CITROEN BIG 15
10'-1½"
15'-6½"
SCALE 1:50
FLOOR TO ROOF 48"
SEAT TO ROOF 35"
SEAT TO ROOF 34½"
SCREEN FRAME TO FLOOR 42½"
14"
14"
24"
23"
10½"
43"
74"
10"
16"
19"
12½"
14
19½"
5'2"
47
52
15"
17½"
15"
19
35½"
33½"
WIDTH OF FRONT DOOR
REAR DOOR
NOT TO SCALE

Test Conditions

Mild, rainy weather with strong wind. Surface, smooth tarmac, wet during tests. (Brakes tested on dry road). Fuel, Pool petrol.

Test Data

ACCELERATION TIMES on Two Upper Ratios

	Top	2nd
10-30 m.p.h.	11.6 secs.	6.5 secs.
20-40 m.p.h.	12.9 secs.	7.4 secs.
30-50 m.p.h.	14.6 secs.	10.4 secs.
40-60 m.p.h.	20.2 secs.	—

ACCELERATION TIMES Through Gears

0-30 m.p.h.	6.9 secs
0-40 m.p.h.	10.9 secs.
0-50 m.p.h.	16.4 secs.
0-60 m.p.h.	29.0 secs.
Standing Quarter Mile	23.7 secs.

MAXIMUM SPEEDS
Flying Quarter Mile

Mean of four opposite runs	70.9 m.p.h.
Best time equals	72.6 m.p.h.

Speed in Gears

Max. speed in 2nd gear	53 m.p.h.
Max. speed in 1st gear	31 m.p.h.

FUEL CONSUMPTION

23.5 m.p.g. at constant 30 m.p.h.
29.5 m.p.g. at constant 40 m.p.h.
25.5 m.p.g. at constant 50 m.p.h.
20.5 m.p.g. at constant 60 m.p.h.
Overall consumption for 459 miles, 17.8 gallons =25.8 m.p.g.

WEIGHT

Unladen kerb weight	22¾ cwt.
Front/rear weight distribution	55/45
Weight laden as tested	26¼ cwt.

INSTRUMENTS

Speedometer at 30 m.p.h.	5% fast
Speedometer at 60 m.p.h.	10% fast
Distance recorder	5% fast

HILL CLIMBING (at steady speeds)

Max. top gear speed on 1 in 20	56 m.p.h.
Max. top gear speed on 1 in 15	49 m.p.h.
Max. gradient on top gear	1 in 11.7 (Tapley 190 lb./ton)
Max. gradient on 2nd gear	1 in 6.4 (Tapley 345 lb./ton)

BRAKES at 30 m.p.h.

0.95 g retardation (=31½ ft. stopping distance) with 100 lb. pedal pressure.
0.74 g retardation (=40½ ft. stopping distance) with 75 lb. pedal pressure.
0.44 g retardation (=68½ ft. stopping distance) with 50 lb. pedal pressure.
0.17 g retardation (=177 ft. stopping distance) with 25 lb. pedal pressure.

In Brief

Price £740 plus purchase tax £412 12s. 3d. equals £1,152 12s. 3d.

Capacity	1,911 c.c.
Unladen kerb weight	22¾ cwt.
Fuel consumption	25.8 m.p.g.
Maximum speed	70.9 m.p.h.
Maximum speed on 1 in 20 gradient	56 m.p.h.
Maximum top gear gradient	1 in 11.7
Acceleration	
10-30 m.p.h. in top	11.6 secs.
0-50 m.p.h. through gears	16.4 secs.

Gearing 17.4 m.p.h. in top at 1,000 r.p.m. 66.3 m.p.h. at 2,500 ft. per min. piston speed.

Specification

Engine

Cylinders	4
Bore	78 mm.
Stroke	100 mm.
Cubic capacity	1,911 c.c.
Piston area	29.6 sq. in.
Valves	Pushrod o.h.v.
Compression ratio	6.5/1
Maximum power	55.7 b.h.p.
at	4,250 r.p.m.
Piston speed at max. b.h.p.	2,790 ft. per min.
Carburetter	Solex 32 PBI downdraught
Ignition	Lucas coil
Sparking Plugs	14 mm. Champion H10
Fuel Pump	AC Mechanical

Transmission

Clutch	Single dry plate
Top gear (s/m)	4.3
2nd gear (s/m)	7.3
1st gear	13.1
Reverse	17.5
Propeller shaft	nil (front wheel drive)
Final drive	9/31 Spiral bevel

Chassis

Brakes	Lockheed hydraulic
Brake drum diameter	Front 12 in., rear 10 in.
Friction lining area	80½ sq. in.
Suspension:	
Front.	Torsion bar and wishbone I.F.S.
Rear.	Torsion bars and dead axle beam
Shock absorbers	Newton telescopic
Tyres	Michelin, 165/400

Steering

Steering gear	Rack and pinion
Turning circle	45 ft.
Turns of steering wheel lock to lock	2⅓

Performance factors (at laden weight as tested)

Piston area, sq. in. per ton	22.6
Brake lining area, sq. in. per ton	61.5
Specific displacement, litres per ton/mile	2,510

Fully described in "The Motor," October 15, 1952.

Maintenance

Fuel tank: 15 gallons. **Sump** 8 pints, S.A.E. 20 **Gearbox and differential:** 3½ pints, S.A.E. 90 E.P. gear oil. **Steering gear:** Grease lubricated **Radiator:** 14 pints (2 drain taps). **Chassis lubrication:** By grease gun every 1,000 miles to 17 points. **Ignition timing:** 12° B.T.D.C. **Spark plug gap:** 0.025-0.028 in. **Contact breaker gap:** 0.015 in. **Valve timing:** I.O., 3° B.T.D.C.: I.C., 45° A.B.D.C.: E.O., 45° B.B.D.C.: E.C., 11° A.T.D.C. **Tappet clearances (Hot):** Inlet 0.006 in. Exhaust 0.008 in. **Front wheel toe-out:** ⅛ in. **Camber angle:** 1° ± ½°. **Castor angle:** 1¾° ± ½°. **Tyre pressures:** Front 20-22 lb., Rear 22-24 lb., **Brake fluid:** Lockheed. **Battery:** 12-volt 57 amp. hr. **Lamp bulbs:** 12 volt. Headlamps, 36/36 watt. Sidelamps and number plate lamp, 6 watt. Tail/stop lamps 6/24 watt.

Ref. B-F/20/52

The CITROEN Big Fifteen

Spaciousness and Rugged Strength in a Usefully Lively and Economical Car

ROOMIER than the Light Fifteen model which uses an identical power unit, this model offers very generous rear seat legroom.

WHEN the front-wheel drive Citroen was first introduced, it was a boldly unconventional design. That, however, was 18 years ago, and although owners of other types of car still tend to look on the design as "unorthodox" it must in fact now be regarded as one of the most thoroughly proven layouts existing anywhere in the world.

Understanding of this background is essential to the full appreciation of the Citroen. At first acquaintance, it is easy to decide that whilst there is a useful combination of roominess and good performance with moderate fuel consumption, the standard of refinement offered does not justify the rather high price of the car: it takes time to realize that the roughness evident on a short demonstration run will soon be forgotten as other virtues are appreciated, and to realize that for all its moderate weight the Citroen is one of the sturdiest cars which it is possible to buy. These comments probably are true of any Citroen model, but seem especially to apply to the "Big Fifteen" model which combines the power unit of the familiar "Light Fifteen" with the larger body used on the "Six Cylinder" cars.

Quite a simple balance sheet can be prepared, listing the respects in which the Citroen Big Fifteen is better or worse than "ordinary" cars. On the debit side, it must be recorded that under certain operating conditions the engine is not up to 1952 standards of smoothness and silence, the gearchange is not above reproach, the springing is decidedly firm by current standards (firmer also than on other Citroen models), the steering is inclined to be heavy, and the bodywork looks rather old-fashioned. To the credit of the "Big Fifteen" go an over 70 m.p.h. top speed (with quiet cruising at up to 60 m.p.h.), accommodation for 6 people and a fair amount of luggage, an overall fuel consumption of around 25 m.p.g., superb cornering qualities, good ventilation inside the body, an excellent lighting installation, and extremely sturdy construction. Quite bluntly, not everyone will like this car, but there remains a hard core of big-mileage motorists who understandably keep returning for another new Citroen every few years.

Good Accessibility

Straightforward in layout, and fitted with renewable wet cylinder liners, the 4-cylinder overhead-valve engine with its economizer-type Solex carburetter is a very willing worker. At tick-over it is completely inconspicuous, and at speeds between 30 and 60 m.p.h. in top gear it is certainly the quietest Citroen which we have yet tested: there remains, however, a moderate amount of noise at high r.p.m., and a very slight irregularity of running under light load in town traffic. Cold starting proved easy on frosty mornings, and with some use of the choke the car was immediately willing to idle or pull quite well. For some unknown reason the radiator filler and the oil dipstick are under opposite sides of the bonnet, but all such items as the battery, fuel pump, distributor, dipstick and tappets appear very easily accessible.

ACCESSIBLE and visible in this view are the sparking plugs, fuel pump, oil filler and dipstick, screen wiper motor, ignition distributor and coil, and the sturdy remote gear-control linkage.

Three forward ratios are provided in the gearbox, the upper two of them having synchro-mesh engagement: the gear lever projecting from the facia panel is unconventional,

The Citroen Big Fifteen - Contd.

but is soon accepted as perfectly convenient. Quite well chosen for all-round use, the gears are not objectionably noisy, but the synchro-mesh mechanism is not potent enough to prevent slight clashing if a gearchange is hustled: unusual is the positive locking of the gear lever until the clutch pedal is depressed, and almost equally unusual is the commendable sturdiness of the remote-control gearchange linkage.

The firmness of the suspension on this model rather surprised us, although it is perhaps in character that torsion bar flexibilities should be chosen especially to suit fully loaded conditions. Unladen, the car rides rather jerkily, although it is of course quite free from pitching and has ample spring travel available to absorb the worst shocks of "colonial" potholed gravel tracks. The pressed steel body which serves as the car's frame is, to all appearances, totally rigid.

Very positive and rather heavier than is nowadays fashionable, the steering is of a kind which might tire some lady drivers who need to manoeuvre into or out of parking places very frequently; but, it is steering which gives great confidence at all times. The turning circle is not very compact, but apart from imposing a limitation in this respect the use of front-wheel drive has very little evident effect on the steering : only by deliberate experiment is it realized that to turn on power when cornering slightly increases both castor action and the under-steer characteristics of the car, or vice versa. In ordinary driving, one is merely conscious of steering which feels very solid and can be implicitly trusted to take the car where the driver wishes.

Excellent Stability

Cornering qualities also are outstanding, on this as on other Citroen models. The ultimate limit of adhesion of the Michelin tyres which are mounted on wide-base rims cannot in fact differ greatly from that of other tyres, but the low build of the car, its firm springs, and the complete absence

DOWNWARD moving pedals are an unusual Citroen feature. As may be seen here, the instruments are on a metal panel, set into the polished wood facia.

LUGGAGE is provided for in a rear locker of useful size, the spare wheel being secured automatically by the sprung pad inside the locker lid.

of vicious handling characteristics combine together to make abnormally high cornering speeds seem completely normal.

Very generously-dimensioned and rigid brake drums make for pleasantly consistent brake performance : it may be noted that although relatively moderate areas of friction lining material are used, the drums are of 12 inch diameter at the front and 10 inch diameter on the less heavily loaded rear axle. All the pedals are similarly pivoted, to move downwards rather than forwards when pressed, but despite alterations compared to early models the accelerator and clutch are not ideally comfortable.

Excellent night-driving vision is provided by a pair of external headlamps which, whatever their effectiveness as air brakes, are mounted far enough above road level to give a long-range main beam and a useful but non-dazzling dipped setting. The instrument illumination is not ideally glare-free, but having once discovered how to use it we were immensely pleased by the one finger-tip switch which actuated the lights, the headlamp dipping, and one or two horns as required. The new luggage locker is, incidentally, effectively illuminated at night from the number plate lamp.

Cheap Heat

Costing only £1, the Citroen interior heater comprises nothing more than an air intake behind the engine radiator and a duct leading the warmed air into the body. For extreme climates such a system may be inadequate, but even with atmospheric temperatures just above freezing a small amount of radiator blanking produced really good warmth inside the front compartment of the car : as it is fresh air which is being heated, the atmosphere does not become tiring. We were not able to try the car in hot weather, but noted a windscreen which will open a short distance, a scuttle ventilator, front windows which disappear completely into the doors, and as a catalogue option not fitted to the test car a sunshine roof for an extra £10.

Rather unusually firm springs have been mentioned already, the telescopic dampers having appropriate settings so that the car is at its best when travelling fast or carrying a full load. A bench-type front seat has a rather low folding central armrest, as has the rear seat, and the upright seat high above a very low floor proved to need unusually little adjustment for drivers of varied heights. For passengers, however, neither the front seat nor the extremely roomy rear seat seemed to be ideally shaped to give support laterally and under the knees.

As has been indicated, the test car proved able to exceed a genuine 70 m.p.h. on level road, in unhelpful weather conditions and with the engine evidently still gaining performance after a running-in mileage of rather under 1,500. A manual ignition timing control is provided on the facia panel, and whilst this needs no attention in normal driving, the extent to which incorrect settings would produce loud pinking in one direction or a loss of performance in the other hinted at a preference for something better than Pool petrol. Comparison of tests figures which appear on the data page with those recorded 1½ years ago on a similarly powered and similarly geared Light Fifteen show less than 2 m.p.h. drop in maximum speed with the larger "Big Fifteen" body, and losses in acceleration and fuel consumption which only become appreciable in the upper ranges of speed. Roomier bodywork has increased the weight by very little indeed, but extra air resistance does apparently penalize performance and economy to a measureable extent—not that an overall fuel consumption of 25.8 m.p.g. for 450 miles, about half of them driven really fast, is anything but creditable for a car of this roominess.

Sensible Equipment

The finishing details of this model are very sensibly planned, there being for example a forward-sloping cubby hole, pockets in the front doors and behind the front seat, ash trays front and rear, a well-placed rear-view mirror, and a useful interior light. The new luggage locker gives useful capacity and incorporates a neat spare-wheel stowage, and although not beautiful it seems to be effectively weatherproofed.

The word "functional" is apt to be overworked nowadays, but it summarizes this Citroen very well indeed. A car is a device for transporting people, and the Big Fifteen does just this thing with the minimum of danger, delay or extravagance, and gives promise of continuing to do it dependably for an unusually long time.

CONTINUED FROM PAGE 60

employed. Lecot's car used 125 grammes of Mobiloil for every 62 miles. Michelin tyres were employed throughout and the life of a tyre varied from 12,500 miles to 18,600 miles, according to weather and road conditions encountered. The lowest mileage covered by any one tyre in really terrible conditions, during heavy winter floods, was 11,100 miles, but, for the most part, tyres were changed after 18,000 miles.

I spoke just now of the car being driven over ordinary roads. So they were, and the French route nationale has an excellent surface as a rule. All the same, they were not quite the kind to which English motorists are accustomed, for Lecot's itinerary included mountain work and we have no mountains in this country, only hillocks. For instance, on the Lyon-Monte-Carlo-Lyon section of Lecot's everlasting shuttle journey, the Esterel mountains provided 185 hairpin turns in 30 miles or less.

During November, 1935, Lecot found his usual route barred by great floods in the south of France. He turned back and continued his daily driving on the Lyon-Paris-Lyon road only, with a wiggle to make up the daily schedule of 1,100 kilometres.

By way of variety, he entered for the Monte Carlo Rally of 1936 and went off to Portugal to find a starting place at Valenca, still keeping all the while to his regular 1,100 kilometres per day. He completed the Rally without a penalty mark, but lost a few points in Monte Carlo through the mistake of a mechanic who poured in some oil without authorization.

M. Lecot started on July 22, 1935, and by July 22, 1936, he had covered exactly 396,888 kilometres, an average of 65 k.p.h., or 40.38 m.p.h. His goal had, however, been fixed at just over 400,000 km. in order to make the 250,000 miles, and he continued driving for four more days until July 26 to complete the full distance, which was 400,134 kilometres.

DISTINCTION
with a Difference

The individual and handsome styling of CITROEN gives a distinction that marks it apart from all other cars. But in addition to its external beauty, CITROEN design . . . still the most modern and most successful over the last eighteen years . . . sets it in a different class. Only the fortunate possessors of CITROEN cars are able to appreciate the surging power—the superb, safe road-holding—the effortless high cruising speed—and the wonderful ability that belongs to CITROEN alone to make a smooth run on any road.

FRONT WHEEL DRIVE • INDEPENDENT FRONT SUSPENSION • TORSION BAR SPRINGING
DETACHABLE CYLINDER BARRELS

CITROEN

PRICES
"LIGHT FIFTEEN"
SALOON
from £1,067 . 1 . 2
"SIX CYLINDER"
SALOON
from £1,525 . 18 . 11
(inc. Purchase Tax)

CITROEN CARS LTD   SLOUGH   BUCKS
Telephone: Slough 23811   *Telegrams: Citroworks, Slough*

Road Testing
The Front-Drive CITROEN

The road-testing crew found this most popular European car to have performance qualities which make it unique among family-type automobiles. In handling and stability it is superior to many sports cars, yet it offers dependable transportation and long service to the car owner

AFTER putting the Citroen through its paces, the road testing crew came up with a conclusion and a question: never try to follow a Citroen around a fast corner; and why don't auto builders make more front wheel drive cars?

The Citroen is a perfect answer for those who want a family car and still seek an auto that handles better than any other sedan and is superior to many sports cars. Its front wheel drive virtually enables a novice to drive like a budding Ascari. And the more one drives the car, the more amazing are the handling characteristics and cornering ability.

As a matter of fact, the body sway

Dyno test at Nicson Engineering showed unusual hp delivered to front wheels.

and tire squeal are practically nil. The driver can just put his foot in it, point the car where he want to go and he gets there—swiftly and effortlessly. These features, plus others, undoubtedly are responsible for the car continuing to head the best seller list for European cars the world over, despite the obvious fact that the body styling has been unchangd in 20 years.

Along with the unusual handling qualities, the Citroen owner has a car that can carry five people, in addition to ample trunk space and many innovations that few other cars have.

THE desirable mechanical features include adjustable spindle bushings which preclude replacement of the front end bushings; a manual spark setting coupled with a normal vacuum advance that enables one to change spark to suit altitude and fuel conditions; a manual windshield wiper which operates independently of the normal electric wiper; and a manual primer on the fuel pump.

All engine parts are constructed of heavy duty materials and will stand up under long use. Maintenance on the car is inexpensive, partly due to use of wet sleeves which can be removed from the cylinders and replaced without going through the trouble or expense of reboring. A typical overhaul job on the engine, including new sleeves, rings, pistons and pins, labor and miscellaneous items, will run about $75—well under the cost for a similar work and parts on the average automobile.

Another point which makes the car a god buy is the fact that because changes in the body styling have been few, depreciation is extremely low.

THE Citroen is unique from the standpoint of engine power delivered to the front wheels. With the average car, hp delivered from engine to the ground is between 50 and 60 per cent. Dyno tests made at Nicson Engineering, in Los Angeles, showed that 87.5 per cent of the power from the four-cylinder engine was delivered to the road. One of the factors responsible for this is absence of a driveshaft.

The Citroen was tested over an officially measured mile on concrete on a warm, windless day when humidity was low. It had performance that was not sensational, but entirely adequate for the type of car. Repeated tests provided the following averages:

Citroen has not changed body styles since the 30's, but car remains biggest selling European auto in the world.

0 to 30 mph—five seconds in low gear.
30 to 60 mph—16.8 seconds in high gear.
0 to 60 mph—19.8 seconds all three gears.

Top speed was 76.59 mph clocked through the flying mile. The speedometer, which registers in kilometers, had the following corrections: 20 mph (18.3 mph); 40 mph (37.7 mph); 60 mph (56.1 mph). New models of the car will have speedometers that read in mph.

The car has three speeds forward and one reverse. Gear shift is located on the dash.

Other test data included fuel consumption. Averages were 31.5 miles per gallon in city driving, 24.1 miles per gallon for highspeed highway driving.

Steering radius, with 2¼ turns from lock to lock, is 44 feet.

ENGINE of the Citroen is four cylinders, in line, with overhead valves and pushrods. The maximum hp is 55.7 at 3800 rpm and torque is 90.4 foot pounds at 2200 rpm. Bore is 3.07 inches, with stroke of 3.9 inches. Total displacement is slightly under two liters, 1911 cc or 117 cubic inches. Compression ratio used is 6.5 to 1. Engine is mounted in rubber with coil spring stabilizers.

Steering gear is rack and pinion The front suspension is independent with torsion bars and rear suspension is trailing link with torsion bars. The three-speed transmission has the following gear ratios: first gear, 13.1; second gear, 7.3 to 1; and third gear, 4.3 to 1.

Factory lists the dry weight at 2350 pounds, with the front wheel load being 1278 pounds and the rear wheel load at 1072 pounds. Wheelbase is 114.5 inches, overall length is 168 inches. Height from the ground is 59.5 inches with ground clearance being measured at seven inches.

The listed price for the car is $2,275, with Campbell Motors of South Pasadena, Calif., being the distributor in the United States. In

Citroen engine is ohv 4-cylinder that develops 55.7 hp at 4250 rpm, has displacement of 117 cubic inches.

Transmission is located in front of radiator and can be quickly serviced by removing grille. Front suspension is independent, torsion bar.

Europe, the car is available in only one color (black), but the cars sold in the U. S. give the buyer an optional of having body painted any color desired.

To sum up the general overall value of the car, it is a good buy. It possesses many features which recently have been incorporated in U. S. production autos, and many others which have not been copied on any large scale elsewhere.

The owner can be satisfied that he has dependable transportation with the prospect of long life without excessive servicing. These qualities, in addition to the performance characteristics mentioned, make the Citroen a good standard for other maunfacturers to meet.

Wet cylinder liners are removable, permit easy maintenance on rugged Citroen engine.

No. 1491 : CITROEN BIG FIFTEEN SALOON

Low build, the wide track and the minimum of front and rear overhang promise good road holding, which is fully achieved in practice. The car is very similar in appearance to the six-cylinder model.

The Autocar ROAD TESTS

THE Citroen Big Fifteen is a model which was well known before the war and has recently been reintroduced to the British market. It combines the economy of the 2-litre four-cylinder engine with the large six-seater bodywork used for the six-cylinder Citroen and, besides the detail improvements which have been made in recent years, it has the newly introduced luggage locker at the rear. Like all the Citroens assembled at the British factory at Slough, it has real leather upholstery and is available in a variety of colours, whereas the cars from the French factory are supplied only in black. Most of the electrical equipment and accessories are of British manufacture.

It needs only a short period at the wheel to act as a reminder that the occasional run on a front-drive Citroen is really an obligation for anyone whose business it is to pass judgment on cars; it is a powerful help in maintaining a sense of proportion about more recent developments. The fact is, that in the vital qualities of roadholding, stability and safety, this eighteen-year-old design is still away ahead of most post-war cars. Part of the test was carried out on roads covered in snow and ice, which allowed a full appreciation of the car's unique qualities, where other cars were sliding and slithering about. If one takes a corner too fast on ice or snow the Citroen will slide like any other car, but if the throttle is opened the wheel eventually find a grip and pull it back to the required line. If the brakes are applied too suddenly under such conditions the tail may slide, but a cautious opening of the throttle quickly draws the Citroen straight again. Even on wet roads astonishing liberties can be taken, more especially as this model has not generally a sufficient reserve of power to break down the wheel adhesion, and it will take corners in a way which is a revelation even to those who are used to riding in fast sports cars. The superb cornering ability greatly reduces the effort of judgment required from the driver; he simply turns the wheel and the car takes gentle curves or sharp corners on a level keel without tyre squeal or protest. This undoubtedly contributes to the lack of fatigue which is experienced after long hours at the wheel.

Despite the large bodywork, the Big Fifteen cruises happily at any speed up to 60 m.p.h., taking most main road hills in its stride. In hilly or mountainous country it is rarely necessary to come down below second gear in the three-speed box, as this gives an adequate reserve for pulling away from sharp and steep corners. Top gear acceleration is not startling, but the engine is sufficiently flexible to allow one to pull away from speeds below 30 m.p.h. without feeling that it is essential to change down. By using the gears, quite brisk acceleration is possible from a standstill up to 50 m.p.h., but the rate at which speed builds up beyond that point is influenced to a fairly marked degree by wind and gradient. With the ignition setting used during the test, scarcely any pinking was experienced when using first-grade fuel.

The four-cylinder engine will not appear particularly smooth to drivers whose ideas have been influenced by the

Front-wheel drive allows Citroens to have a completely flat floor. Prominent on the facia are the gear lever and the control for the opening windscreen. The pendant pedals are also visible.

Rear leg room is generous, and there is a foot rail for short-legged passengers. Folding centre arm rests are provided at front and rear. Wide doors hinged on the centre pillar allow easy access to all seats.

The exposed head lamps are an unusual feature today, but they are set high and give a powerful beam. The fog lamp is included in the standard equipment. An external luggage locker, hinged at the top, constitutes the first major change in appearance which Citroen have made for nearly eighteen years. There are twin tail and stop lamps in the wings and semaphore direction indicators in the centre door pillars.

ROAD TEST

latest flexible mountings, but it is a robust, untiring power unit with a fine record of reliable service, and stands up to long periods of full-throttle running without producing disagreeable symptoms such as auto-ignition. When the throttle is fully opened there is a certain amount of roar from the engine, and a slight whine is audible from the transmission, becoming more prominent at speeds above 60 m.p.h., but the general level of mechanical noise is not excessive. The engine started instantly, even after standing out all night at freezing temperatures, and warmed up quickly.

The gear change, by the characteristic Citroen facia lever operating in an open gate, is unusually smooth and light in action. A fairly deliberate movement is imposed upon the driver, because the selector locks are interconnected with the clutch pedal, and it is therefore impossible to move the gear lever until the pedal is fully depressed. The synchromesh between second and top gears is effective in normal use, but can be beaten by an excessively vigorous hand on the lever. The clutch takes up the drive smoothly and is quite light in action.

The suspension merits special praise. The formula of a very low, wide body, between wheels which are set out near to the maximum width and length of the car, produces a natural stability and resistance to roll to be found on few other vehicles. At normal cruising speeds it ensures a level, comfortable ride, and the car can be driven fast over rough surfaces with virtual disregard of the most severe bumps and potholes. At speeds below 30 m.p.h. the ride does appear somewhat firm by modern standards and a fair amount of sharp vertical movement is experienced, especially in the rear seats, but on long, fast journeys the Citroen remains one of the least tiring cars to ride in.

Steering is rather heavy at low speeds, and when turning sharp corners some slight snatch is experienced from the transmission, but as the speed rises the steering becomes lighter and is admirable for main road motoring. It does not transmit road shocks, its accuracy is exemplary, and there is fairly pronounced understeer. Castor action is always adequate. The turning circle is rather noticeably large and a certain amount of manœuvring is necessary when parking in confined spaces. Good braking power is produced with a moderate pedal pressure, and the brakes stood up well to hard use without deterioration, although there appeared to be a temporary increase in pedal travel in the course of the braking tests.

An upright driving position gives a good forward view, with a clear sight of the bonnet, head lamps and both front wings. The windscreen pillars are relatively thick, but the screen is sufficiently close to the driver to prevent their being felt as an obstruction in normal driving and the shallow near-vertical pane throws back no distracting reflections. The gear lever is well placed and the pendant pedals, a feature pioneered by Citroen which is now coming into wider use, are conveniently arranged, permitting simultaneous use of the brake and throttle pedals when changing down. The pull-out hand brake proves adequate in practice. Under the steering wheel is an excellent multi-purpose French electrical unit with one finger-tip lever which acts as light switch, head-lamp dipper and horn control. The big head lamps give an entirely adequate beam.

The facia on the British-finished cars is of polished wood and the interior surrounds to windscreen and windows are grained in imitation of wood to suit the English taste. There is a fair-sized glove box without a lid, and the simple instrument panel comprises a speedometer with total and trip mileage recorders, clock, ammeter, fuel gauge and warning light for low oil pressure. Facia illumination is discreet, but the intensity is not variable. The twin screenwipers, which can be switched on together or separately, sweep a good area; they are parked by turning the control knobs. The rear window is somewhat smaller than current fashion dictates, but the mirror gives an adequate view for normal requirements, and the large blind rear quarters do not specially impede the driver when parking the car.

A central folding armrest is provided for the front seat as well as the rear, but in spite of the car's cornering abilities it rarely seems to be required, for it becomes apparent that what causes passengers to slide about is roll rather than cornering speed, and on this car the roll is negligible. Some

The bonnet is hinged on the centre line. One side opens to reveal the simple heater duct collecting warm air from behind the radiator. Oil and water fillers, fuel pump, distributor, plugs, coil, battery and wiper motor are all accessible. On the opposite side, carburettor, dynamo, starter, water pump, electrical junctions, brake reservoir and master cylinder can be reached with unusual ease.

Behind the spare wheel a space capable of carrying considerable luggage is enclosed by the new locker lid. The interior is illuminated by the number plate lamp at night. Tools are carried in the well alongside the wheel.

wind noise is heard at speeds near the maximum, but the interior sound level is normal, and there is less road noise transmitted on rough surfaces than on some post-war unit construction cars.

The heater, which is a simple duct collecting warm air from behind the radiator, now has an orifice controlled from the facia panel and after five to ten miles from a cold start it delivers a sufficient flow of warm, fresh air to keep the interior pleasantly comfortable, even when it is freezing outside. The lack of demisting and defrosting equipment is noticeable, especially when the car is held up in traffic blocks, but it is mitigated to some extent by the fact that the windscreen can be opened, a rare feature nowadays.

In the new luggage locker there is space for several suitcases, a distinct improvement on the previous arrangement. The extra space is not as great as it first appears, however, because the spare wheel, previously carried externally, is now housed inside the locker; usable locker space is actually four inches longer and four inches higher than before. It was felt that in some aspects, particularly the sealing round the lid, the new locker compares unfavourably with the proprietary luggage trunks which have been offered in France for these cars during the past year or two.

The old style two-piece bonnet, with its central hinge, gives such unimpeded access to all the main mechanical items that it comes as something of a shock to realize how much has been sacrificed by modern styling. Points requiring regular attention with the grease gun number 17.

Its strictly practical layout, coupled with the big body space and moderate operating costs, must have an appeal for many people, quite apart from the car's road behaviour. Outmoded though it may be in appearance, and crude by current standards in some of its details, it is easy to see why a large number of people still refuse to buy anything else.

CITROEN BIG FIFTEEN SALOON

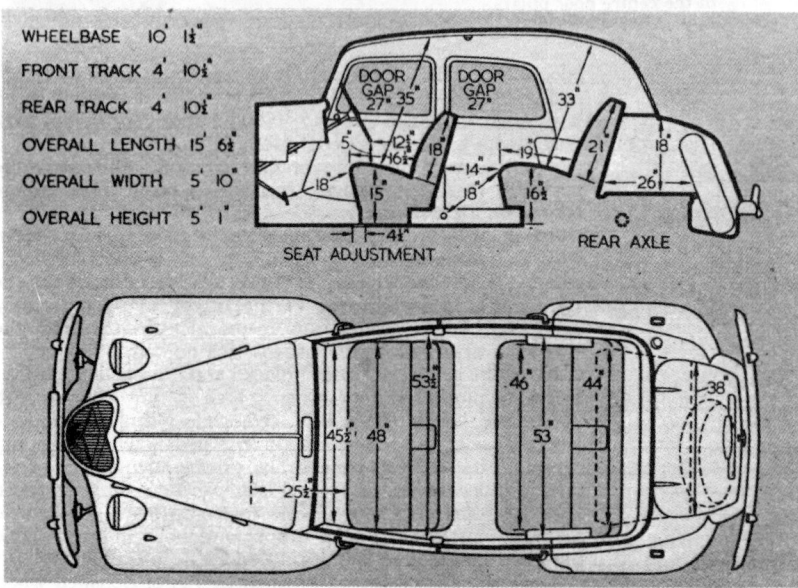

WHEELBASE 10' 1¼"
FRONT TRACK 4' 10½"
REAR TRACK 4' 10½"
OVERALL LENGTH 15' 6¼"
OVERALL WIDTH 5' 10"
OVERALL HEIGHT 5' 1"

Measurements in these ⅛in to 1ft scale body diagrams are taken with the driving seat in the central position of fore and aft adjustment and with the seat cushions uncompressed.

PERFORMANCE

ACCELERATION: from constant speeds.
Speed, Gear Ratios and time in sec.

M.P.H.	4.3 to 1	7.3 to 1	13.1 to 1
10—30	11.9	6.7	5.7
20—40	11.6	6.9	—
30—50	12.8	9.8	—
40—60	18.4	—	—

From rest through gears to:

M.P.H.	sec.
30	7.3
50	17.9
60	29.1

Standing quarter mile, 23.3 sec.

SPEED ON GEARS:

Gear		M.P.H. (normal and max.)	K.P.H. (normal and max.)
Top	(mean)	70.3	113.1
	(best)	72.0	115.8
2nd		36—52	58—84
1st		16—32	26—51

TRACTIVE RESISTANCE: 23.7 lb per ton at 10 M.P.H.

TRACTIVE EFFORT:

	Pull (lb per ton)	Equivalent Gradient
Top	200	1 in 11.2
Second	347	1 in 6.4

BRAKES:

Efficiency	Pedal Pressure (lb)
88 per cent	100
79 per cent	50

FUEL CONSUMPTION:
23.6 m.p.g. overall for 806 miles (12.0 litres per 100 km).
Approximate normal range 22–26 m.p.g. (12.8–10.9 litres per 100 km).
Fuel, First grade.

WEATHER: Dry, cool, medium breeze.
Air temperature 43 degrees F.
Acceleration figures are the means of several runs in opposite directions.
Tractive effort and resistance obtained by Tapley meter.
Model described in *The Autocar* of October 17, 1952.

SPEEDOMETER CORRECTION: M.P.H.

Car speedometer	10	20	30	40	50	60	70	80
True speed	10.5	18	26.5	35	44	52	62	72

DATA

PRICE (basic), with saloon body, £740.
British purchase tax, £412 12s 3d.
Total (in Great Britain), £1,152 12s 3d.
Extras: Heater £1 11s 2d.
Sliding roof £15 11s 1d.

ENGINE: Capacity: 1,911 c.c. (116.6 cu in).
Number of cylinders: 4.
Bore and stroke: 78 × 100 mm (3.07 × 3.93 in).
Valve gear: overhead, push rods.
Compression ratio: 6.5 to 1.
B.H.P.: 55.7 at 4,250 r.p.m. (B.H.P. per ton laden 40.9).
Torque: 90.4 lb ft at 2,200 r.p.m.
M.P.H. per 1,000 r.p.m. on top gear, 17.4.

WEIGHT (with 5 gals fuel), 23½ cwt (2,632 lb).
Weight distribution (per cent) 53.4 F; 46.6 R.
Laden as tested: 27.2 cwt (3,046 lb).
Lb per c.c. (laden): 1.59.

BRAKES: Type: F, Leading and trailing shoes. R, Leading and trailing shoes.
Method of operation: F, Lockheed hydraulic. R, Lockheed hydraulic.
Drum dimensions: F, 12in diameter, 1.38in wide. R, 10in diameter, 1.38in wide.
Lining area: F, 49.62 sq in. R, 40.08 sq in (66 sq in per ton laden).

TYRES: 165 × 400 mm.
Pressures (lb per sq in): 20–22 F; 22–24 R.

TANK CAPACITY: 11 Imperial gallons.
Oil sump, 8 pints.
Cooling system, 14 pints.

TURNING CIRCLE: 45ft (L and R).
Steering wheel turns (lock to lock): 2¼.

DIMENSIONS: Wheelbase 10ft 1¼in.
Track: 4ft 10½in (F); 4ft 10½in (R).
Length (overall): 15ft 6¼in.
Height: 5ft 1in.
Width: 5ft 10in.
Ground clearance: 7in.
Frontal area: 20 sq ft (approx).

ELECTRICAL SYSTEM: 12-volt; 57-ampère-hour battery.
Head lights: Double dip; 36–36 watt.

SUSPENSION: Front, Independent, torsion bars and wishbones. Telescopic dampers. Rear, Torsion bars and trailing arms with solid axle beam. Telescopic dampers.

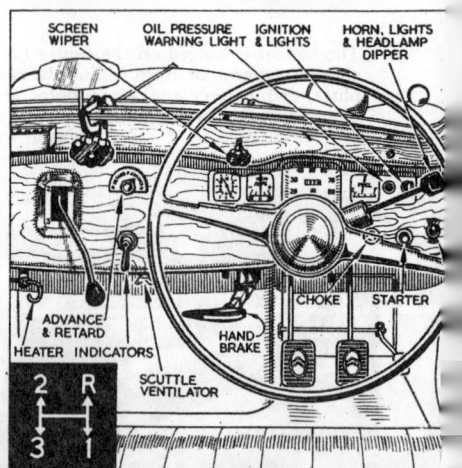

The Citroen Six retains the classic lines associated with the marque *for many years. The styling is simple and in many ways very functional, with a complete absence of frills. Front and rear doors are hinged from a central pillar. In addition to the mud flap on the front wings, splash plates are fitted to the rear wings.*

CITROEN SIX SALOON

THE products of the Citroen company are outstanding in many ways and the reaction to this make is usually quite strong. In this country there are, as it were, two camps, the Citroen enthusiasts and the rest. This state of affairs is all the more interesting when it is realized that the car occupies a very different position in its country of basic origin across the Channel. Although these cars have been assembled in England for many years, the front wheel drive Citroen is considered as something rather special and a car for the enthusiast or connoisseur, while in France it is one of the most popular makes and is very much a car for everyday transport.

Why should this difference in outlook or reaction exist? This is a question that can best be answered by considering the various qualities built into the design and also in some degree by considering the differences in operating conditions between the two countries. The Citroen is perhaps unique inasmuch as its basic design has been in existence for almost two decades. Various engine sizes as well as several body sizes have been produced, but the main layout has remained unchanged; yet in a number of ways it is still not outdated.

Weight distribution with a bias towards the front wheels, rack and pinion steering and wishbone and torsion bar independent front suspension, are all features that were well ahead of their time when they were first introduced, and are still current today.

The Citroen Six was last tested by *The Autocar* over four years ago and during that time very few modifications have been made, the most noticeable being to the appearance of the body by the addition of a luggage locker lid that encloses the spare wheel. The general proportions and rakish lines of the Six suggest that this is a car for "real motoring," and right from the start the general stability and fine roadholding qualities are very noticeable. Although the engine has a capacity of almost 3 litres its power output is fairly low—76 b.h.p. at 3,800 r.p.m., running on a compression ratio of 6.5 to 1. However, it should not be thought that the car is in any way sluggish; on the contrary it is very lively, with a mean maximum speed of a genuine 80 m.p.h. together with particularly good top gear acceleration around the 50-70 mark. Although its ultimate maximum speed may be less than that of some cars of a similar engine capacity, it is the overall time of a journey that is the important factor. On one run of 100 miles over a main road route that contained a reasonable amount of traffic, and under conditions of fairly heavy rain, the Six recorded almost 45 miles in the hour without exceeding a maximum of much over 70, which is a comfortable cruising speed for the car. On another occasion 60 miles were covered in 66 minutes.

As with the Light Fifteen, the transmission and final drive unit is placed in front of the engine, which extends right back to the vertical bulkhead in front of the driving compartment, and although the engine is "just the other side of the wall" it is well insulated. The engine is quite smooth and has good bottom end power. It is free from pinking with the manual overriding ignition lever in the fully advanced position when running on first grade fuel, while on ordinary grade fuel it is necessary to retard the

There is a distinct step down into both compartments of the Citroen and because of the front wheel drive the floor is completely flat. Both front and rear seats have folding central armrests; there are pockets in the front doors and on the back of the front seat. A foot rail is provided in the rear compartment.

117

A recent change in appearance has been made by modifying the luggage locker, which now encloses the spare wheel. Combined stop and tail lights are mounted in the rear wings and a separate central light illuminates the number plate.

setting slightly to obtain smooth operation free from pinking. The clutch is smooth and quite light in operation, with just about the right amount of pedal travel. An unusual feature is the interlock mechanism between the clutch and the gear lever, so that it is not possible to change gear (or for that matter for the gears to jump out of engagement) while the clutch is engaged. On the particular car tested the change was light and smooth on all gears except second, which was sometimes a little awkward to engage. The gate for the change is mounted in the facia panel, the gear lever extending back and to the right in a somewhat unusual but convenient position. Synchromesh is provided on second and top gears of the three-speed gear box; it is effective for normal operation but can be beaten if snappy changes are made. Some slight transmission gear noise can be heard, as the drive is through a gear train even in top gear.

Perhaps the finest features of this car are the cornering and general roadholding qualities, which are of a very high order and immediately inspire confidence. Over all types of road surface the ride in both front and rear seats is very comfortable, a little hard by present-day standards, but not unduly so. The car does not pitch, nor is the suspension harsh, and there is a negligible amount of roll on corners, which can be taken very fast. There is a marked understeer tendency and this further increases the general feeling of stability. Rack and pinion steering produces a very accurate method of control. It is extremely accurate and there is a complete absence of any feeling of vagueness; on the other hand with $2\frac{1}{4}$ turns from lock to lock it is heavy at low speeds, although at high speeds this is not so noticeable. The self-centring action is good and quite marked if the throttle is opened while the car is being cornered. Although the steering is fairly well insulated from normal road shocks, a certain amount of vibration was transmitted back through the steering wheel, and this was particularly noticeable when the car was driven fast; this fault was later rectified by changing the tyres.

Twelve-inch diameter hydraulic brakes with leading and trailing shoes are fitted to all wheels. Under normal conditions the braking is well up to the car's requirements, and even under the severe conditions of performance testing, which call for brake applications at much more frequent intervals than is necessary under normal operation, no brake fade was noticed. With two up, however, there is a tendency for the rear wheels to lock first if a violent stop is made. It is interesting to note that the total lining area for the front brakes is a little *less* than that provided at the rear.

The general arrangement of the three pedals is very good, and the throttle is in a convenient position to permit the heel and toe type of gear change if required. The position of the large two-spoke steering wheel is also very comfortable and well arranged in relation to both the seat and the pedals. Because of the low build of the car and the flat floor, made possible by the absence of a propeller-shaft tunnel, the seat is relatively high off the floor, with the result that a normal sitting position is provided. The front bench type seat is well proportioned and gives ample support, although a short driver might well find the cushion a little long. The seat itself is well sprung and comfortable after long periods at the wheel. There is plenty of vacant floor space for the driver's left foot.

Driving Vision

From the driving seat there is a good view of the road ahead and both front wings and side lamps can be clearly seen. The flat windscreen is wide and although the screen pillars are relatively thick they do not obstruct the driver's vision unduly. The rearward view is also satisfactory, and the mirror is placed in a position where it does not mask a useful forward area of the windscreen, although it does cover the left wing. All the instruments, which include a clock, ammeter, fuel gauge and speedometer, are mounted immediately in front of the driver, who has an unobstructed view of them through the steering wheel. A green light is used to indicate low oil pressure. The instrument lighting is satisfactory and does not reflect in the windscreen at night.

Minor controls are conveniently grouped on the right of the facia, with the exception of the combined light and horn switch, which is mounted on the steering column and looks rather like a small edition of a steering column gear-change lever. If the knob is pressed inwards towards the centre of the steering column the horns are sounded (a light push produces a soft note while a harder push increases the volume); if it is twisted to 90 degrees the side lights are switched on, while a further 90-degree twist switches on the head lamps. By moving the lever up and down from the dip to the straight-ahead position the lights can be

The large six-cylinder engine nearly fills the space under the bonnet. A sheet metal cowl surrounds the fan, which is mounted on the end of the dynamo in front of the cylinder block.

A strut holds the locker lid in the open position to facilitate loading. The spare wheel and tools are carried in the rear section of the locker, leaving the flat floor for luggage. Deep overriders and a wide bumper protect the rear of the car.

flashed. The Citroen still retains an opening windscreen. The wiper blades are well placed and cover an effective area of the screen. They also overlap, and so avoid an unwiped portion in the centre of the glass.

The rear seats are comfortable but a little hard, there is a satisfactory amount of head room, while the leg room is well above average. There is a small tray for parcels behind the rear seats as well as an open glove box in the facia panel. Ashtrays are fitted on the facia panel and at the back of the front seat. A simple form of heater consisting of a duct which collects warm fresh air from the back of the radiator and conveys it to the passenger compartment is fitted as standard. It is usual practice to dismantle this duct during warm weather.

Either single- or double-dip head lamps can be provided, and the car tested was fitted with the single dip and switch arrangement. This provided a good spread of light, but a longer range would be advantageous for fast night driving. The horns had an unusual note and from inside the car they appeared to be fairly quiet; however, in operation they proved to be quite effective. The suspension system has eight grease nipples which require attention at intervals of 600 miles. Starting from cold was very good and very little use of the mixture control was required.

The Citroen Six is a car that has retained what may be called good vintage qualities, while at the same time possessing modern features such as torsion bar suspension. It is a robust, rugged car designed for fast journeys, such as a day trip from Paris to the Riviera, or London to Scotland. It does not have the smoothness or some of the refinements possessed by a large number of cars today; on the other hand, it does have outstanding roadholding qualities that enable long distances to be covered at quite high average speed; also, for its size it has a modest thirst for fuel.

CITROEN SIX SALOON

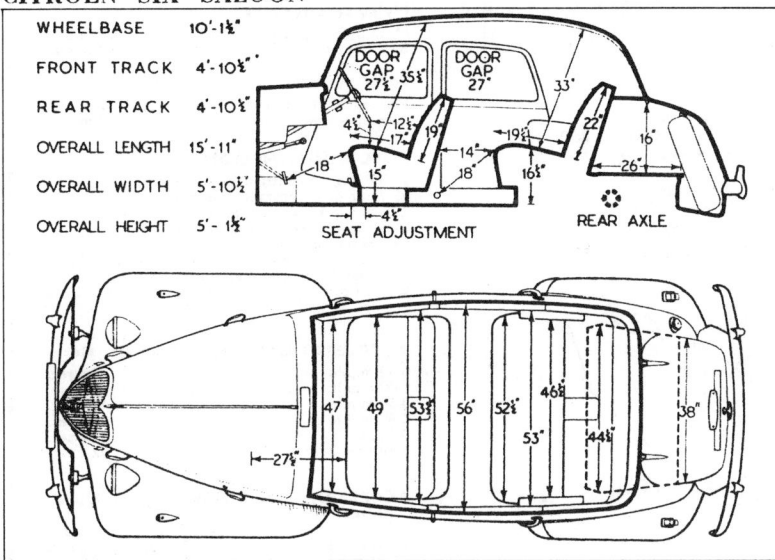

WHEELBASE	10'-1½"
FRONT TRACK	4'-10½"
REAR TRACK	4'-10½"
OVERALL LENGTH	15'-11"
OVERALL WIDTH	5'-10½"
OVERALL HEIGHT	5'-1½"

Measurements in these ¼in to 1ft scale body diagrams are taken with the driving seat in the central position of fore and aft adjustment and with the seat cushions uncompressed.

PERFORMANCE

ACCELERATION: from constant speeds.
Speed, Gear Ratios and time in sec.

M.P.H.	3.875 to 1	5.62 to 1	13.24 to 1
10—30	10.2	6.0	—
20—40	10.1	6.2	—
30—50	10.7	7.1	—
40—60	12.0	—	—
50—70	17.2	—	—

From rest through gears to:

M.P.H.	sec
30	5.4
50	12.5
60	19.3
70	31.2

Standing quarter mile, 21.4 sec.

SPEED ON GEARS:

Gear		M.P.H. (normal and max.)	K.P.H. (normal and max.)
Top	(mean)	80	128.8
	(best)	84	135.2
2nd		50—58	80—93
1st		20—26	32—42

TRACTIVE RESISTANCE: 42.5 lb per ton at 10 M.P.H.

SPEEDOMETER CORRECTION: M.P.H.

Car speedometer	10	20	30	40	50	60	70	80	88
True speed	12	21	30	38	48	57	67	77	84

TRACTIVE EFFORT:

	Pull (lb per ton)	Equivalent Gradient
Top	205	1 in 10.8
Second	410	1 in 5.4

BRAKES:

Efficiency	Pedal Pressure (lb)
74 per cent	100
58 per cent	80
31 per cent	50

FUEL CONSUMPTION:
18.5 m.p.g. overall for 252 miles (15.3 litres per 100 km).
Approximate normal range 17-20 m.p.g. (16.6-14.1 litres per 100 km).
Fuel, first grade.

WEATHER: Damp surface; wind fresh.
Air temperature 50 degrees F.
Acceleration figures are the means of several runs in opposite directions.
Tractive effort and resistance obtained by Tapley meter.
Model described in The Autocar of August 22, 1952.

DATA

PRICE (basic), with saloon body, £940.
British purchase tax, £392 15s 10d.
Total (in Great Britain), £1,332 15s 10d.

ENGINE: Capacity: 2,867 c.c. (174.9 cu in).
Number of cylinders: 6.
Bore and stroke: 78 × 100 mm (3.07 × 3.937 in).
Valve gear: Overhead; push rods and rockers.
Compression ratio: 6.5 to 1.
B.H.P.: 76 at 3,800 r.p.m. (B.H.P. per ton laden 49.7).
Torque: 138 lb ft at 2,000 r.p.m.
M.P.H. per 1,000 r.p.m. on top gear, 19.95.

WEIGHT (with 5 gals fuel), 27 cwt (3,027 lb).
Weight distribution (per cent) 58.7 F; 41.3 R.
Laden as tested: 30.5 cwt (3,427 lb).
Lb per c.c. (laden): 1.2.

BRAKES: Type: F, leading and trailing. R, leading and trailing.
Method of operation: F, Hydraulic. R, hydraulic.
Drum dimensions: F, 12in diameter, 1¾in wide. R, 12 in diameter, 1¾ in wide.
Lining area: F, 68 sq in. R, 72 sq in. (89 sq in per ton laden).

TYRES: 185-400 mm.
Pressures (lb per sq in): F 20; R 23 (normal).

TANK CAPACITY: 15 Imperial gallons.
Oil sump, 12½ pints.
Cooling system, 21 pints.

TURNING CIRCLE: 45 ft 3 in (L and R).
Steering wheel turns (lock to lock): 2¼.

DIMENSIONS: Wheelbase, 10ft 1½in.
Track: (F) 4ft 10½in; (R) 4ft 10½in.
Length (overall): 15ft 11in.
Height: 5ft 1½in.
Width: 5ft 10½in.
Ground clearance: 7 in.
Frontal area: 28.5 sq ft (approx.)

ELECTRICAL SYSTEM: 12-volt; 57 ampère-hour battery.
Head lights: Single or double dip; 36 watt.

SUSPENSION: Front, independent; wishbone links and torsion bars.
Rear, Dead axle; trailing arms and torsion bars.

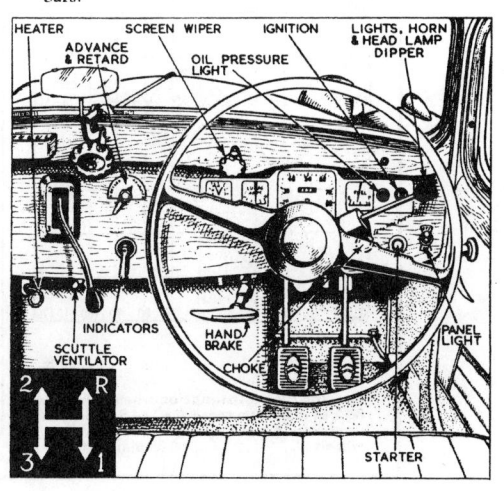

The Motor Road Test No. 10/54

Make: Citroen
Makers: Citroen Cars Ltd., Trading Estate, Slough, Bucks
Type: Six-cylinder

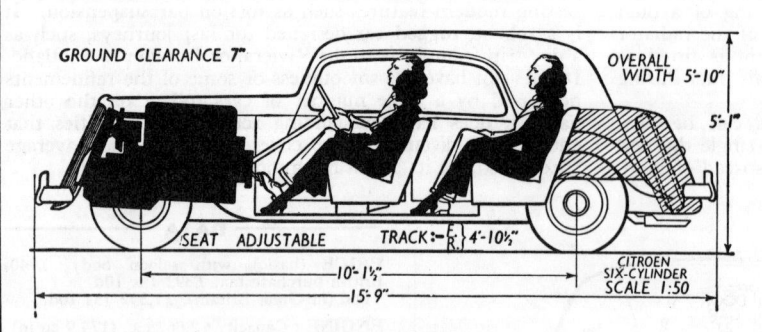

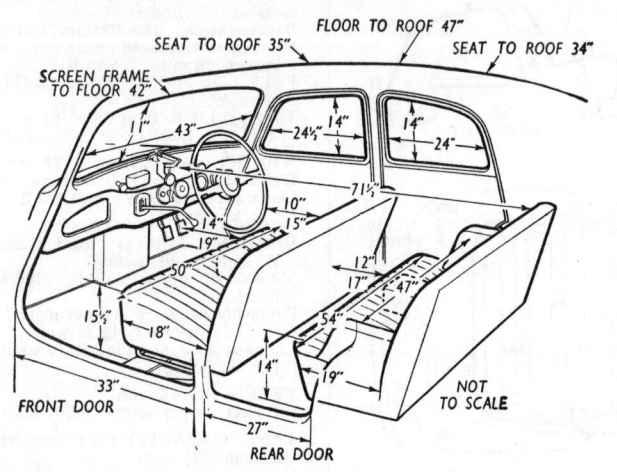

Test Data

CONDITIONS: Dry with strong wind. Smooth tarred road surface. Premium grade pump fuel.

INSTRUMENTS
Speedometer at 30 m.p.h.	9% fast
Speedometer at 60 m.p.h.	5% fast
Distance recorder	2% fast

MAXIMUM SPEEDS

Flying Quarter Mile
Mean of four opposite runs	81.1 m.p.h.
Best time equals	85.7 m.p.h.

Speed in Gears
Max. speed in 2nd gear	60 m.p.h.
Max. speed in 1st gear	27 m.p.h.

FUEL CONSUMPTION
- 26.0 m.p.g. at constant 30 m.p.h.
- 25.0 m.p.g. at constant 40 m.p.h.
- 23.5 m.p.g. at constant 50 m.p.h.
- 21.5 m.p.g. at constant 60 m.p.h.
- 19.0 m.p.g. at constant 70 m.p.h.
- Overall consumption (driven moderately hard) for 495.8 miles, 26.6 gallons = 18.6 m.p.g.
- Overall consumption (driven very hard) for 824.2 miles, 49 gallons = 16.8 m.p.g.

ACCELERATION TIMES Through Gears
0-30 m.p.h.	6.6 sec.
0-40 m.p.h.	9.9 sec.
0-50 m.p.h.	14.5 sec.
0-60 m.p.h.	21.2 sec.
0-70 m.p.h.	31.6 sec.
Standing Quarter Mile	22.1 sec.

ACCELERATION TIMES on Two Upper Ratios
	Top	2nd
10-30 m.p.h.	13.0 sec.	7.6 sec.
20-40 m.p.h.	11.6 sec.	6.8 sec.
30-50 m.p.h.	12.2 sec.	8.3 sec.
40-60 m.p.h.	14.3 sec.	—
50-70 m.p.h.	18.8 sec.	—

WEIGHT
Unladen kerb weight	26¼ cwt.
Front/rear weight distribution	62/38
Weight laden as tested	30 cwt.

HILL CLIMBING (at steady speeds)
Max. top gear speed on 1 in 20	70 m.p.h.
Max. top gear speed on 1 in 15	64 m.p.h.
Max. top gear speed on 1 in 10	45 m.p.h.
Max. gradient on top gear	1 in 9.5 (Tapley 235 lb./ton)
Max. gradient on 2nd gear	1 in 5.5 (Tapley 400 lb./ton)
Max. gradient on 1st gear	1 in 3.9 (Tapley 545 lb./ton)

BRAKES at 30 m.p.h.
- 0.92 g retardation (=33 ft. stopping distance) with 120 lb. pedal pressure.
- 0.86 g retardation (=35 ft. stopping distance) with 100 lb. pedal pressure.
- 0.77 g retardation (=39 ft. stopping distance) with 75 lb. pedal pressure.
- 0.50 g retardation (=60 ft. stopping distance) with 50 lb. pedal pressure.
- 0.23 g retardation (=131 ft. stopping distance) with 25 lb. pedal pressure.

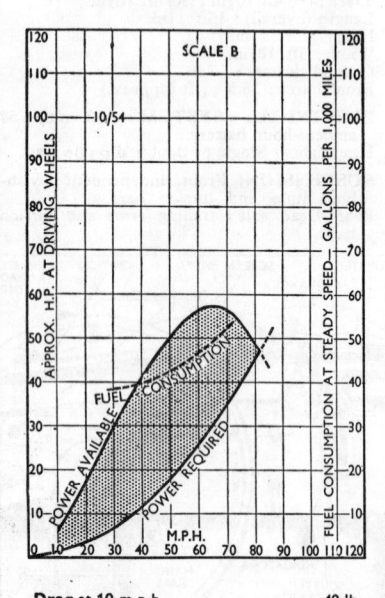

Drag at 10 m.p.h.	40 lb.
Drag at 60 m.p.h.	163 lb.

Specific fuel consumption when cruising at 80% of maximum speed (i.e., 64.9 m.p.h.) on level road, based on power delivered to front wheels .. 0.85 pints/b.h.p./hr.

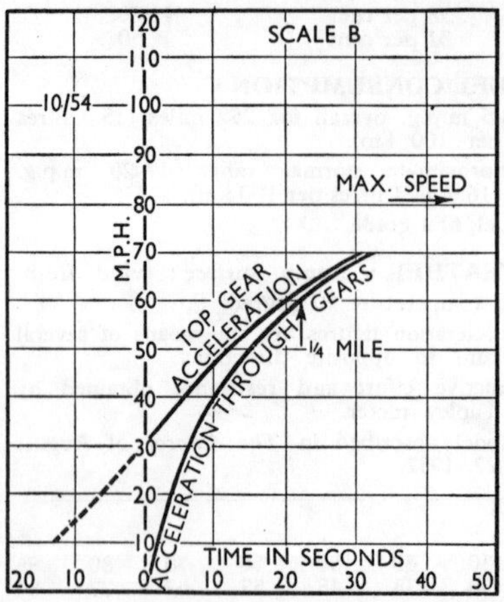

Maintenance

Sump: 12 pints, S.A.E. 20. **Gearbox & Differential:** 4¾ pints, S.A.E. 90, extreme pressure. **Steering gear:** Shell Retinax A. **Radiator:** 21 pints (1 drain tap, 1 drain plug). **Chassis Lubrication:** By grease gun every 1,000 miles to 8 points. **Ignition timing:** 8° initial advance. **Spark Plug gap:** .016-.020 in. **Contact Breaker gap:** .016. **Valve timing:** I.O. 3° B.T.D.C. I.C. 45° A.B.D.C. E.O. 45° B.B.D.C. E.C. 11° A.T.D.C. **Tappet clearances:** (Hot): Inlet .006 in. Exhaust .008 in. **Front wheel toe-out:** 0 to 3/32". **Camber angle:** 1°. **Castor angle:** 0. **Tyre pressures:** Front 20 lb. Rear 23 lb. **Brake fluid:** Lockheed. **Battery:** 12 volt, 57 amp.-hr. **Lamp Bulbs:** Headlamps, double-filament, 36/36 watt; Sidelamps 6 watt.; Tail lamps 6 watt; Stop-lamp 18 watt; Rear No. plate lamp 3 watt (2 bulbs).

The CITROEN Six

Increased Refinement shown by a Roomy and Fast Anglo-French Car

In Brief
Price £952 (plus purchase tax £397 15s. 10d.) equals £1,349 15s. 10d.
Capacity ... ... 2,866 c.c.
Unladen kerb weight ... 26½ cwt.
Fuel consumption... 16.8 to 18.6 m.p.g. (driven hard)
Maximum speed ... ... 81.1 m.p.h.
Maximum speed on 1 in 20 gradient... ... ... 70 m.p.h.
Maximum top gear gradient 1 in 10.4
Acceleration:
 10-30 m.p.h. in top ... 13.0 secs.
 0-50 m.p.h. through gears 14.5 secs.
Gearing: 20 m.p.h. in top at 1,000 r.p.m.; 76 m.p.h. at 2,500 ft. per min. piston speed.

CHARACTERISTIC.—The six-cylinder Citroen makes no concession to ultra-modern styling, and preserves its traditional appearance while remaining essentially good-looking.

ALTHOUGH built up from bare body pressings in England, and incorporating a large proportion of entirely British components, the Citroen Six is French in conception and in many of its mechanical parts. Sampling recently an example of this model, which has not been the subject of any major design change announcements since it was last the subject of a Road Test Report 4½ years ago, we found ourselves inverting a familiar French quotation and saying "Plus c'est la meme chose, plus ca change." Ostensibly almost the same as its predecessors, this latest Citroen Six is in fact a very much more refined and more versatile car.

Powered by a 3-litre 6-cylinder engine, and able to seat six passengers, this largest Citroen model is a big car with quite high performance on the road. It stands on its own because, whereas most cars currently in production are essentially post-war designs, the Citroen Six is the result of about 20 years of detailed development work applied to a design which was boldly new long before the war.

Adherence to a general layout concerning which little can now remain unknown has brought both advantages and disadvantages. There should be a very high standard of reliability, there is certainly roominess of a kind which is not altogether common, and there is a remarkable absence of departure from an even keel during vigorous cornering or braking. On the other hand, the traditional form of bodywork limits maximum speed to some extent, riding comfort does not come up to the highest 1954 level, and standards of silence also lag slightly behind current practice.

Externally, the principal change which has been made since we last tested a Citroen Six is the provision of a much enlarged luggage locker, with a lift-up lid and internal illumination from the number-plate lamp. Inside the car, a new and better-looking arrangement of the instruments and controls on the polished wood facia panel is immediately observed.

It is in driving the Citroen Six or riding in it, however, that subtle yet significant changes become apparent. Formerly, for example, there were grounds for criticism in respect of noise from the all-indirect gearbox, but this fault was not evident on the new test car which came to us at an indicated mileage of 5,000.

From the owner-driver's point of view the 6-cylinder Citroen possesses a number of admirable characteristics and several which are tiresome. On the credit side there is good engine accessibility, and a floating oil level indicator which cuts out the messy part of checking the sump. The combination of opening windscreen (which provides ventilation but not, unfortunately, enough travel for use in fog) and sunshine roof will please many motorists, and the general finish throughout the car is of a high order. The headlamps produce a beam equal to maximum speed at night and the well-known fingertip control which combines dip switch, horn and illumination control is a feature not bettered by any other manufacturer.

Less satisfactory is the heater which takes the form of an air pipe running from the radiator block into the car and, although adequate in temperate zones, cannot be said to be equal to the worst that

OPENING screen and sunshine roof are a combination seldom found these days on a production car. Separate headlamps are another feature which many motorists still admire.

The Citroen Six

FRONT-WHEEL DRIVE means an absolutely flat floor for driver and passengers. The gear lever, protruding through the facia panel, offers no obstruction. Despite a low roofline the seating position is pleasantly upright.

winter weather can produce in Europe. Nor is there any provision for a much-needed demister, which is unforgivable on a comparatively expensive vehicle. The mirror is badly sited in that it causes a blind spot, and if the window winder is left in one position it is possible to remove the skin from the driver's right hand while gripping the wheel. Armrests are provided front and rear, which is desirable on so fast-cornering a vehicle, but the front one could be better sited.

A simple and yet fundamental merit of the Citroen Six is a driving position providing physical and mental comfort, the bench seat being upright and well shaped, the view forwards over a long but low bonnet excellent, and the controls generally very well placed. Upwards vision is none too good, however, except when the sliding roof panel is open, and rather blind rear quarter panels do not facilitate reversing in confined spaces. Familiarity with the car minimizes but does not eliminate the nuisance of an abnormally large-radius turning circle, the result to some extent of front-wheel drive being used on a long-wheelbase car.

This same combination of front-wheel drive and a long wheelbase also provides, of course, for a completely flat floor inside the car, a floor the height of which above road level is barely more than the ground clearance. Also, for greater space than is nowadays usual between the front and rear seats, without sacrifice of the advantages of inter-axle seating. Long wheelbase, wide track, and low build are, of course, the basic ingredients for stability on the road, and the Citroen Six is indeed an exceptionally stable and controllable car.

At low speeds the steering remains distinctly heavy, and when the car is being backed the normal castor action becomes reversed so that the wheel has to be pulled back to the central position. Once the car is rolling at 10 m.p.h. or more, however, the steering becomes quite acceptably easy, rather firmer than is fashionable but also rather quicker in response.

Stable on the straight, this car can be swung rapidly along winding roads with exceptionally little fuss of any kind. Front-wheel drive does not disclose any snags, although handling qualities are appreciably influenced by throttle opening, the castor action and under-steer characteristic which are evident when accelerating diminishing quite noticeably if the accelerator pedal is released. We drove many miles on icy surfaces during this test, and the only effect of reaching the limiting speed on a slippery corner was a tendency for the front of the car to run wide. One point noted when driving on roads coated with snow and ice was momentary tugs at the steering if one front wheel spun suddenly on a patch of ice, but this was not in fact at all disconcerting. Passengers

FORWARD VISIBILITY is good for all but the very tall. The lighting of the instruments at night is clear and unobtrusive and the standard of finish, in general, is high. A flap which controls the warm-air intake can be seen below the cubbyhole. The mirror gives a good view rearward but is inclined to impede forward vision. New instruments are a great improvement.

could ride for surprising distances without ever suspecting that wet roads had frozen to ice, there being no swaying to disturb them even at quite fast cruising speeds.

Torsion bar springs are used at both front and rear of this car, springs which were unusually flexible when this design was introduced but are nowadays regarded as inclining towards firmness. The moderately firm springs in themselves should suit many people's tastes, but unfortunate shock-absorber settings seemed to be in use on the test model which did not provide sufficient control over persistent small-amplitude bouncing of the car on its springs: no doubt this is a matter which will soon be rectified. Apart from this unwanted motion, the riding is pleasantly flat, and low unsprung weight on the undriven rear axle allows rear-seat comfort to be at least equal to that enjoyed in the front of the car. Legroom and headroom in the rear compartment are very ample, so that a special footrest is provided for rear-seat passengers.

THE ABSENCE of a propeller shaft enables a low floor line to be used and passengers step down into the car. A footrest is provided and the armrest folds away.

ACCESSIBILITY is good. The petrol pump can be seen just behind the metal tube which carries warm air from the radiator block into the car when desired. Just behind the petrol pump is a floating dipstick. Distributor, windscreen wiper motor and battery may all be reached with the minimum trouble.

During the frosty weather which formed much of our test period, the six-cylinder engine was by no means a first-touch starter after overnight parking in the open air, but once it fired it behaved very well indeed. Operation of the starter produced sound effects reminiscent of a peal of bells, but in contrast the engine idled so quietly and smoothly as sometimes to arouse false suspicions that it had stalled. Flexibility at low speeds is up to the standard expected with six cylinders, and the engine also runs quite freely up to fairly high r.p.m. when high performance is wanted.

The three-speed gearbox has unfortunately never been the most endearing feature of front-wheel-driven Citroen models. In this instance, second is a very useful ratio providing reasonably quiet and very potent acceleration between 5 and 50 m.p.h., and the unusual facia-mounted gear lever is quite convenient, but the synchromesh system was none too powerful and clutch-controlled locking of the gearchange mechanism seems to prevent smooth feel being obtained.

As has been said already, the car subjected to this test was very much quieter than the model driven four years ago. Improved standards of workmanship probably account for the fact that transmission noise has now been largely suppressed, and silencing has been very effectively applied to the carburetter air intake and also to the exhaust system. What remains, it appears, is mainly tyre noise, this being more than usually evident at cruising speeds up to 55 m.p.h., but seeming to fade away in the 60-70 m.p.h. speed range which is customarily used on clear roads.

ONE of the few alterations in the appearance of the Citroen during recent years is the adoption of a larger boot lid which improves luggage accommodation without sacrificing the accessibility of the spare wheel.

Mechanical Specification

Engine
Cylinders	6
Bore	78 mm.
Stroke	100 mm.
Cubic capacity	2,867 c.c.
Piston area	44.42 sq. in.
Valves	Pushrod o.h.v.
Compression ratio	6.5/1
Max. power	76 b.h.p.
at	3,800 r.p.m.
Piston speed at max. b.h.p.	2,493 ft. per min.
Carburetters	Solex downdraught
Ignition	Lucas coil
Sparking plugs	K.L.G. FA 50
Fuel pump	AC Mechanical
Oil filter	Gauze on pump

Transmission
Clutch	Single dry plate
Top gear (s/m)	3.87
2nd gear (s/m)	5.62
1st gear	13.24
Propeller shaft	Nil (front-wheel drive)
Final drive	Spiral bevel

Chassis
Brakes	Lockheed hydraulic, 2 LS Front
Brake drum diameter	12 in.
Friction lining area	162 sq. in.
Suspension:	
Front	Torsion bars and wishbones
Rear	Torsion bars and dead axle
Shock absorbers	Telescopic
Tyres	Michelin 185—400

Steering
Steering gear	Rack and pinion
Turning circle: Right	46 feet
Left	42 feet
Turns of steering wheel, lock to lock	2

Performance factors (at laden weight as tested):
Piston area, sq. in. per ton	29.6
Brake lining area, sq. in. per ton	108
Specific displacement, litres per ton mile	2,865

Fully described in *The Motor,* September 9, 1948.

Coachwork and Equipment

Bumper height with car unladen:
Front... (max.) 16 in., (min.) 12½ in.
Rear ... (max.) 18½ in., (min.) 15 in.

Starting handle	Yes
Battery mounting	On scuttle
Jack	Harvey Frost screw type
Jacking points	Under axles

Standard tool kit: Tyre pump, grease gun, wheelbrace, adjustable spanner, screwdriver, pliers, plug box spanner and tommy-bar, tool roll and holdall.

Exterior lights: Two headlamps, two side lamps, two stop/tail lamps, number-plate lamp.

Direction indicators ... Semaphore type, self-cancelling

Windscreen wipers ...Lucas two-bladed, electric

Sun Visors ... Two

Instruments: Speedometer (with trip), fuel contents gauge, ammeter, electric clock, manual ignition control.

Warning lights ... Oil pressure

Locks:
With ignition key ... Ignition; o/s front door; luggage boot
With other keys ... Nil

Glove lockers ... One on facia panel

Map pockets... Two on front doors; one on scuttle side, one behind front squab.

Parcel shelves One behind rear seat squab

Ashtrays One on facia; one behind front squab

Cigar lighters ... Nil

Interior lights ... One in centre of roof

Interior heater ... Fresh-air type (without demisters or fan)

Car radio ... Optional

Upholstery material ... Leather and plastic leathercloth.

Carpets ... Pile carpet

Exterior colours standardized: Green, red, grey and black.

Alternative body style: Fixed-head saloon, £940 plus purchase tax £392 15s. 10d., equals £1,332 15s. 10d.

Wheels Road Tests the CITROEN

In the Light Fifteen the motorist has a car which will motor far, fast and safely, and is good fun to drive

Above: This is quite plainly a motor car. It can not be mistaken for a gin palace or drawing room on wheels. The styling is old-fashioned, but highly functional. From the front the car does not betray its front-wheel drive. Notice the cut-away wings to let plenty of air to cool the brake drums.

OUR outstanding impression of the Citroen Light Fifteen was its roadholding and handling.

This front-wheel drive four or five passenger car gave us no anxious moments on our fast and arduous test drive, and displayed hairline response and extraordinary adhesion.

It is French in origin and characteristics.

It has the typical French virtues of robust construction, a suspension which soaks up extraordinarily rough surfaces, a direct feeling of control and confidence-inspiring roadworthiness.

Like other French cars, the vision from the driver's seat is not good, engine noise inside is too loud and the body picks up a high level of wind noise at speed.

The car is a five-passenger all-steel saloon of unitary body-and-chassis construction with the engine, gearbox, final drive, front wheels and universals mounted in one unit at the front of the car.

To service any major component the entire unit is rolled out of the car — on its own wheels — and presents no repair problem.

There is no difference in the control of the car from a normal model except that the gear lever protrudes from the facia.

Cruising speed is from 60 to 70 miles per hour, with a top speed of 80 m.p.h.

TECHNICAL DETAILS

ENGINE:
4 cyl., o.h.v., wet liners, bore 78 mm. x 100 mm. stroke, capacity 1,911 c.c. Comp. ratio 6.1 to 1. Develops 56 b.h.p. at 4,200 r.p.m. Road speed at 2,500 ft./min. piston speed, 663 m.p.h., 17.4 m.p.h. per 1,000 r.p.m. in top gear. Single downdraught Solex carburetter, A.C. mechanical fuel pump and fan cooling. Lucas 12 volt ignition, 57 amp. hour battery.

TRANSMISSION:
Single dry plate clutch, F.W.D. Three speed gearbox, synchro on top two speeds. Ratios 4.3, 7.3, 13. Reverse 17.5 to 1. Spiral bevel final drive through universals and half shafts.

SUSPENSION:
Lockheed hydraulic, two-leading shoes in front, Independent at front by torsion bars; torsion bars and head axle at rear. Telescopic shock absorbers front and rear.

BRAKES:
12" drums in front, 10" at rear. Friction area, 97.3 sq. in. Parking brake mechanically linked to rear wheels.

CHASSIS:
Chasis and body combined in stressed-skin pressed-steel construction.

MEASUREMENTS:
Wheelbase, 9' 6"; track, 4' 6"; length, 14' 2"; height, 5'; width, 5' 5"; ground clearance, 7". Unladen weight, 21 cwt.

TANKS:
Petrol, 10 gallons; sump, 8 pints; radiator 14 pints.

STEERING:
Rack and pinion, turning circle 45 feet, 2 1/3 turns from lock-to-lock.

WHEELS:
Disc. Michelin 165 x 400 tyres.

PERFORMANCE:
Acceleration from rest through gears:
From rest to 30 m.p.h., 5 secs; rest to 40, 7 secs.; rest to 50, 12.2 secs.; rest to 60, 18.2 secs.; rest to 70, 22 secs.
Acceleration in top gear:
From 10 to 30 m.p.h., 8 secs.; 20 to 40, 10 secs.; 30 to 50, 10.2 secs.; 40 to 60, 12 secs.
Standing quarter mile: 22 secs.
Maximum speeds:
In first, 30 m.p.h.; 2nd, 58 m.p.h.; top, mean of three runs 77 m.p.h.; best time 81 m.p.h.
Speedometer correction: 1 m.p.h. fast at top speed.
Fuel Consumption: 33 m.p.g. on test.
Laden weight as tested: 22½ cwt.

Above: Engine accessibility is very good for maintenance and minor repairs. Any major overhauls can be tackled by undoing a few bolts and connections and rolling engine, gearbox, differential and front wheels out of the car. Right: The interior was highly functional, but we didn't approve of the close spacing of the clutch and foot pedals. Awkward placing of the clutch against the bulkhead is another bad point, too

The Citroen has been made in its current form for nearly 20 years; although the chassis is still modern the body is outdated in appearance by current trends. However it is highly functional in all respects and for that reason is respected by the owner who wants a motor car to motor and not to be a pretty ornament.

There are bench seats front and rear and a flat floor in both compartments. General finish throughout is up to a high standard and the recent addition of a larger luggage locker eliminates the only deficiency which the car possessed.

This is the first car we have tried with an opening screen. We found it very useful for ventilation on hot days, particularly when driving off after the car had been parked in the hot sun. Other ventilation is provided by an opening scuttle vent.

Driving position is very comfortable, with soft seats and plenty of cushion under the knees to support a long-legged driver. Wheel rake and positioning of controls is very good, but the brake and clutch pedals are too close together for comfort. The clutch is positioned rather awkwardly in relation to the bulkhead, and attention here would make a world of difference.

Both controls are light and positive and there is no noise or shudder from the clutch even when it is used brutally. It gives the feeling of strength in manufacture.

Driver's vision is limited because the windscreen pillars cut noticeably into the line of sight. Vision to the rear is satisfactory.

After short acquaintance with the three-speed gearbox control on the facia we found that the selection of gears was positive and easy, with a pleasant firm feel. If used very quickly the synchomesh can be overridden.

We found the ratios well chosen. An additional feature is that the lever cannot be knocked out of gear, as the clutch must be used before the gearshift can be moved.

After our short acquaintance we set off to enjoy some fast motoring. The engine responds instantly to the driver's wishes, and although it is very noisy inside the body it is smooth, free from transmission snatch and takes hard work without effort.

It will rev without fuss and combined with the road manners and steering, gives a genuine cruising speed of 70 m.p.h.

This is the highest cruising speed yet encountered by "Wheels." The Citroen is the fastest car we have tested from point-to-point over winding highways.

Acceleration is quite good, the figure of 22 seconds for a standing quarter mile placing the car between the strong getaway cars and the more leisurely family saloons.

At top speed of 80 m.p.h. the engine is noisier, but still not straining, but the wind noise is very high. The car's stability at top speed is unimpaired.

Rack-and-pinion steering is fitted and gives hairline accuracy under all conditions, with a definite feeling of response which inspires confidence. An accurate line can be taken on a corner or straight regardless of the surface.

Like all rack-and-pinion layouts, it is heavy to use at low speeds and when parking, but there is considerably less shock transmitted to the driver's hands than we expect from a rack-and-pinion layout. The damping in this car is very good.

There was slight understeering characteristics at speed.

Under all conditions the suspension behaves well. On very tight corners it is hard to provoke any sideways wheel movement at all. At the limit, despite the front wheel drive, the back slips in readily controllable fashion. The steering is geared fast enough to give quick correction.

The car's adhesion to the road and sheer ability to corner fast impressed us. The car would be very hard to beat on a testing road.

The suspension is firm, and bumps can be felt individually by the passengers. There is a definite sensation of there being a wheel at each corner, and over atrocious surfaces the car did not pitch, bounce, or skitter. Altogether a most roadworthy design.

Braking is very good, and we could not produce any fading after prolonged hard use. Pedal pressure is light, and there is no side pull under hard stops. The parking brake is a pistol-grip lever which is easy to use.

Road noise through the body is very low, but it is nullified by the engine and wind noise.

The instrument panel is not worthy of the car, and the instrumentation is complete. The affect is jumbled and we found the individual dials hard to read. The facia is in well-finished polished wood.

The instruments embrace a water temperature but no oil pressure gauges. There is an ammeter, fuel gauge, clock and speedometer.

Since our test, a newer model has appeared with modifications to the instrument panel and an oil pressure gauge.

At 45 feet, the lock is very limited, a result of the front wheel drive. For a car of this size it is not very easy to turn in city streets or park.

The electric windscreen wipers have manual controls so they can be used if the motor fails. The headlights are operated from a knob on a lever projecting underneath the steering wheel.

Engine accessibility through the side-opening bonnet panels is excellent, and for major repairs the whole gearbox, engine and transmission is rolled out of the car on the front wheels.

Furnishings are of good quality, and the car is very comfortable to travel in over long distances. The carpet on the floor is neatly fitted, and there are pockets in the front doors.

Luggage space is not substantial, but enough gear can be stowed in the car to deal with most trips.

Starting is by press-button solenoid.

The lacquer finish is clear and free from orange peel, there are ash trays front and rear, and a glove box in the facia.

We left the Citroen with very satisfied feelings. Here is a car which is meant to be driven far and fast, and a car which rewards a good driver.

It has shortcomings in refinement, and a little more thought to a couple of small matters outlined would remove the sources of what are, after all, carping criticism.

For business or private uses we consider the Citroen to be admirably suited, and has a sufficiently robust construction to handle hard work.

Right: Underside view of the suspension at the wheel. The piston chamber and gas sphere can be seen, also a conical rubber rebound stop to the left of the end of the piston rod. Above the stop can be seen the anti-roll bar linking the trailing arms.

AN entirely new type of suspension, using neither laminated nor coil springs, and having inherent spring damping, has been produced by Citroen and is already being offered in France on the six-cylinder saloon model. Although full facilities for an external examination and for practical road tests have been offered, Citroen engineers show considerable reserve as regards the technical features of this revolutionary hydro-pneumatic system of suspension.

It is applied to the rear only, the front suspension being by torsion bars lengthened and of greater flexibility than those on the standard model. The some-

Revolutionary Rear Suspension

Hydro-pneumatic System Devised by Citroen in France

what complicated installation can be divided into two parts: an oil pressure system under the bonnet, and the suspension unit proper at the rear. The independent rear wheels are mounted on trailing arms, these arms being linked by a transverse anti-roll bar. The suspension unit is a metal sphere divided into two parts by a diaphragm and containing a gas, the nature of which is not revealed. The weight of the car moves a piston which compresses the gas by means of a liquid similar to the fluid used for hydraulic brakes. "Compression" of this fluid takes place under the bonnet, a small seven-piston pump being belt driven from the camshaft and running at camshaft speed. The pump draws the fluid through a flexible pipe from a tank of about one gallon capacity mounted on the scuttle by the side of the brake fluid tank. The fluid is delivered to a container, or storage chamber, where constant pressure is maintained by a make-and-break mechanism. The degree of pressure maintained is not indicated. From the container the fluid is carried by piping to the piston compressing the gas contained in the wheel sphere.

It might be thought that with the engine stopped pressure would drop and the rear of the car would settle. This is avoided by a cut-off which insulates the fluid container and the gas-filled sphere.

The new Lagonda for Le Mans. Here David Brown, head of the organization which controls Aston Martin and Lagonda, is sitting at the wheel of the 4½-litre Lagonda sports car, two examples of which are entered for the Le Mans 24-hour race in June. The engine is an entirely new V-twelve unit (bore 82.55 mm, stroke 69.85 mm) with twin o.h.c. on each bank of cylinders; carburation is by three four-choke Weber carburettors. The chassis is a development of that of the famous Aston Martin DB3S This car will make its debut in the sports car race at the Silverstone meeting on May 15, in the hands of Reg Parnell.

If, however, all pressure should be lost, it will be built up again and maintained within 12 to 15 seconds of the engine being started.

The immense advance embodied in this type of suspension over anything yet produced was amply demonstrated by a test over roads comprising every type of surface—rough paving stones, smooth paving stones, good tarmacadam, a surface dug up by a scarifier, humps which could be taken at speed, a serpentine section with bad pot-holes, and a fast straight. Whatever the surface, the car kept on a perfectly level keel, and on no occasions were the passengers bounced off their seats or subjected to lateral displacement. The Citroen is remarkable for its road-holding ability, but this is still further enhanced by the new suspension. Two pronounced humps right across the road were taken at speed and had the car's occupants not been notified they would hardly have been aware of their presence. As a demonstration, a bottle of champagne was uncorked and three glasses were filled at between 45 and 50 miles an hour on a winding section, without a drop being spilled. The only difference that could be felt between smooth *pavé* and the scarified road surface was in the hum of the tyres on the former.

An auxiliary advantage of this hydro-pneumatic suspension system is that it can be used for raising the car for wheel changing. This can be done with the engine running or, rather less rapidly, by the fluid in the storage chamber.

In reply to objections raised against the system, such as leaks, the breakage of a fluid pipe, and so on, Citroen state that the worst which could happen would be the breakage of a fluid pipe, in which case (if a spare was not carried) the car could be driven home slowly. It is claimed that this system has been experimented with and thoroughly tested over a period of several years, and it is now being offered to the public with a trouble-free guarantee. It is already available as optional equipment at a cost of about £70. There is no intention of fitting this to the Fifteen, for the two-fold reason that space is not available for the extra equipment and the increased cost would not be justified on what—in France at any rate—is considered a "popular" car.

It should be noted that for this suspension Michelin X tyres—steel and rubber carcase—are recommended. The rear seats have been made softer by the use of sponge rubber and there is a layer of sponge rubber under the floor mat.

The Motor Road Test No. 7/55

Make: Citroen. **Type:** Six (Hydro-pneumatic suspension)
Makers: Citroen Cars Ltd., Slough, Bucks, England.

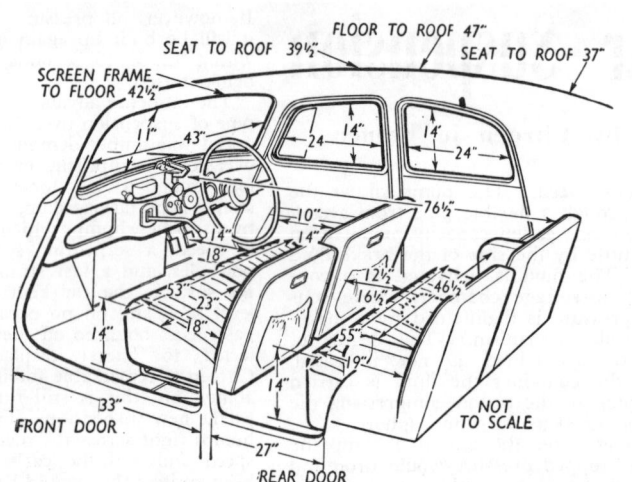

Test Data

CONDITIONS: Cool, dry weather with 5-12 m.p.h. E.N.E. wind (temperature 40-45° F., barometer 30.45-30.48 in.). Dry tarred road surface. Intermediate-grade pump fuel.

INSTRUMENTS

Speedometer at 30 m.p.h.	14% fast
Speedometer at 60 m.p.h.	10% fast
Speedometer at 80 m.p.h.	9% fast
Distance recorder	4% fast

MAXIMUM SPEEDS

Flying Quarter Mile
Mean of four opposite runs .. 83.5 m.p.h.
Best time equals .. .. 84.1 m.p.h.

Speed in Gears
Max. speed in 2nd gear .. .. 63 m.p.h.
Max. speed in 1st gear .. .. 25 m.p.h.

FUEL CONSUMPTION

26.5 m.p.g. at constant 30 m.p.h.
25.5 m.p.g. at constant 40 m.p.h.
23.5 m.p.g. at constant 50 m.p.h.
21.0 m.p.g. at constant 60 m.p.h.
17.5 m.p.g. at constant 70 m.p.h.
Overall consumption for 665 miles, 36.8 gallons equals 18 m.p.g. (15.5 litres/100 km.)
Fuel tank capacity 15 gallons

ACCELERATION TIMES Through Gears

0-30 m.p.h	5.2 sec.
0-40 m.p.h	8.8 sec.
0-50 m.p.h	13.0 sec.
0-60 m.p.h	18.6 sec.
0-70 m.p.h	25.9 sec.
Standing Quarter Mile	21.4 sec.

ACCELERATION TIMES on Two Upper Ratio

	Top	2nd
10-30 m.p.h	9.7 sec.	5.6 sec.
20-40 m.p.h	9.9 sec.	5.7 sec.
30-50 m.p.h	10.2 sec.	7.2 sec.
40-60 m.p.h	11.3 sec.	—
50-70 m.p.h	14.3 sec.	—

WEIGHT

Unladen kerb weight .. 27¼ cwt.
Front/rear weight distribution .. 61/39
Weight laden as tested .. 31 cwt.

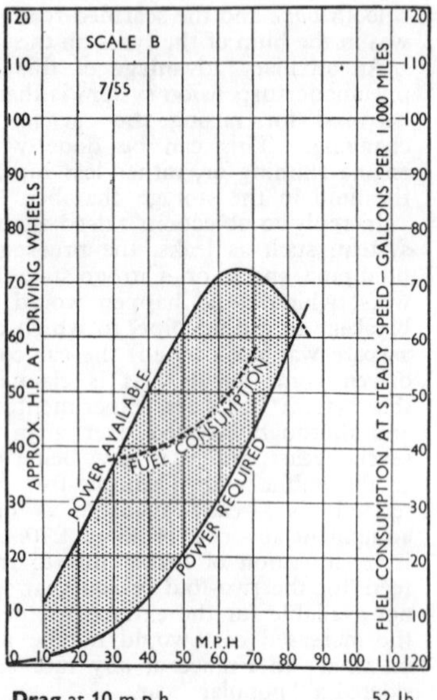

HILL CLIMBING (at steady speeds)

Max. top gear speed on 1 in 20 .. 70 m.p.h.
Max. top gear speed on 1 in 15 .. 66 m.p.h.
Max. top gear speed on 1 in 10 .. 52 m.p.h.
Max. gradient on top gear .. 1 in 9.3 (Tapley 240 lb./ton)
Max. gradient on 2nd gear .. 1 in 5.3 (Tapley 415 lb./ton)

BRAKES at 30 m.p.h.

0.95g retardation (=31½ ft. stopping distance) with 120 lb. pedal pressure
0.80g retardation (=37½ ft. stopping distance) with 100 lb. pedal pressure
0.51g retardation (=59 ft. stopping distance) with 75 lb. pedal pressure
0.31g retardation (=97 ft. stopping distance) with 50 lb. pedal pressure

Drag at 10 m.p.h. .. 52 lb.
Drag at 60 m.p.h. .. 191 lb.
Specific Fuel Consumption when cruising at 80% of maximum speed (i.e. 66.8 m.p.h.) on level road, based on power delivered to rear wheels .. 0.75 pints/b.h.p./hr.

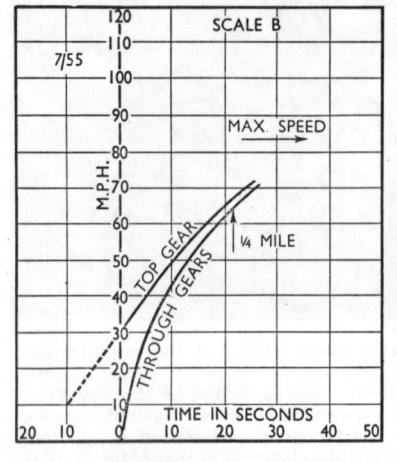

Maintenance

Sump: 12 pints, S.A.E. 20. **Gearbox and differential:** 4¾ pints, S.A.E. 90, extreme pressure. **Steering gear:** Shell Retinax A. **Radiator:** 21 pints (1 drain tap, 1 drain plug). **Chassis lubrication:** By grease gun every 1,000 miles to 8 points. **Ignition timing:** 8° initial advance. **Spark plug gap:** .016-.020 in. **Contact breaker gap:** .016. **Valve timing:** I.O. 3° B.T.D.C.; I.C. 45° A.B.D.C.; E.O. 45° B.B.D.C.; E.C. 11° A.T.D.C. **Tappet clearances:** (Hot) Inlet .006 in. Exhaust .008 in. **Front wheel toe-out:** 0 to ⅛ in. **Camber angle:** 1°. **Castor angle:** 0. **Tyre pressures:** Front 23 lb. Rear 24½ lb. **Brake fluid:** Lockheed. **Battery:** 12 volt, 57 amp.-hr. **Lamp bulbs:** Head lamps, double-filament, 36/36 watt; side lamps 6 watt.; tail lamps 6 watt; Stop lamp 18 watt; rear number-plate lamp 3 watt (2 bulbs). **Rear suspension reservoir:** 3.7 pints brake fluid.

The CITROEN Six

(Hydro-pneumatic)

New Standards of Riding Comfort in a Car Noted for Roadholding and Roominess

INHERENT STABILITY.—Three factors combine to give the Citroen a unique combination of stability on main roads and comfort over rough tracks. These are the wide track, long wheelbase, and low centre of gravity shown above and the soft, servo-levelled pneumatic rear suspension system pressure-controlled from the hydraulic pump and reservoir seen beneath the bonnet.

THE point of cardinal importance in the new Citroen Six is the hydro-pneumatic suspension; what it is, still more important, what it does.

The main trend in private car suspension over the past 20 years has been to secure a progressively softer ride, so that bumps are absorbed, and to eliminate fore and aft pitching which can be an intolerable nuisance to anyone occupying the rear seats. The means adopted to achieve these ends have been uniformly to soften the front springs and simultaneously to increase the weight carried by the front wheels. The result of this forward concentration of mass has been two-fold. The car has been endowed with an arrow-like stability which promotes under-steering characteristics, and with the front end tied down, as it were, a really flat ride can be induced over comparatively poor surfaces. Nevertheless, the modern flat ride has been achieved by a sacrifice of stability when cornering, of driveability on wet or slippery roads, and it exists only so long as the hydraulic damping units are fully effective.

The idea of using air as a suspension unit is as old as the industry, for an air spring has, or can have the important quality of a rapidly rising rate: that is to say it can be given the equivalent flexibility of a very soft mechanical spring at the beginning of its travel and will rapidly stiffen in resistance as the wheel rises to full bump. This gives in some way the equivalent of a built-in damper, but it has been found from experience that additional fluid damping is required. The main problem of the air spring is how to retain the air within the working chamber, for very high pressures are realized, and the lubrication of moving parts is difficult. There has, therefore, been considerable interest in the possibility of containing the air within rubber bags so that the leak problem would not arise, and the possibility of so doing has been increased in recent years by the introduction of improved rubbers both natural and artificial.

The rear springs on the new Citroen are made in this manner and they are supplemented by hydraulic dampers. This, however, is but the beginning of the story. The damper units are of piston type with two-way valves, and the damper fluid is used as the reacting medium between the air spring and the frame of the car. That is to say, as each wheel swings on a trailing arm the piston of the damper moves in and out of the cylinder, and after driving oil through a restricting valve compresses the gas in the sealed rubber container. The damper is kept constantly supplied with oil under pressure from a separate pump and, most important of all, the supply of oil is controlled by a valve which is responsive to the angle of the car in relation to the road. This in effect changes the fulcrum about which the spring unit operates, the result of which can be seen in most dramatic form when the engine is started with the car at rest.

As one approaches the new Citroen it is seen to be very much down by the stern. The engine drives a seven-piston pump supplying brake fluid to a pressure accumulator and from this it is bled off to the suspension system. So a few moments after starting the engine the tail of the car comes up until it reaches the designed position. This will now hold irrespective of load, and if the tail of the car be depressed by an outside force the car will be forced back to the level not only by the return action of the springs but also with the additional effect of the oil servo.

When the car is driven away on an ordinary southern English road the benefits of this apparently complicated arrangement are not immediately apparent. The driver certainly and the passengers probably are more immediately impressed with the high standard of stability which arises from some of the other novel features of the design. Having by modern standards an exceptionally long wheelbase, a track of 4 ft. 10½ in., and an unusually low centre of gravity, the car is almost wholly free from roll. This in conjunction with the remarkable cornering power inherent in the Michelin X tyres gives a feeling of safety when sweeping through quite severe bends at between 60 and 70 m.p.h., which must be experienced to be believed.

These qualities are partly the product of front-wheel drive, but a perhaps more im-

---

In Brief

Price: £1,040 plus purchase tax £434 9s. 2d. equals £1,474 9s. 2d.
Capacity ... ... 2,886 c.c.
Unladen kerb weight ... 27¼ cwt.
Fuel consumption ... ... 18 m.p.g.
Maximum speed ... ... 84.1 m.p.h.
Maximum speed on 1 in 20 gradient ... ... 70 m.p.h.
Maximum top gear gradient 1 in 9.3
Acceleration:
  10-30 m.p.h. in top ... 9.7 sec.
  0-50 m.p.h. through gears 13 sec.
Gearing: 19.9 m.p.h. in top at 1,000 r.p.m.; 76 m.p.h. at 2,500 ft. per min. piston speed.

The Citroen Six Hydro-pneumatic

CONVENTIONAL PATTERN.—The control gear of the Citroen (including the finger-tip light switch and horn button) has not been greatly changed and the separate seats are outstanding for their exceptional comfort.

portant aspect of this arrangement is that the 60% of car weight carried by the front wheels which is required for a pitch-free ride is coupled with drive through the front wheels. Thus, the smaller proportion of weight carried by the rear wheels does not affect traction. This more than compensates for transfer of weight backwards as the car accelerates or climbs a hill. Hence, under conditions of snow and ice, speed can safely be maintained, especially once the driver has mastered the technique of "power on" cornering.

The old criticism that cornering characteristics of front-wheel drive cars change completely as between power on and throttle closed, does not apply in this instance, for the difference, which admittedly exists, is of a very small order. There may, however, be some reaction into the steering if wheel-spin develops when accelerating on the lower ratios. The turning circle is definitely on the large side, and the steering, which needs only $2\frac{1}{2}$ turns from lock to lock, continues to be heavy in comparison with most modern cars, although it is substantially lighter than we have experienced on any earlier Citroen.

As on all earlier examples of this rack and pinion system, it is highly positive, free from backlash, and by virtue of the stabilizing effect of driven front wheels, free from road reaction except on the worst of surfaces.

Up to now this latest version of a basically 1939 motorcar has produced few surprises. As compared with the model tested by *The Motor* in 1954, the mean maximum speed has risen from 81.1 m.p.h. to 83.5 m.p.h., there is no change which is significant in the acceleration or hill climbing factors, and but little difference in fuel consumption. The car feels smoother and quieter and there have obviously been improvements in finish and equipment of which mention will later be made. But let us now continue this journey in the mind's eye on the assumption that a normally smooth main road is succeeded by a secondary road which is pronouncedly wavy; one which will limit the comfortable speed of a normal car to, say, 50 m.p.h., or reduce the average vehicle which has run 30,000 or 40,000 miles to 30 or 40 m.p.h. The Citroen continues (*ex hypothesi*) at its normal cruising speed (bends included) of 70 m.p.h. and the driver and passengers immediately notice—nothing! The increased flexibility of the torsion bar suspension at the front absorbs the punishment handed out to that end of the vehicle, and the combination of air spring characteristics plus the ride leveller ensure that every upward or downward movement of the body is limited to one or two oscillations, after which the back end is as it were clamped into its normal position.

Let us continue on to even worse surfaces. When the driver chooses to demonstrate by taking the car on to unmade roads he instinctively braces himself for the shock which will attend the striking of 4 or even 5-in. obstacles at speeds of between 60 and 70 m.p.h. But if the passengers can be persuaded to read a book or newspaper they will find that they continue to do so and have little or no knowledge of the road surface, appalling as it is, over which the car is being driven.

The Citroen hydro-pneumatic suspension, in fact, sets standards of ride which are without parallel in our experience, and it is a particular merit of the car as a whole that the superlative comfort for the passengers (and the back seat passengers in particular) is coupled with the high-speed stability which has for so long been the hallmark of the products of Quai de Javel.

Our Roads Too Good?

It may be unfortunate for the sales prospects of this model that the roads of England as a whole are too good to show off the new system to its best except on special occasions. Also such bad roads as do exist are mainly to be found in the northern and colder sections of the country and although the suspension of the Citroen gives it particular appeal as a chauffeur-driven vehicle, it suffers rather badly in respect of interior ventilating and heating arrangements. Warm air is piped from behind the radiator to a point just above the front passenger's feet, but in cold weather the temperature of the delivered air is little above the ambient and this benefit is dissipated before the incoming air reaches the back seat passengers. Moreover, as there is no air demisting, in humid conditions it is almost essential to drive with a window open. For these reasons the habitability of the car falls considerably below the standards one may reasonably expect for a car of this class and price and unfortunately, this is not the only anomaly perpetuated in this pre-war design. The enclosed luggage locker is largely occupied by the spare wheel, and the theoretical carrying capa-

HYDRO-PNEUMATICS.—A sphere containing compressed gas supports the weight carried by each rear wheel, a fluid column forming the reaction member between the spring and a piston on the suspension arm attached to the frame. The oil is restricted as it is moved by the piston and it is held permanently under a pressure, which is varied by a device sensitive to angle, so that there is both damping and a "power push" to restore a level ride after an external disturbance. The system can also be used to help wheel-changing (*right*).

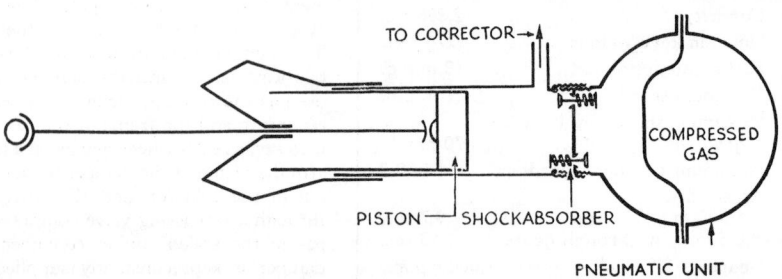

130

- - - - - Contd.

BACK SEAT COMFORT.—The exceptional rear-seat ride of the Citroen is matched by deeply upholstered seats, and detail amenities such as ash trays, map pockets and a foot rest are provided.

city is limited further by an awkward shape. Both the front windscreen and the rear window seem shallow in comparison with later designs, and the conformation and styling of the body makes no pretence to modern standards.

As a by-product of the unique suspension system a rear wheel change becomes an exceptionally easy operation. Beneath the luggage locker is a three-position tap. If this is turned to "high" additional oil is fed between the exceptionally soft pneumatic spring and the damper piston, and this raises the carcass of the car considerably above normal. Chocks may then be fitted beneath the hull and if the tap be turned to the "low" position, oil will be abstracted which will result in the rear wheel rising clear of the ground. After changing the wheel the reverse procedure can be ended by returning the tap to "normal."

Top-gear Performer

Turning now from the dominant features of the car—many good, and some bad—to the more ordinary, an outstanding feature is the extreme flexibility of the low output, high torque six-cylinder engine. The top-gear acceleration from 10 m.p.h. upwards is both smooth and rapid, but in addition speed can be reduced to a slow walking pace and the car then smoothly accelerated by placing one's foot hard down on the accelerator. Unless a gradient sensibly stiffer than 1 in 10 is met the indirect gears can be regarded as for emergency use only, especially in the case of bottom gear, as with a little practice smooth starts can readily be made in 2nd. It is perhaps just as well, for the interlock which prevents a gear being released until the clutch pedal is fully depressed will not be found to everyone's taste. Nor does the facia-mounted gear lever really lend itself to a right-hand drive car.

As frozen roads were experienced through much of this test it is impossible accurately to report on the endurance and resistance to fade of the braking system, but previous models of this make have always shown up excellently in these respects.

As indicated above, the type now tested compares very favourably with any earlier model in respect of smoothness. This is partly due to the provision of sponge rubber under the carpets, and there is generally a high level of furnishing, the seats being large and softly sprung; in fact, for a woman the length of the rear seat is if anything excessive. The cubbyhole on the facia has a small opening, but is deep and inclined so that anything put into it will not emerge as the consequence of violent manoeuvres, and it is supplemented by large pockets in each of the front doors and two on the seat squabs. The instruments are grouped neatly in front of the driver and are illuminated at fixed intensity through a separately controlled switch. One may perhaps question the policy of providing a speedometer which records up to a speed nearly 50% higher than the car is capable of realizing, and in addition to this the instrument provided on the test car was unusually erratic in its promise of more than could really be performed. The combined lighting and horn switch is a feature so useful that one is surprised that it is not universal. A recessed square on the end of a stalk is rotated clockwise to switch on successively the side plus dipped beam and side plus full beam lighting systems. Flicking the stalk away from the steering wheel will change from high beam to dipped beam, or alternatively from side lamps only to dipped beam. Pressing the stalk inward sounds the horn which can be worked at two noise levels.

Superb Roadholding

The car tested was fitted with two fog lamps which are optional extras; an additional charge is also made for two Lucas de-icers which can be fitted at the base of the windscreen, and it is worth noting that the latter is almost unique amongst present-day cars in that it can be opened at its base to improve hot weather ventilation.

To sum up, here is a car that blends superb roadholding with soft, yet controlled, suspension, that can be driven hard without anxiety concerning mechanical failure, is robust to the equivalent of A1 at Lloyds, but which is in certain respects lacking in modern amenities.

Mechanical Specification

Engine
Cylinders	6
Bore	78 mm.
Stroke	100 m.m
Cubic capacity	2,866 c.c.
Piston area	44.4 sq. in.
Valves	O.h. push rods
Compression ratio	6.5/1
Max. power	76 b.h.p.
at	3,800 r.p.m.
Piston speed at max. b.h.p.	2,620 ft. per min.
Carburetter	Solex twin D.D.30 PAAI
Ignition	Coil
Sparking plugs	Champion H.10
Fuel pump	AC mechanical
Oil filter	Gauze on pump

Transmission
Clutch	Twin dry plate
Top gear	3.88
2nd gear	5.62
1st gear	13.29
Propeller shaft	Nil
Final drive	Spiral bevel
Top gear m.p.h. at 1,000 r.p.m.	19.9
Top gear m.p.h. at 1,000 ft./min. piston speed	30.3

Chassis
Brakes	Lockheed
Brake drum diameter	12 in.
Friction lining area	143 sq. in.
Suspension: Front	Torsion bar and wishbones
Rear	Hydro-pneumatic and trailing arms
Shock absorbers: Front	Telescopic
Rear	Nil
Tyres	Michelin "X" 165-400

Steering
Steering gear	Rack and pinion
Turning circle (between kerbs): Left and right	43¼ feet
Turns of steering wheel, lock to lock	2¼

Performance factors (at laden weight as tested)
Piston area, sq. in. per ton	28.7
Brake lining area, sq. in. per ton	92
Specific displacement, litres per ton mile	2,790

Fully described in *The Motor*, Sept. 29, 1948

Coachwork and Equipment

Bumper height with engine running:
Front (max.) 16 in., (min.) 12¼ in.
Rear (max.) 18¼ in., (min.) 15 in.
Starting handle ... Yes
Battery mounting ... On dash beneath bonnet
Jack   Automatic rear; bevel and screw type
front jacking points... 4
Standard tool kit: Tyre pump, grease gun, wheelbrace, adjustable spanner, screwdriver, pliers, box plug spanner and tommy bar, tool roll, rear jacking chocks and supports, Harvey Frost screwjack.
Exterior lights: 2 headlamps, 2 sidelamps, 2 tail lamps, 2 stop lamps, Number plate lamp.
Direction indicators: Semaphore self-cancelling
Windscreen wipers ... 2
Sun vizors ... 2
Instruments: Ammeter, fuel gauge, clock, speedometer, manual ignition control.
Warning lights ... Oil and charge
Locks: with ignition key ... 2
Glove lockers ... One on facia
Map pockets... 2 in front doors, 2 behind front seats
Parcel shelves ... Behind back seat
Ashtrays ... 1 on facia, 2 behind front seats
Cigar lighters ... Nil
Interior lights ... 1 in roof
Interior heater ... Duct from radiator
Car radio ... Optional
Extras available: Fog lamps, windscreen de-icers, sunshine roof.
Upholstery material ... Leather
Floor covering ... Pile carpet over rubber
Exterior colours standardized: Green, red, grey, black.
Alternative body styles ... Nil

SUSPENSION UNDER

Good forward visibility and excellent controllability make the Citroen Six a car that can be taken through narrow lanes with confidence

WHATEVER other comments are made on the new Citroen suspension system, extraordinary must be the first. It is no ordinary system which hisses and as a result makes a garage mechanic ask if the tyres are going down, that causes the sprung mass to sink at the back and rise at the front with an almost ballet-like grace when a rear passenger climbs aboard, and that leaves the empty car to restore its own level as you walk away from it. And if you roll back and brake, the six stops with a clonk and the rear rises while the nose falls. These are the oddities associated with the suspension, though its behaviour on the road still calls for the unusual adjective.

The first impression is the close identity between the feel of the six-cylinder on its suspension and that of the 2 c.v., in spite of the differences of the systems. A blindfold driver of any perspicacity would have little difficulty in deciding that both cars came from the same factory. Both have made a virtue, it might be said, out of what used to be regarded as a vice of suspension—"float." For that is the predominant sensation, that of floating along at a certain height above the ground, only remotely affected by what is happening below. Float as a vice results in amplification of road irregularities.

The six-cylinder was given a very thorough test indeed. All aspects of normal motoring were covered, and in addition two special routines were imposed: one was the obvious one of taking the car over a rough stone track, deeply rutted and pot-holed, the other a fast late night drive over a minor road thoroughly known since childhood, the twelve miles of which ought not, in more normal circumstances, to find a car going at over 40 m.p.h., owing to the wave in the surface and the fairly abrupt ups and downs of the way. These tests were taken with four persons aboard, of rather less than average weight, and while the rough track provided a more or less expected performance following experience on the main road, causing little vertical movement of the body, the night drive was impressive. None of the passengers complained of anything untoward in the car's behaviour although they almost certainly would have done so had it been a comparable "normally" sprung car; they would, in fact, have been inclined to hang on and wonder if the driver was trustworthy.

Low speed surface unevenness gives about the same result as for any suspension, the slight pattering of the wheels over stone setts, say, being felt; this is neither surprising nor disappointing, for the tyres used in conjunction with the suspension are Michelin X, which is a tyre to pick up an uneven surface with its steel-reinforced tread; minor waves, however, are not felt, and major waves result in a slow, well-damped rise of the body and what seems to be an immensely leisurely restoration or rebound, so long drawn out as to result in the damping-out of the oscillation during the single rebound. The car, therefore, is deflected vertically and then restores its level; no more, it seems.

The sensation is one of airiness, akin to the feeling within a very softly sprung car with a high ratio of sprung to unsprung weight. So far, so good, though until the derestriction signs are reached a certain misgiving is felt in case the Citroen suspension should have the defects of this type of ride when it comes to fast, open road, motoring. Roll is the worst of them, but they sum up to a feeling of insecurity brought about by excessive movement of the sprung mass upon its unsprung basis.

There is no comparable reaction to the Citroen suspension. In the straight line, speed makes very little difference to the feel. Still the seats maintain their steady float in mid-air, apparently unrelated to the movement of the wheels over the road contour. Still it remains possible to read a newspaper or study the small type of a map without difficulty; for some minutes one of the passengers read a book. The driver gains confidence, for not only is there no intrusion of vagueness into the heavy—at low speeds, at least—and high-geared steering of the six, but also he quickly becomes aware of the absence of any roll tendency on corners.

How far this is the system, and how far the front anti-roll bar, is analysed on the following pages. The importance lies in the result, and that is quite remarkable. In only one respect at speed does the new system seem as if it might compare unfavourably with its more metallic counterparts, and that is in the reaction to a rapidly taken hump-backed bridge. The sensation of "aircraft stomach" is a little prolonged as the restoration of ground clearance takes place. However, there are few suspensions that can cater properly for the really nasty hump-backed bridge. Even if the springing is nearly "solid," as in an old-fashioned sports car, all that happens is that solid objects in the vehicle which are unattached tend to hit the ceiling. Too much rebound damping, on the other hand, means that the wheels themselves easily leave the ground.

Cornering

Complete confidence having been gained on the straight, the driver begins to acquire the feel of fast corners. Of course, anyone who does not normally drive a front-wheel-drive car has a double task in this respect. He has first to overcome his diffidence at handling a vehicle with the power laid on at the wheels that steer, and secondly to go on up the speed curve as a suspension test.

This driver had had a certain amount of f.w.d. experience, enough for him to scoff at the "experts" who prophesy doom if a fast corner is taken on the over-run. When rear-driving engines were set about nine inches abaft the front axle the point could be made that an f.w.d. vehicle introduced a certain strangeness by its high concentration of weight for'ard and the fact that its front wheels were subject to the application of power, but now that rear-drive engines have gone ahead of the axle, so that we have become used to understeer characteristics, there is nothing in it. In fact, I would prefer to corner fast in a Citroen to many of the contemporary vehicles in which a six-cylinder engine overhangs the axle out front.

Lay on the power and Citroen cornering is almost phenomenal; yet one utters the warning that is necessary with a combination of design virtues and X tyres. At some speed, obviously, tyre adhesion must be lost, and the speed is very high indeed in terms of centrifugal force on this car. If, therefore, a driver does take the car beyond breakaway point, he will have quite a handful to contend with. The faster the cornering potentialities of a car, the more strictly should safety margins be maintained.

Anyway, on the six-cylinder corners were several times taken so fast that the X tyres squealed in protest, and still the car behaved itself admirably. It is not easy to make X tyres squeal. The car does take up an angle, as any car is bound to that has its roll centre below its centre of gravity, but the angle is very small indeed, and the only discomfort

SCRUTINY: *The Citroen Hydraulic System on the Road*

to passengers is caused, not by roll but by the sheer intensity of the centrifugal force which the speed brings to bear upon their bodies.

The Citroen Six has a range of cruising speeds from about 45 m.p.h. to a recorded 80 on the speedometer. This suspension fully does its share in making them possible. If circumstances force one to cram on the brakes at a very high speed nothing untoward results.

Suspension is chiefly relevant, in terms of comfort, to the back seat. The driver was driven for some distance while he occupied the back seat, from which he navigated in spite of his normally queasy stomach. Citroens have always been numbered amongst those cars whose back seats are as good as those in the front—mostly because they have the wheels at the corners—and the six has almost improved upon that claim. In certain circumstances, notably for those who want forty winks, the back of the six-cylinder is almost better than the front, especially as the upholstery of the six is luxurious and the carpet has a sponge rubber underlay.

It is odd that this car should arouse so much enthusiasm 23 years after it was conceived. Back in 1932, André Citroën standardized on torsion bar independent front suspension, integral construction, rack and pinion steering, a gear change off the floor and front-wheel drive. In 1955 his products continue to earn approbation and to excite the keen motorist who comes in contact with them. The six-cylinder—known as the Fifteen in France, but not to be confused with the Light Fifteen over here—is the biggest of the range and is a quite exceptional car. The merits of its design are reflected in the ability that the car has to make long, tearing journeys over the roads of its homeland and to do so with the minimum of mechanical trouble and attention.

The design has a few disabilities (the steering lock is poor, for instance, owing to front-wheel drive and the very long wheelbase), but in the main it commands the respect of all who handle Citroens. Some of the incidentals are notable; the dipstick is a slender, flexible affair like a stay-bone, which weighs little and which can be threaded into its accessible hole at almost any angle. The sunshine roof, an optional fitting, is much appreciated by the now long boxed-in British, and the mere depositing of the spare wheel in the locker where the closing lid secures it is a masterly way of saving trouble for manufacturer and owner. As a corollary, the wheels can be changed without manual jacking effort, for the idling engine does the lifting hydraulically (as described later) when the necessary controls are operated, until the offending wheel is clear of the ground.

When Citroen ultimately produces a new car, the start on the rest of the world will still be considerable. If the contemporary virtues are held and just a few modern ones (mainly in passenger space and looks) are introduced, the result will make the motoring world sit up.

Citroen lines, virtually unchanged since the early 'thirties, are still such as to appeal for their rakishness. The car looks well in any surroundings

SUSPENSION UNDER SCRUTINY

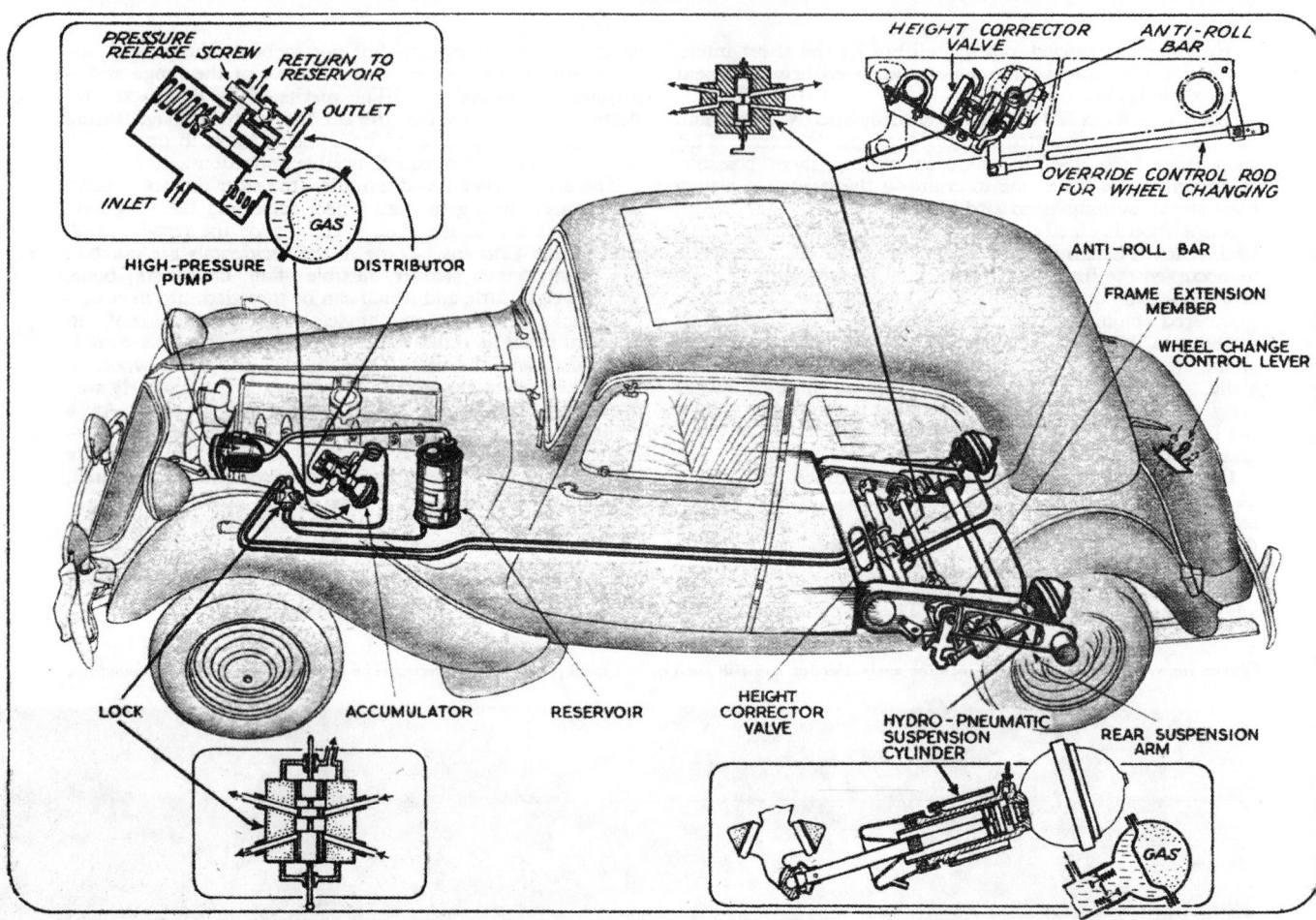

Layout of the Citroen hydro-pneumatic rear suspension system. The main functional components are the pressure pump, accumulator and distributor valve, isolation cock, height corrector valve and wheel suspension cylinders. The front suspension continues to employ torsion bars in conjunction with wishbones, but the bars have been lengthened to give a considerable reduction in rate

THE road-holding of the Citroen has always been good, due in no small measure to its basic features of long wheelbase (10ft 1½in), wide track (4ft 10½in) and a very low centre of gravity. Lack of fore and aft pitch indicates that the centre of oscillation of the car is forward of the front wheels, as the vertical movements experienced appear to be of less magnitude in the front seats than in the rear. The worst possible condition of ride, incidentally, is when the centre of oscillation lies within the wheelbase near to the car's centre of gravity, which, in most designs, occurs around the front seat position. With such a combination the occupants of the front seats are subjected to quite violent forward impulses, while the rear passengers receive a mixture of vertical movement and fore and aft motion, a most unpleasant combination.

Another feature which is of some importance in relation to the Citroen's road-holding is the high percentage of weight carried by the front wheels. Unladen, with five gallons of fuel in the tank, the weight distribution is 60 per cent front and 40 per cent rear. There is thus a high ratio of sprung to unsprung weight at the front, which gives a high inertia value against which the springs can react; similarly, with no axle the ratio is also high at the rear end.

This means that the wheels are in contact with the ground for longer periods than if the ratio of sprung to unsprung weight were of a lower order. The concentration of weight forward in combination with the inherent characteristics of front-wheel drive, give the car its uncanny directional stability and precision of handling.

The six has now been equipped with hydro-pneumatic suspension at the rear, which has enabled a softer ride to be provided; at the same time, the front suspension has been softened. The general arrangement of the front wishbones is retained but the torsion bars have been considerably lengthened and the rate reduced. In addition, an anti-roll bar, connected to the lower wishbone arms by drop links, has been added.

Air springing is not new and it has always attracted designers. Its property of a rapidly rising rate gives low flexibility at small deflections, with increasing resistance to bumps at high deflections. One application was seen in the suspension struts used for B.R.M. racing cars. Air suspension is also becoming popular on commercial vehicles in the U.S.A., the air being in rubber containers placed between the axle and frame. By using such containers the difficulty of preventing the escape of air under compression is overcome, and a rubber bag is also used in the Citroen system. But although an air spring has this desirable quality of increasing stiffness to load, it has been found necessary to supplement it with hydraulic damping.

The point of real importance in the Citroen rear suspension is that it incorporates a constant level device which returns the rear of the car to the same static position irrespective of the load carried. This is achieved by adjusting the length of the oil column forming the reaction member between the air spring and the piston of the hydraulic struts.

With the normal type of suspension system, using mechanical springs, it is necessary to have a spring rate higher than is desirable for the best ride conditions, to avoid too much change in spring deflection between the laden and unladen conditions. If, as in the Citroen, the static position remains constant irrespective of load it permits a reduction of spring rate, thereby giving a softer ride. By present standards the Citroen rear suspension is very soft, having a frequency of approximately 40 oscillations per minute; an average figure for the equivalent modern British saloon would be in the region of 70 to 75 per minute. Soft springing can, of course, cause excessive roll, but, as explained above, the inherent roll resistance of the Citroen is high and the car is almost completely free of this undesirable feature.

The rear of the car has been modified by mounting the new suspension units on a sub-frame extending backwards from the main body sills. This extension piece is based on a steel tubular cross-member bolted to the main frame with end attachment plates. Two box-section side-members are welded to this cross-tube and terminate at another large-diameter tube welded between their rear ends. The trailing links of the suspension are each mounted on this extension with the aid of two opposed taper roller bearings; an anti-roll bar is connected between the pivot points. A further arm extends downwards from the pivot point of each trailing link to carry the reaction point of the hydro-pneumatic jack and the bump and rebound rubber stops.

The hydro-pneumatic spring unit is attached to the rear cross-member of the frame extension and the piston rod is actuated by the lower end of the trailing link drop arm.

The suspension unit consists essentially of a sphere screwed into the end of the hydraulic cylinder. The gaseous mixture (the nature of which is not revealed, but it is probably one of the inert gases) is separated from the hydraulic damper fluid by a rubber diaphragm. Two-way damper valves are carried in the neck of the spherical component where it is screwed into the cylinder.

Hydraulic pressure is supplied by a seven-cylinder pump driven by a V-belt from the crankshaft pulley and supplied from a reservoir mounted on the scuttle. The hydraulic fluid is delivered to a hydro-pneumatic accumulator incorporating a distributor valve which pressurizes the system from the accumulator when the pump is inoperative. With the pump delivering, this valve feeds the fluid under pressure to recharge the accumulator and to feed the suspension system. Thence the circuit passes through the isolation cock to an automatic height corrector valve, and from this a single pipe connects to each hydro-pneumatic spring unit. This corrector valve regulates the pressure in the spring units and thus allows for variations in the load carried. The hydraulic fluid is the same as that used in the braking system.

The accumulator screwed into the end of the distributor valve body is, like the suspension units, a sphere containing gas on one side and the hydraulic fluid on the other, separated by a flexible rubber diaphragm, the gaseous mixture being the same as that in the suspension unit.

The isolation cock fitted between the accumulator and height corrector enables the rear portion of the circuit to be isolated from the front, so that the system is locked to maintain the suspension at its static height when the engine is stopped. This cock is opened automatically by the first application of the clutch pedal after it has been closed.

Height Correction

The automatic height corrector is a slide valve actuated by a tongue attached to the anti-roll bar. Thus, as the suspension arms rise and fall this slide is moved up and down. As the arms rise the slide uncovers the delivery port and allows hydraulic fluid to pass under pressure into the suspension cylinders. As the arms fall the delivery port is closed and the fluid passes from the springs back to the reservoir. So long as the suspension unit remains at normal height the delivery and return ports are closed.

An overriding control is attached to the slide valve so that it can be operated manually by a lever in the boot of the car, used when changing the rear wheels. This control has three positions—normal, high and low. A stand is inserted under the vehicle on one side or the other, immediately in front of the rear wheel. With the control placed in the low position the suspension arm is raised, bringing the wheel clear of the ground by action of the hydraulic fluid, and the weight of the car is taken on the stand so that the wheel can be removed.

Doubts may be raised as to the possibility of failure, or leaks, in the hydraulic system. The worst which could happen is the fracture of the fluid pipe, in which case the car could be driven home slowly with the suspension arms resting against the rubber bump stops.

Undoubtedly this system sets a new standard in ride comfort in an ingenious manner and may well promote a new line of thought for suspension systems of the future.

Left: With the aid of a stand provided in the tool kit the hydraulic system of the rear suspension can be utilized for easy wheel changing Right: The hydraulic pressure pump is belt-driven from the crankshaft and, with the accumulator, is mounted on the cylinder block. A reservoir for the hydraulic fluid incorporates an external glass tube level gauge

Progressive Maintenance of t...

Dismantling the Engine: Timing Chain Replacem... Adjustment and ...

By C. ...

THERE is no reason why the presumed complication of front wheel drive should deter Citroën owners from attempting progressive maintenance at home.

Mention will not be made of operations detailed in the instruction book issued with the car but many other quite substantial home jobs are easily possible and can be tackled single-handed. Although this should result in substantial saving, work best left to Citroën specialists includes torsion bar adjustment, front wheel bearing renewal, differential settings and camshaft and crankshaft repairs, including bearing remetalling.

Components needing maintenance or renewal may be grouped as those requiring prior removal of (a) radiator, (b) engine, (c) bonnet only.

Group (a) includes gearbox, front engine mounting and pulley drive from camshaft front end. Under group (b) come clutch and camshaft drive. Group (c) comprises starter, pistons and cylinder barrels.

Good quality metric spanners are essential —open-ended, ring and socket. Approximate equivalents waste valuable time. Numbered consecutively, those most frequently needed are Nos. 8-14 and 16-18 inclusive. The remainder up to No. 25 are useful purchases but those above are disproportionately expensive and seldom used.

The whole engine design is, of course, "back to front," the gearbox-differential unit being carried in the nose. Then follow clutch bell-housing, crankcase and finally timing case which carries the rear engine mounting.

Dismantling

The key to major dismantling lies in rapid disconnection of the inner drive shaft couplings from the gearbox after removal of the radiator. The latter rests on a pressing which combines the function of support for the front engine mounting and upper cross-bracing to the cradle forming the front "axle."

Proceed by raising both front wheels fairly near the hubs so that the drive shafts can be rotated at the normal angle. Next, remove the nuts from the four cross-bracing bolts, set a jack under the gearbox drainplug, remove the studs supporting the front engine mounting and drive out the bolts from the ends of the cross-bracing which can then be detached. Raise the gearbox half an inch by the jack and wedge the engine firmly against the body shell on one side. Then remove the four nuts from the differential drive flange on the opposite side.

Free the drive coupling from three of the four studs, turn the drive shaft through 90 deg. and leave one stud in the coupling recess. Next, wedge the engine against the body shell on the opposite side whereupon the second coupling can be dismantled and will fall free. Either the gearbox or the engine can then be removed.

Gearbox Removal

Whether the car is in gear or in neutral the gear selectors (shown in Fig. 1) are locked in position. They are freed by depressing the clutch pedal which has the effect of drawing back the "T" locking rod against its spring, so releasing the pressure of two small balls in contact with notches in the selector rods.

The gearbox-differential unit (shown in Figs. 1, 2 and 3) draws forward off the front of the clutch bell-housing, the castings parting in the plane of the drive shaft axes.

Leaving the front wheels blocked up and free to rotate, transfer the jack from the gearbox to the sump drain plug. While the gearbox is draining, withdraw the speedometer drive, remove the gear control pillar (Fig. 1) undo the nine nuts holding the gearbox in place and unscrew the gear lock adjusting plate.

Drawing the unit forward is a combined operation of progressive raising of the engine by jack and rotation of the roadwheels in turn to free the drive shaft studs and clear them above the "axle" cradle.

Two safeguards are worth mentioning. Dismantling should begin with freeing of the clutch rod or cable, for if the clutch is freed with the gearbox withdrawn the clutch centreplate may move out of place. Also, the clutch bell-housing aperture should be stuffed with rag, for if anything drops in by mistake further dismantling will be necessary.

Engine Removal

As the weight to be lifted is some six cwt. a light chain block is essential, as is adequate strength of garage roof frame.

First, disconnect the exhaust downpipe, clutch cable or rod, and all essential fuel, electrical, throttle and gear connections; then jack down the front wheels. Sling carefully under the water pump casing. Slight raising will first clear the drive couplings and allow the rear engine mounting to come forward free of its socket. The engine weight rests mainly on two transverse helical spring supports. As the compression goes from these they can be slipped out by hand. With further lifting the car can be pushed back and clear. The engine is then lowered on to blocks or, for preference, on to a simple wooden cradle.

Timing Chain Replacement

The timing case is now free for removal. The camshaft is driven from its rear extremity by a double roller chain which for preference should not have a free link. If it does, see that it is securely fixed. The sprockets are keyed and an easy sliding fit on their shafts. For the first chain renewal the original sprockets can reasonably be retained, but should be renewed the second time. Retention of correct valve timing presents no problem if the shafts remain stationary. A tooth on each sprocket is centre-punched. First lay the sprockets on a flat surface with the punchmarks innermost and in line with the sprocket centre-lines, then fit the chain in that position and slide the complete assembly on to both shafts in one movement, watching that the keys remain in position. If either shaft has moved, rotate the crankshaft until both keyways come into the fitting position. Then thoroughly tighten the sprocket ring nuts and secure firmly with new locking washers.

Fitting New Pistons

The new assemblies come from the makers complete, comprising pistons with four rings fitted, gudgeon pins and cylinder barrels (with their base gaskets).

Having removed the cylinder head as for normal decarbonising, take off the sump and note that the big-end caps and connecting rods are centre-punched numerically at their forward ends. Then cut a hardwood block to fit across the cylinder barrel skirt and tap sharply upwards to break the barrel gasket seal. Remove No. 1 big end cap and draw the whole assembly—piston, connecting rod and barrel—up by

Fig. 1.—(A) Gear-locking "Tee" rod. (B) Gear selector rods. (C) Front engine mounting. (D) Gear control pillar (displaced).

Fig. 2.—The gea... Bellhousing ap... cradle. (C) Drive... bellhousing. (E)... drive housing.

Fig. 3.—A further view of the gearbox-differential housing. (A) Differential drive flange. (B) Gear lock ball housing. (C) Gear locking spring and rod. (D) Main shaft.

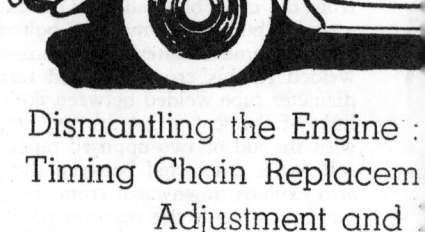

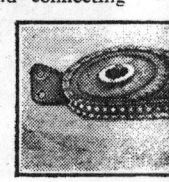

Fig. 4.—Timi...

CITROËN "Light Fifteen"

...x and Engine Removal :
...ing New Pistons : Brake
... of Starter Motor

...ONG

hand and out of the cylinder block. Repeat for each cylinder, finally stripping the connecting rods from the pistons and trying one of the new gudgeon pins through each little end. Unless they are a very good fit, have the latter rebushed, sending one of the new gudgeon pins as a pattern.

Check Big Ends

Before despatch check the big ends for lift and soundness; also measure up the crankpins. If ovality is negligible, lift not excessive and the bearing surfaces even, most of the clearance may be taken up very carefully on both cap and connecting rod equally without remetalling. If in doubt, however, have the big ends remetalled along with the rebushing of the little ends, bearing in mind that slackness at either end will be more audible with the new pistons.

To begin assembly of the replacement parts, fit one circlip in each new piston and get them fairly hot. See that the piston skirt slot will be fitted on the nearside of the car, i.e., on the same side as the pushrods. The cold gudgeon pin is a sliding fit in a hot piston. Follow up with the second circlip and position the ring gaps alternately opposite one another. Then stand the barrel upside down on a flat surface and work the piston slowly down into it with a slight rocking motion, easing the rings past the barrel chamfer one by one with the fingers. Keep the piston well within the barrel, for if the top ring escapes free at the other end it will have to go back in at that end with no chamfer to help because the connecting rod will not follow through the barrel.

Cylinder Block Seatings

Each barrel has an external flat from end to end and these have to be set flush with one another in pairs. Carefully clean the seatings in the cylinder block on which the barrel base gaskets are to rest, applying jointing material and placing each pair of gaskets in position. These gaskets will have cooling water above them and engine oil below, so careful fitting is important.

Turn the crankshaft until No. 1 crankpin is in the lowest position, push the new No. 1 piston half way down its barrel with the skirt slot on the correct side, lower the whole unit on to the gasket, press on the piston top to seat the big end, and wedge a block of wood between all the cylinder head studs to prevent the fitted barrel from rising again when the others are placed in position. Connect up and secure each big-end cap, using new bolts and new locking washers. Finally refit the cylinder head after checking that all the barrels are standing 2-4 thou. proud of the cylinder block. Fill the radiator and check for watertightness of the barrel gaskets *before* replacing and topping up the sump.

It is not worth considering reboring; replacement is easier and more satisfactory.

Camshaft Extension Drive

This forward extension shaft carries a belt pulley driving the fan, water pump and dynamo. It is carried in the clutch bellhousing on two roller bearings and has an oil-thrower and a driving tongue at its rear extremity. The forward end of the camshaft carries a similar tongue. These tongues, which should clearly be allowed to suffer as little wear as possible, are connected by a cylindrical slotted dog coupling. Replacement of the dog at, say, yearly intervals is the best safeguard against wear. Withdrawal of the shaft is straightforward and a piece of wood cut to fit the slot firmly will retrieve the dog. The same piece can be reused for replacement if its end is first pared down to be a slack fit in the slot. There are eight alternative positions in which the new coupling can be fitted and it is worth while checking which position results in the least backlash.

Lubrication

A simple wire-gauze filter strains the engine oil and should be taken apart and washed in petrol whenever the sump comes off. A supplementary bypass filter is easily fitted and is a good investment, as is a spare oil-pressure indicator. In course of time the pressure indicator will cause the warning light to flicker on in normal use. Unless the oil level is too low this shows the need for replacement of the indicator. If topping up the sump makes no difference, nor does a new indicator, have the lubrication system checked before running the engine again.

Brakes

The handbrake cables to the rear wheels are unenclosed at the front and are of Bowden type towards the rear. In time the cables tend to rust a few inches inside the flexible casing near its forward end, thus annual attention at this point greatly prolongs cable life.

First disconnect the cables at their forward ends and clean them well. Then remove each brakedrum. Unhook the rear end of the cable from its operating arm in the brake mechanism and withdraw it backwards to the fullest extent. Clean the withdrawn length of cable thoroughly, treat well with graphite grease, slide forward again and reconnect.

Relining

The rear brake shoes slide readily off their pivots for relining, but before ordering new linings check whether the rearmost lining covers the whole shoe or only half. Also the linings for the front and for the rear brakes are not interchangeable; they differ both in curvature and in rivet hole spacing.

Front brake relining is equally straightforward once the wheel hubs have been removed (Figs. 5 and 6). A simple hub-puller is needed, but can be made inexpensively and is a useful, if seldom used, addition to the tool kit. The hub, with the brake drum, is keyed to the tapered stub axle and must be a tight fit with the key properly positioned on reassembly. The

Fig. 6.—The hub puller assembled in position.

lock-nut must on no account be moved once the new securing pin has been fitted.

The starter is secured to the clutch bellhousing by a single stud and lock-nut. Having slackened these undo the lock-nut holding the offside helical engine support and screw the spring down to its fullest extent. Then undo the cable to the starter solenoid at its upper end, whereupon the body of the starter can be drawn back, rotated through about 180 deg. and the rear end lifted in one continuous operation. By so doing it can be lifted clear and reinserted in the reverse manner.

Fig. 5.—Hub puller dismantled. (A) Collars on hub and puller. (B) Hub nut to be removed.

...tial unit. (A) Front axle ...gs. (D) Clutch ...ward extension ...ial housing.

...sprockets.

NOW AVAILABLE...

A Sports Car For the Family!

CITROEN

FRONT WHEEL DRIVE   TORSION BAR SUSPENSION

OUTSTANDING ROADABILITY
Front wheel drive. Torsion bar suspension.

FLASHING PERFORMANCE
Up to 90 M.P.H. Two engine sizes to choose from: 4 and the 6.

GREAT ECONOMY
Up to 48 miles per gallon.

AMPLE ROOM FOR FIVE
Two body sizes to choose from: 114½" and 121½" wheelbase.

MONOBUILT BODY
Solidly built for safety.

CAMPBELL MOTORS
United States Distributor
Dealer inquiries invited
818 Fair Oaks Ave., South Pasadena, Calif.   •   SY 9-1544

USED CARS on the Road

No. 144 — 1955 CITROEN LIGHT FIFTEEN

PRICES: Secondhand £595; New—basic £750, with tax £1,064.

		Acceleration from rest through gears:	
to 30 m.p.h.	8.5 sec	20 to 40 m.p.h. (top gear)	16.1 sec
to 50 m.p.h.	21.3 sec	30 to 50 m.p.h. (top gear)	16.0 sec
to 60 m.p.h.	32.0 sec	Standing quarter-mile	23.3 sec
Petrol consumption	20-23 m.p.g.	Mileometer reading	49,556
Oil consumption	negligible	Date first registered	March 1955

Provided for test by C. N. K. Motors, 353, Finchley Road, London, N.W.3.   Telephone: HAMpstead 5712/8532.

The sliding roof opens readily, and for additional ventilation there is an opening windscreen. Anti-draught panels have been added to the front windows. Tiny arcs of the flat windscreen are cleared by the wipers

AFTER a run lasting 23 years, production of the Citroen Light Fifteen ended in August 1957. Throughout this time the basic design specification remained largely unaltered, and because of the lack of change from year to year, and of the excellent reputation for durability which the car established, well-kept used examples of the model held their prices well. The one which is the subject of this test certainly comes in the "well-kept" category, and in terms of condition related to age it is among the best cars so far tested in the series.

It is finished in black, and the only blemishes on the cellulose are one or two long but shallow scratches. There is evidence of carefully executed retouching in some areas, but practically no rust is to be seen anywhere on the bodywork. Although admittedly the car is only four years old, it creates the impression that proper care has been given to the finish, to preserve it from corrosion and the progressive decay which is too readily accepted by used car buyers as "normal." This thought is confirmed by the above-average appearance of the chromium which, with one or two exceptions—notably the bonnet and boot handles, is unmarked. The same absence of rust was noticed when the underneath of the car was inspected.

Inside the Citroen, the slightly grey appearance of the roof linings and mild creasing of the brown leather seat upholstery are the sole indications of the considerable use which the car has seen. In almost all other respects the interior is practically up to new car standards, and is abnormally good in relation to a mileometer reading of nearly 50,000. Even the brown floor carpets are virtually as new.

When the Citroen was selected for test, C.N.K. Motors explained that the engine was to be overhauled before sale. The figures recorded above for acceleration and petrol consumption were measured before the engine was taken down. It is a four-cylinder o.h.v. unit of 1,911 c.c.

Starting is good, but the engine warms up slowly, and needs the choke if it is to pull away smartly during the first one or two miles after a cold start. It is quiet at all speeds, but is not really smooth, some degree of lumpiness and lack of torque at low revs being noticeable. The manual ignition control is in need of adjustment, but even without its use the engine runs happily on commercial petrol—a point to bear in mind when assessing the car's fairly high petrol consumption.

After new pistons, rings and exhaust valves had been fitted, and the big end bearings had been remetalled, the car was tried again at "running-in" speeds. It was then appreciably quieter, and noticeably smoother. The buyer of this Citroen obviously will kick off with a virtually new engine.

Protruding from the facia is the lever for the three-speed gear box; cranked towards the driver, it proves extremely convenient and light to operate. The synchromesh is weak, and changes cannot be hurried without grating the gears—a fault which was criticized in our Road Test of the model when new in 1952. There is a positive lock arrangement which prevents the gear lever from being moved unless the clutch pedal is depressed. The clutch take-up is slightly juddery when reversing, and it is necessary to pause for a second or two with the pedal fully depressed to engage bottom or reverse silently from rest.

One of the best features of the Citroen is its suspension by torsion bars all round, with trailing arms for the rigid axle at the rear, and independent wishbones at the front. It is comparatively firm, yet has a remarkable ability to absorb road irregularities at speed almost without any jolt at all to the car. At certain speeds there is a limited degree of firm vertical movement, but the dampers are still effective and the ride is pleasantly taut. There is complete absence of body rattles.

With its relatively wide track and long wheelbase, the Citroen holds the road remarkably well, and the car can be driven with confidence really fast on corners; there is a minimum of roll, and no tyre squeal occurs until the very high limit of adhesion is being approached. With these commendable safety factors goes rack-and-pinion steering which is light and extremely positive. Only at low speeds is there occasionally a slight tug at the steering wheel, reminding the driver that the car has front wheel drive.

Effective and dependable braking is available in return for reasonably light pedal pressures; and it is rare indeed to come across such a powerful hand brake as the one fitted on this Citroen. It is controlled by an umbrella-type lever below the facia.

Four practically new Michelin X tyres have been fitted. The spare wheel—which stands at the front of the luggage locker—has a little-used Regent Remould tyre. The toolkit handle is confined to a jack and wheelbrace, and there is a starting handle in the luggage locker.

A flat-beam Marchal fog lamp has been added to the car, and a windscreen washer, two spring-loaded wing mirrors and an Ekco radio are the other extras. The radio has good tone, but some attention is required to the volume control, which gives rise to crackling. The windscreen washer is also out of order, but this is the only other faulty item of the equipment.

This Citroen Light Fifteen leaves a satisfying impression of efficiency with those who drive it. It makes no pretence to be a stylish car; it scores simply by being comfortable and convenient for normal motoring on short or long journeys, with a praiseworthy ability to cover the ground rapidly, yet safely. Visibility is good in spite of the shallow windscreen, and a pleasing array of bonnet, head lamps and front wings is within the driver's forward view, making it easy to place the car.

There has been no change in the system for selecting the subjects of this test series; as previously, this car was picked at random from *The Autocar* classified advertisements in our issue of 12 June.

The Citroen's comprehensive equipment includes a simple form of heater which collects fresh air warmed by the radiator, and passes it by ram effect to the interior where it is controlled by a rubber flap valve

TRENDSETTERS

FRENCH WITHOUT TEARS

To the average Frenchman the Traction Avant Citroens which ran from 1934 until 1957 were synonymous with gangsters and policemen. No better recommendation could be given.

CITROEN HAS almost specialised in dropping motoring bombshells. The dust is still clearing after the fabulous little GS which last year brought everyone in Europe into agreement over the "Car of the Year".

In 1955 the shock was even greater when the first streamlined, disc-braked, hydropneumatically-sprung DS Goddess dawned, probably 20 years ahead of its time. And after World War II the 2CV was greeted with derisive laughter — but over 2 million examples later it is still a best selling car in France.

Andre Citroen dropped the biggest bombshell of them all in France some 21 years earlier when he unveiled the car which paved the way for front wheel drive in Britain and Europe — the TA or "traction avant" model series.

And what a shock it was then to the conservative European motorist reared on a steady diet of running boards, fabric roofs and chassis rails.

The first TA, the model 7A (seven French Treasury horsepower) came with a 1303 cc four-cylinder engine with removable wet liners. The body was low, long and lean, and except for the very front, resembled nothing else before or after.

There was no chassis, and the passengers actually stepped down into the underslung body. This removed the need for running

The "Traction Avant" Citroen became familiar in Europe during the '30s and '40s partly because it was such a successful get-away car for criminals.

boards, a point which was heavily criticised by motoring "experts" of the time.

Four wheel hydraulic brakes, torsion bar suspension, rack and pinion steering, a pressed steel roof (the first on a European car), an opening external rear luggage door with a built-in spare wheel cover, a hot-air heating system and efficient soundproofing were among its unusual features.

But of course the focus for attention was on the front wheel drive system which put the gearbox and differential ahead of the engine in a similar fashion to that used by Renault in its 16 model today.

Of course Citroen was not the first firm to produce a saleable "traction avant" model. Lancia had been at it for years and in America the Miller racing cars and the 1929 Cord L-29 saloon had popularised the system. But a popularly-priced FWD? Never before.

The 7A was created by direction of Andre Citroen himself. In 1931, he called for a light car of revolutionary design intended to succeed the 8, 10 and 15 hp cars whose production was then about to begin. A 65 kph sedan to carry four passengers in comfort and safety, have a fuel consumption between 20 and 28 miles a gallon and combine striking looks with outstanding comfort — that was the design task.

What was revolutionary in 1934 was still years ahead in 1951 when this Light 15 came off the assembly line. The underslung "monopiece" all-steel body without running boards was criticised in 1934 but its advantages were soon realised.

The problems were of course enormous at that stage of automotive history.

Despite the mechanical complications inherent in designing a front axle layout to take care of both drive and steering and of an automatic transmission system which had to be abandoned (Sensaud de Lavaud's progressive and continuous torque variator) the body and motor will always remain two masterpieces of design.

Seldom has any car been subject to such strong criticism as was the traction avant when it was first announced. However during its long and successful career the design ultimately triumphed. Whereas production of the earlier models had been spread over some 15 years, the front-wheel-drive 7s started coming out of the works in 1934 and were still coming out up to the beginning of World War II. The similarly-bodied "11 hp" TA models were in production right up to 1957.

Naturally other manufacturers used the Citroens as design lessons. Some borrowed the engine (Chenard ET Walker), others the complete power complex (Georges Irat, Rosengart in 1939); or more unassumingly used the body alone (Delage, Licorne).

Even after the armistice old Cord cars fitted with either 11 hp or 15 hp engines in place of the original 125 hp eight and its troublesome electrically-operated preselector gearbox, were to be seen on the roads.

The first born of the series, the 7A was soon followed by the 1529 cc 7B, and the 1911 cc 7 "S" which was in fact the first Citroen Light 15 — our feature car this month. A model 7C was also introduced with a running series of modifications from 1934 to 1939. It was a first cross between an "A" and an "S" model having the bore of the former and stroke of the latter to produce 1628 cc.

Although the general design of these models was unchanged, the Citroen company experimented with a number of different transmission and suspension systems. There were two forms of engine mountings, two types of water pump, three oil pumps, three types of transmission, three types of front axles, three types of rear axles and three types of steering gear.

Three months after the introduction of the "7" models in May, 1934, the company released the 11A and 11AL models. The "11" referred to 11 French Treasury horsepower which represented 15 RAC horsepower, hence the reference to the cars in British journals as Light and Normal 15 models.

The 11 is remembered as the vehicle in which Francois Lecot covered 400,000 km (about 250,000 miles) in 400 days in 1935-36.

The 11AL (L for Light) was the successor to the short-lived 7S and was identical to that model save for a 1 centimetre greater rear track, making it 2 cm wider than the other 7 models.

The 11A or Normal model came with a bigger-all-over, more-spacious body. There were also two alternative body styles on this model — a five-door berline and a nine seater family version.

Both cars shared the same 7S engine which produced 46 brake horsepower at 3800 rpm on a modest compression ratio of 5.9:1. They were the beginning of the most famous line of the TA models which ran through into the middle '50s although the versions and variations on their design in the interval were numerous.

The engine itself was perhaps the least changed of any item and apart

A boot with a rear door and a spare-wheel cover were unique in 1934. Relatively slippery shape added to car's impressive performance and top speed. Lack of vent windows is a supposedly new feature but the Citroen went without nearly 40 years ago.

from a "Performance" version with 6.2:1 compression ratio introduced in March, 1939, and a further boost to 6.5:1 in 1950, the engine was largely untouched until it was replaced by the 59 brake horsepower, 6.8:1 11D engine in 1955.

Body changes were continuous. In 1936 the 7s got new instruments and a new one-piece bonnet. The following year there was a two-seater fixed head coupe version of the 7C. In 1938 all models received Michelin's new Pilote tyres and an 11B sports model was shown at the same time. In 1939 a new fresh-air heater was introduced and in 1940 the 7C went to 6.2:1 compression and was renamed the 7E.

War halted production although a few "11" models slipped through before the Germans stripped the Paris factory. There was not much left when peace came, but some of the "11" moulds were salvaged and production resumed in 1947.

However, the family and commercial models were dropped as the moulds were beyond repair. These were not to re-appear again until late 1953 in the case of the family and 1954 in that of the commercial version.

Meanwhile a new six cylinder model, the 15 Six D with a 77 hp 2867 cc engine appeared on the market.

This model was developed from the 15 Six pre-production model of 1938 which in turn became the 15 Six G — so named because its engine turned to the left — gauche.

The 1947 example had clockwise crankshaft rotation and was hence called the 15 Six D — D standing for Droit, or right. The model continued in production up to the time of the revolutionary DS and in 1953 introduced one of the latter model's unusual features — the hydropneumatic self-levelling rear suspension.

These cars also had lengthened forward torsion bars visible on either side of the radiator grille. An anti-roll bar set at the front connected the front lower axle arms.

Amid the vast array of similar models produced by the Paris firm, two highly unusual bybrids were conceived.

One was the TA 22CV. This was an eight-cylinder front wheel drive version which was intended for production in 1935. Development work took place on the car at the same time as the "11" models were being readied. The brief called for a car with all possible refinements — "the safest and fastest standard production car in the world".

A range of beautiful bodies was to have included a cabriolet, faux cabriolet, berline, coupe de ville, coupe long amd familiale. The engine was to be a 100 horsepower unit comprised of two "11" blocks mounted on one crankcase to give a 90 degree V8. Altogether 20 prototypes were built using 21CV Ford V8 engines for testing. However, the model plans were shelved in 1937.

It was a Citroen bombshell which never went off.

Another unusual Citroen called the UA was also built in the 1930s. This was a car with rear wheel drive to cater for customers who were scared by the revolutionary front wheel drive principle. Using "7" and "11" mechanical components and bodies, the models were built between January 1935 and July 1938, some with 1766 cc diesel engines. In all only 15,000 of the cars were built — certainly one of the most exclusive Citroens ever.

But let's get back to the most popular TA Citroen and our feature car, the Light 15.

This was the model to capture the public imagination for a number of reasons. For a start it was well-priced, it had enough room for four — although the 15 Normal had more — it had a good turn of speed, was economical and of course it had the excellent front wheel drive system.

An English Motor test of a 1951 example showed it to be capable of 79 miles an hour, 25 miles a gallon and 23.2 seconds over the standing quarter mile. It praised the car's stability, handling and reliability.

Perhaps the best Citroen Light 15 in Australia is the 1951 model you see on these pages which is owned by 24-year-old Gerald Propsting of Box Hill, Melbourne.

It is one of three Citroens he owns, the others being a 1939 11A model under restoration and a 1965 DS19 Pallas which serves as hack transportation.

And if that is not enough, Gerald's brother has a 1956 Light 15, his mother has a similar 1954 model and his father has a 1953 15 Normal.

The family Citroen tree reads like this: Mr Propsting senior's first car was a 1926 Lancia Lamda which was later swapped for a 1927 Lancia. In 1954, his taste for traction avant whetted, he bought a 1938 Light 15, but a gearbox problem blew the motor apart.

In 1960 Mr Propsting tried again with the 11 Normal and this car has so far chalked up 105,000 miles, many of them pulling a caravan. His wife's current Light 15 was bought just two years ago.

Gerald Propsting's history of Citroen ownership also reads like a column of the classified pages. His first Citroen was a 1952 standard model Light 15. This was replaced by a $1495 38,000 mile 1954 Light 15 but was recently sold to buy the ex-European Nato forces Pallas ($2400).

Meanwhile the current deep maroon 1951 Light 15 was bought with 145,000 miles on the clock. And, has undergone full and very complete restoration.

Included has been a careful re-paint — although Gerald maintains it is due for another — a fully reconditioned engine, gearbox and front-end plus new brakes.

The interior trim is also new — cream Deerhide PVC sadly — which although excellent does not look as nice as the original leather and vinyl.

The dashboard has been carefully re-polished and the cloth roof-lining is still in excellent original condition. However the carpets are due for replacement and are the first item now on the owner's perpetual list.

The car went back on to the road in January 1970 and has since spent a lot of its time in shows, picking up concours awards with monotonous regularity.

But such quality does not come cheaply. Gerald originally paid $235 for the car unrestored and now it owes him over $2000.

Driving Gerald's 15 you are immediately reminded that although the car itself is just 20 years old, the design is now nearly twice that.

The layout of the controls, the style and the initial feel of the car verify that. Yet in almost every department other than straight-line performance and braking, the Light 15 is a surprisingly modern machine. It corners in approved FWD fashion with very little body roll and complete stability; its steering is heavy but very direct; it handles road-ripples with a smooth, firm action.

In fact it would take surprisingly little to bring the car up to date. I look forward to saying the same things about the new Citroen GS in the year 2000. *

CITROEN
A TRACTION AVANT

BY DAN DE PARTO

CITROEN CABRIOLET — The 1934 Citroen 7 S Cabriolet was a classic design which still looks attractive today almost 40 years later. Note close resemblance the Ford of the same period.

Commenting on Facel Vega, automobile historian Richard Langworth describes the six-cylinder Citroen "TRACTION-AVANT" in so many words: "one of the most innovative and exportable models France was producing." He was commenting on the French government decision to tax high powered cars off the road, a policy that, together with astronomical taxes on gasoline, drove such French greats as Delahaye and Talbot into oblivion.

The last Citroen with that six-cylinder *"petard"* under its hood came off the line in 1957.* It's referred to as the 15-SIX-H which means 15 *cheveaux* (French taxable horsepower: 16CV), six cylinders, and hydropneumatic suspension. That's the fairly well tooted suspension system used on the first ID, the system that made those "innovative" products of French genius squat when the driver ordered to sit and rear up on their haunches while standing perfectly still.

The introduction of the new, rather far-out body style in 56 (DS-19)* required that the already famous and certainly well-loved "tractions" be put out to pasture. And they were literally put out to pasture because virtually every other French farmer or his son came into possession of one type or another of these practical cars for a generation.

They were used for another decade or more as all around farm vehicles, when they weren't being used as ATVs. In fact, some are still used to charge across plowed fields or, in rare cases, still serve in their original, highly polished, robust condition as "Sunday-go-to-meetin' " transportation, *a la francaise*.

There's where you find them. But just try to pry one loose from its sentimental, ultra conservative owner. Even a fist-full of twenty dollar gold pieces, the French farmer's first love, wouldn't make many of them bat an eye, although you would catch a twinkle, now and then.

As for good restorable *"tractions"*, that's another story. They're not as readily available as Model A Fords, but their parts are! Some have been stocked for a rainy day. Others belong to members of the up and coming generation of Frenchmen who have formed small clubs. And the remainder are scattered about the world, places like Turkey, Morocco, Indochina, where they still do a day's work and are maintained as any everlasting automobile should be.

So, depending on how lucky you are, it's not all that unusual to find one of the first models to come off the line in 1934. That would be the 7A with a four-cylinder, 1303 cc engine, 72cm bore, 80cm stroke, rated at 7 CV. Citroen built something like 7000 of them. Of course, you're more likely to run into a 7C (1622cc, 9CV) or the 11 AL (1911cc, 78 cm bore, 100 cm stroke, 11 CV) or 11 B, 11 BL, 11 C, all with the 1911cc engine.

Telling the above mentioned apart, without looking at the name plate under the hood or checking serial numbers for the year and model, is quite a chore. From 1934 thru 1957, at first glance, they look like they all came from the same mold. Your best clue, if you're watching one go down the road, is the spare. If you see it, mounted in the rear, you can be sure the car is an older type. None of the

FUNCTIONAL STYLE — The Citroen sedan combined a nice blend of flowing styling and roomy interior space. Note the louver doors in place of cut louvers which appeared later. The model who is probably a great-grandmother by now, is not identified.

prewar models had the trunk-bulge with spare inside.

To become an aficionado, you'll need a copy of the "Revue Technique", Reedition Archives R.T.A., entitled CITROEN TRACTION-AVANT, published by R.T.A., 20-22 rue de la Saussiere, 92 Boulogne sur Seine, France. You'll discover on page 3, for example, that those very special wheels, called *"roues pilotes"* were first used in 1938. This publication, by the way, lists serial numbers from 1934 thru 55.

The evolution of this car, a car a good 20 years ahead of the industry the day it was put on the market, is about as spectacular as that of Germany's VW. Of course, there were thousands of minor modifications, over the years, but few major ones. Not only did Citroen stick with that classic body style for close to a quarter of a century, but they (and everyone else) knew when they had a good thing. Check the post war, international rally

---

For answers to your questions, the best source of information is Citroen Public Relations, 133 Quai Andre Citroen, Paris 15, 75 France. But I hasten to point out that Mademoiselle Jacqueline Dupont, who heads up that service, does not have the help she needs to handle any volume of correspondence. It would be preferable to contact the fellow who is organizing the A B C Club (AMERICAN BOOSTERS OF CITROEN), a club with an international flavor, geared to the interests of the buff inclined to hop a charter to far away places. His name is BOB HARDICK, P.O. Box 86, SEASIDE HEIGHTS, N.J. 08751, (201) 830-1592. He can point you in the right direction and, at the same time, save Mlle. Dupont a lot of frustration. I also want to point out that this article could not have been written without Mlle. Dupont's friendly cooperation, above and beyond the call of duty.

CITROEN PROTOTYPE — The 1934 Citroen might have looked like this if the design had been accepted. This car was shown at the 1934 Paris Salon De L' Automobile, but it never went into production. One wonders if it still survives.

RACING TRACTIONS — Many of the "11 Legere" Tractions were fitted with extra driving lights and special equipment and used to set records. Shown here is Francois Lecot with his 1935 Citroen 11 AL. Lecot was very successful in breaking records with the car. The other two men are not identified.

CUTE COUPE — Even though the top makes the car look like something akin to a baby buggy, the 11 BL "La Legere" Cabriolet was a popular car in its time. Note the Michelin "Pilote" wheels and tires.

BELLY PAN — The belly pan of the 1934 Citroen is a classic study in design and engineering thought. The pan forms the lower portion of the car's frame and gives added strength to the entire body. Note the following technical advances: torsion bar suspension, "floating" motor, front wheel drive, hydraulic brakes, hydraulic shocks, straight rear axle for simplicity. The transmission was also synchronized in second and third. Compare these features with the 1934 Ford.

CLEAN DASH — *The driver's view of the Citroen was business-like and uncluttered. Note the instruments are in one cluster, although they are a bit hard to read. The windshield swings open at the bottom, and the shift lever is easy to reach. Light controls are under the steering wheel on the right side.*

POWER PLANT — *The heart of the Citroen was this motor and transmission unit. The factory spent its money on engineering and development, not on making the motor look fancy. Note paint runs (concours judges take note!) and other sloppy finishing touches. Even so, this was one of the world's finest motors and had enormous stamina and endurance. Ask any Frenchman.*

CROWDED QUARTERS — *The interior of this Citroen 11B "Familiale" is crowded, but no one could accuse the designers of wasting space! Nine people in a body this size is an exceptional feat — both for the engineers as well as the passengers. We were not informed as to the amount of time needed to seat the passengers, but with a car load of cute passengers — who cares about crowding? (Opposite)*

Hydropneumatic suspension detail  *English-built coupe with rumble seat*  *Post-war Citroen 11 BL*

FAST SERVICE — There are three ways to remove the motor from a Citroen, and this is one of the most effective, although it usually required a few special things such as the body supports shown. According to a French mechanic, the engine could be removed in this manner in less than four hours. (Above, top, right)

149

RESULTATS EN RALLYES DE TRACTIONS-AVANT

9 Fevrier 1947 - Grand Prix d'Hiver de Suede
    1ere au classement general: Citroen 15-six   M. Sjoquist

29 Septembre 1947 - Grand Prix du Motor-Club Norvegien
    1ere categorie 1500-2000 cm3: 11 BL Citroen   M. Finkelberg

Mars 1949 - Rallye des Neiges (France)
    1e, 2e, 3e classement general: 3 tractions-avant Citroen

25 Janvier 1953 - 23e Rallye de Monte Carlo
    3e classement general: Citroen 15-six   M. Charmasson-Marion
    14e classement general: Citroen 15-six
    19e classement general: Citroen 15-six
    sur 10 berlines 15-six au depart (+ 4 berlines 11 CV) 404 concurr.

10/11 Janvier 1953 - Criterium Neige et Glace
    1e classement general: 11 CV Citroen   MM. Ricou-Prestail
    1e categorie + 1500 cc: 11 CV Citroen   MM. Ricou-Prestail
    2e categorie + 1500 cc: 15-six   Funel-Funel

Rallye International Acropole 1953
    1e classement general: 11 BL Citroen   M. etMe Prestai

Rallye International Lyon-Charbonnieres
    1e serie A-1er Groupe: 15-six Citroen   Gautruche-Funel
        (4e classement general)
    2e serie A-1er Groupe: 15-six Citroen   Pouchol-Buisson
    3e serie A-1er Groupe: 15-six Citroen   Bertrand-Francin
    1e serie A-2eme groupe: 11 CV Citroen   Balmy-Rudin

IVe Rallye d'Aix-en-Provence - 4/6 Juin 1954
    1er classement general - 11 BL Citroen   Guigou
    1er categorie serie A-2e classe Citroen   Guigou
    1er categorie serie A-1e classe: 15-six Citroen   Courtes

records. And there were exploits equal to the crossing of the Gobi Desert, like the run from the Cape of Good Hope to London by G.W. Taylor, 20,000 kilometers in 17 days.

Most worthy of mention is the man picked by Andre Citroen himself to make a name for the "TRACTION-AVANT". The man, Francois Lecot, was 56 years old when Andre Citroen started his publicity campagne in March of 1934. He had been an athlete, a champion all-terrain cyclist, and kept himself in shape. He also had a reputation as a rally drive and test driver of 10 years standing, both for Rosengart and Citroen. And he had just competed in the 1934 *Tour de France*.

The proposition entailed an endurance test to be supervised by the Automobile — Club de France: 250,000 miles in one year, that's damn near 700 miles a day, every day, one car, one driver with lunch on the run, consisting of a sandwich. Just the sacrifice of the noon meal alone is a lot to ask of a Frenchman.

But Lecot did it, racking up those 400,000 kilometers between 22 July 1935 and 24 July 1936, running the Paris-Monte Carlo route, day in, day out.

And to top it off, by way of a break in the routine, he participated in the Monte Carlo Rally in January of 1936 which started from Portugal. In June of that year, maybe because the grind was getting to him, he made a tour of European capitols.

Unfortunately for Lecot, his contract with Citroen was verbal. So when the old man died, he had to go begging for backing. A man of determination, he bought the car, an 11 AL (11 CV), 46BHP at 3800 RPM, an assembly line model, and modified only the gas pedal and the windshield so it would open a full 90 degrees to make for better visibility in fog.

He found backers where he could, as you can see by the photo of his car, and did his thing. Rumor has it he never go a *franc* out of Citroen, whose car he was pushing. Even old Henry couldn't have swung a better deal than that for a million dollars of publicity.

No story of these cars would be complete without a remark or two about the roll they played during World War II. Production of the 11 B, with the 1939 "Performance" engine, ground to a halt in 1940 under the German occupation. A few of the 11 BL, the light, short wheelbase model, were turned out during the first half of 1941. And the last 7 C was built in May of 1942.

An interesting sidelight, concerning the gasoline pinch during the war, has to do with the use of substitute fuels. Outfits like Brandt, Imbert, and Facel were turning out *"gasagene"* systems to burn wood, anthracite, even acetylene. Some of the heavier devices were mounted on trailers and towed along. But most interesting was the bottle gas system developed by Citroens' Toulouse plant.

Somewhere between 50 and 60 cars were equipped with long tanks (on the order of large oxygen tanks), mounted on the roof, for use by Citroen personnel. And privileged drivers, such as doctors, particularly in the Pyrenees region, close to the source of natural gas, were able to have their "TRACTIONS" modified to use that relatively abundant fuel.

What with new car production falling far short of their needs, the German Army "requisitioned" all the second hand *"tractions"* their hearts desired. At the same time the FFI acquired as many of these Citroens as they could, often by hook or by crook, showing a preference for the new 15-SIX.

You can just imagine the running gun-battles that took place on narrow, tree lined, country roads. They made the Chicago "Cops and Robbers" shout-outs seem like Sunday-School picnics in comparison. Some of those FFI war-horses had holes cut in their tops over the rear seats and machine-guns mounted on the forward part of the roof. What a way to die!

After the war, the sure footed, powerful 15-SIX became the getaway car of French organized crime. During that period of shortages and up heavel, the papers were full of stories about daredevil shenanigans of self-styled Robin Hoods with fancy names like *Pierrot-le-fou*. And they always got away . . . well, almost always. After all, those Michelin radials weren't quite bulletproof. [CAR]

RALLY CHAMPION — *The Citroen throughout its long life was a rally champion in almost every contest it was entered in. Here a 15-Six is seen charging along an icy piece of frozen asphalt during the Sestrieres "Rallye" in 1953. These workhorses of yesterday can still be seen on the highways and by-ways of France proving that a good idea lasts, and lasts, and lasts.*

25 years in front

Jonathan Wood, in the first instalment of a two part article, chronicles the evolution of the Traction Avant Citroen, the car that taught the world about front wheel drive, back in 1934

"As soon as an idea is good the money is of no importance."—Andre Citroen

THE ultimate international endorsement of front wheel drive has been provided by the Ford Motor Company's latest creation, the Fiesta. Pull instead of push, for so long regarded with suspicion by the public and insurance companies alike has, in the small car field, gained universal acceptance. Thus the once unconventional has become the orthodox.

A crucial milestone along the road to the public's acceptability of front wheel drive was, of course, the British Motor Corporation's Mini Minor of 1959. But it should be remembered that Alec Issigonis's dedication to FWD sprang from his admiration of a car that had appeared 25 years previously. That car was the *Traction Avant* Citroen, a model that with a single deft stroke firmly allied a major car manufacturer to FWD for the first time and although it incorporated a fair degree of transAtlantic thinking has become identifiably French as Gauloises cigarettes or Camembert cheese.

The story of the evolution of this fascinating car has to be seen against the background of the company's increasingly unstable financial position, which resulted in Citroen losing control of his firm in 1934, and was no doubt a factor in his untimely death the following year at the age of 57. The tragedy of the Citroen story is that the car achieved practically everything hoped for (the last car leaving Quai de Javel in 1957), had a profound effect on car design in Britain and throughout the world and convincingly demonstrated that front wheel drive was a mass production reality.

However, it is only in recent years that the names of the talented design team who created the car have emerged, the principal one being that of André Lefebvre, who appears to have been the guiding genius of the Traction's creation. Unfortunately much of the recently published material on the car has proved to be somewhat contradictory, so any further observations on the foregoing story will be much appreciated. I would, however, single out some excellent research by *L'automoblist* which has done much to clear the air.

I am also greatly indebted to the enthusiastic members of the English Traction Owners' Club, namely John Dodson, Graham Sage and Reg and Ginny Winstone for giving up their time to talk to me of Tractions, for the loan of much valuable material and for allowing me to draw freely on the results of their researches.

André Citroen (1878–1935) was, ironically not French, but the fifth son of a Dutch-Jewish diamond merchant, who had set up his business in Paris. Alas, André was not destined to know his father as he committed suicide when the boy was two. However, the indications of his future sagacity are reflected by the fact that he did extremely well at secondary school, thus qualifying him for a place at the élitist Polytechnique in 1898. Although he did not maintain his earlier academic prowess (being placed 159th out of 200), his time at the Polytechnique provided an excellent foundation for his future career, but when he left in 1900 he entered the French army as an engineering officer.

Tradition insists that while he was on leave at Lodz in Poland he spotted a pile of wooden double helical wheels stacked at the back of a local workshop and recognising the potential of the configuration, when reproduced in metal, decided to go into the business of doing just that. Therefore in 1904 André Citroen bade farewell to the army and with two partners, André Boas and Jacques Hinstin, set up a small works in Faubourg St-Denis, though they later moved to Essonnes and finally to the Quai de Grenelle. There seems little doubt that Citroen streamlined his gearwheel production to an impressive extent because in 1908, André was asked by Emile and Louis Mors, who were based in Paris's Rue du Theatre, to modernise their production methods. It took five years for the full benefits of the "Citroenisation" to take effect, by which time production had increased ten fold from 120 cars a year output to 1200 vehicles per annum.

But when the First World War broke out in 1914, Citroen, as a captain in the reserves, was soon spirited off to the front, only to find that his unit was suffering badly from lack of shells. France, like Britain, suffered severe shortages of ammunition during the early years of the war, so when Citroen worked out a plan to mass produce shells, drawing on the methods he'd used in the helical gear and car business, his proposals were greeted with alacrity.

With the active backing of the Armaments Ministry he acquired 30 acres of market gardens in Paris's Quai de Javel and built a mighty armaments factory, which was soon producing 55,000 shells a day. Even at this early stage, Citroen had already made two trips to America, which complemented his own thoughts on mass production and is reflected by his use of American machine tools for shell production. Therefore, when the war came to an end, Citroen had at his disposal a vast factory equipped for mass production and the obvious answer was to manufacture cars for the car hungry post-war world.

There seems little doubt that André Citroen set out to do in France what Henry Ford had succeeded in doing in America. (Paradoxically Ford was far more similar in personality, origins and temperament to Citroen's great rival, Louis Renault.)

Ford, who could arguably have claimed to have put the world on wheels with his famous Model T, first introduced the well established America moving assembly line, principle for flywheel magneto assembly in 1913 the method then being extended to the entire car. Production of this remarkable vehicle soared and prices fell and the Tin Lizzie was soon dominating markets outside America. Such was the penetration that by 1920 every other car *in the world* was a Model T.

After the end of the First World War, the major car producing countries of the Old World began putting their automotive houses in order and no doubts spurred on by Ford's dazzling success, and profits (they were $59,000,000 for 1916 alone), decided to ape his moving track assembly, and one model policy. Unit construction engines and gearboxes, with the manufacturing advantages they offered were also adopted to some degree. Ford's production philosophy was taken up to a greater or lesser extent by Bean, and later Morris in Britain, by Fiat in Italy and Citroen in France. Thus it can be seen that André Citroen was in the forefront of collective thinking at the time, though as can be seen his connections with the car manufacturing business had been confined to his contact with Mors, his undoubted flair being for *production*, be it shells, gears or cars.

Although he had toyed with a Panhard design during the war, the first production Citroen was an extremely straightforward Salomon design, designated, appropriately enough, Model A. The sidevalve 1327cc engined car was a "no frills" vehicle that easily lent itself to mass production. The gear company's well established double chevron *motif*, being a stylised version of the double helical gear layout, soon became as well known in France as Renault's "losange" or Peugeot's lion rampant.

Throughout his car manufacturing career, Citroen maintained his close links with the New World which was, after all, the forcing house of automobile production methods and design innovations. This is reflected by the fact that Citroen adopted the Budd principle of pressed steel body production from 1924 (three years ahead of Morris). The outcome of another American trip was the appearance of the Chrysler "Floating Power" vibration damping engine mounting which was fitted to the Citroen C6 of 1931.

Citroen continued to exploit the formula he established with the Model A throughout the 'twenties and early 'thirties. The cars were not designed to set the world on fire yet were well made and reliable. Like Henry Ford he stuck to the simple but ingenious designs which lent themselves to quantity production.

His aggressive attitude to publicity was in stark contrast to the traditional spirit of his motor cars. He paid to have the Eiffel tower lit up at night with his name and all over France signposts sprang up to well known landmarks, churches and museums bearing the legend "Gift of Citroen".

Citroen was a great believer in providing social services for his workers and this, coupled with his charm and approachable manner, made him extremely popular among the workforce at Quai de Javel. There seems little doubt that the round the clock working conditions necessitated during the development of the *Traction Avant* were made possible by this excellent relationship.

On the debit side, his attitude to money was a study in itself, as the quotation at the begining of this article makes clear. It meant little to him, that was what the banks were for! Much of his spare time was spent at the gaming tables and he made the mistake of extending the unnecessary chance and risk of the green bieze into the more sombre atmosphere of the boardroom. In one light his decision to throw convention to the wind and build the *Traction Avant* can be seen as a courageous move, but to commit his seldom secure company to such an ambitious programme of research and development boarders on the foolhardy.

Before continuing with our story, it is probably an appropriate moment to reflect on the position that front wheel drive had reached in the automotive world by the early nineteen thirties. J. Walter Christie's efforts in America between 1904 and 1910 have been well chronicled for, apart from employing front wheel drive, the massive engine of his car was mounted tranversely, á la Issigonis. Again, in 1924, Jimmy Murphy, the first American to win a European Grand Prix (1921 French GP in a Duesenberg), asked his fellow countrymen and twin cam exponent Harry Miller to build him a front wheel drive car for Indianapolis.

Meanwhile back in France, Jean-Albert Gregoire had been aware of Miller's experiments and from 1927 until 1931 his front wheel drive Tracta cars were made in small numbers, their use of special constant velocity joints, invented by his friend Paul Fenaille, representing a major step forward in European fwd technology. On a front wheel drive car the front wheels, apart from having to carry out their role of steering the vehicle are also responsible for driving it. Traditional universal joints can only be used to transmit the power from the engine providing the steering arc is limited. Too great an arc will

Far left, Andre Citroen. Left, Sensaud de Lavaud. Bottom left, Andre Lefebvre at the wheel of one of the Voisins he co-designed for the 1923 French GP at Tours. Below, Sensaud de Lavaud transmission adapted for front wheel drive, the Bucciali brothers 1927 Buc.

25 years in front

produce unpleasant juddering such as when the vehicle is turned into a sharp corner.

The more expensive constant velocity joint allowed a far greater steering angle to be used, the drive being constant through the prescribed arc. But they are desirable, though not essential, adjuncts of front wheel drive. In Britain the fwd Alvis, and later the BSA, used conventional Hooke joints on their drive shafts, though the shortcomings involved by their fitting (certainly in the latter case) may go some way to explaining the suspicion with which the general public viewed the layout.

But to pick up Gregoire's story again. He actively campaigned his Tractas and soon became a voluable disciple of front wheel drive and it is no surprise to find him cropping up later in the Traction Avant story.

Towards the end of the decade, America again took up the front wheel drive torch with the L29 Cord and to a lesser extent the Ruxton of 1929. With these Trans-Atlantic essays, great advantage was taken of the absence of a propellor shaft to produce a low sleek look, which in these days was the hall mark of FWD. Of course, these last named creations were out of reach of the average motorist, though fwd came smartly within their grasp with Jorgen Rasmussen's DKW two stroke twin cylinder 500cc F1 model, although a similar transverse engine layout had been a feature of the diminutive and obscure 360cc French Micron of 1926! The DKW and its variants gained great popularity throughout the 'thirties. Certainly as far as Germany was concerned FWD had become a production reality. Adler soon followed suit in 1932 with cheap FWD Trumpf model which ex-Citroen man Lucien Rosengart also produced in Paris under licence.

Behind the scenes a number of manufacturers were toying with front wheel drive projects in the early 'thirties, but two in particular, Voisin in France and Budd in America have a particularly important bearing on our story.

The Voisin project was an excursion into the realms of front wheel drive by that idiosyncratic French engineer, Gabriel Voisin, who graduated to building cars after pioneering work in the aircraft field. (In fact, up to his death in 1973 he remained locked in controversy as to whether he or the Wright brothers had been the first to achieve powered flight.) Voisin was one of those brilliant eccentrics, rather like Bugatti (though lacking le Patron's flair for symmetry in the automotive field), who seem destined to plough their lonely furrows with a vigorous, clearsighted determination to do things *their* way. The cars Voisin built during the 'twenties have his intellectual buoyancy stamped indelibly on them, indeed the body work was an almost architectural quality, for Gabriel first trained as an architect before he was lured away from his studies by the romance of aviation. He began building aircraft at a small factory at Paris's Issy-les-Moulineaux in 1904, the concern gradually expanding during the succeeding decade. But it was the demands of the First World War which catapulted Voisin's works into vivid expansions and by the end of hostilities he had produced about 10,700 aircraft, which averages out to about five a day. Wisely (as the postwar aircraft scene was somewhat bleak) he decided to enter the car manufacturing field, entering the motor trade at about the same time as André Citroen, although it is difficult to imagine two greater extremes of the automotive spectrum. Whereas Citroen provided a small mass produced car to sell to a large unquestioning public, Voisins were made in tiny numbers, being powered by sleeve valve engines of high efficiency and remarkable reliability, for an exclusive clientele. It wasn't long, however, before Voisin managed to work his way through the fortune he had accumulated during the War, no doubt a fair proportion being lavished on the beautiful women who enliven the pages of his autobiographies!

I have dealt at some length with Voisin, not because he had any direct influence on the design of the *Traction Avant*, but because André Lefebvre, who he dubbed as his "spiritual son", can be justly regarded as "Le Pere de la Traction" (the father of the Traction). Lefebvre's master minding of the Traction should be enough to guarantee immortality, but he is also credited with the design of the Citroen 2CV and also the DS. All these projects have the hallmark of individuality, defiance and vitality; spiritual heirs of the Voisins from Issy.

Details of Lefebvre's career are tantilisingly brief (Griffith Borgeson, please note!). He was born on 19 April 1894, the son of a book-keeper and like Voisin, he seems to have been first attracted to the aeroplane, receiving his technical training at Paris's Ecole Superieure d'Aeronautique. In 1915, and still in his very early 'twenties, Lefebvre joined Voisin at Issy and soon became involved in the design of a night bomber, and was later responsible for a four wheel undercarriage equipped with brakes and shock absorbers. Voisin must undoubtedly have been impressed with this young man, for when he decided to go into the car business he took Lefebvre and Marius Bernard (his other "spiritual son") along with him. The remaining members of *La Grande Equipe,* as Voisin called them, were Artaud and Dufresne, who, according to Gabriel, defected soon afterwards taking with them plans for a Grand Prix car to Peugeot.

Lefebvre, however, was soon transferring his undoubted talents in the aviation field to Voisin cars. He was partly responsible for the design of the Voisins which were placed first, second and third in the "touring" Grand Prix at Strasbourg that followed the pukka GP in 1922, though the team of four cars he and Voisin designed for the following years French GP were, to quote *The Autocar* "unlike anything yet seen on wheels".

The cars had been built up during the six months prior to the race in July, and one can do no better than quote the same magazine when they said: "the Voisins may be described as having the profile of an aeroplane wing". The cars' outlandish streamlined appearance was accentuated by the fact that although a conventional front axle was used with a 4ft 9in track, the rear one was only 2ft 6in wide in line with the bodywork. And the driver's head was a mere 3ft 4in from the ground! Perhaps the most interesting feature of *Laboratoire* as the cars were named, was the fact that aircraft practice was reflected in the construction of the body and chassis, being a monocoque of ash, pressed steel members, steel tubes and sheet aluminium panelling. Despite the fact that Voisin had claimed before the race that six months work of streamlining was worth six years of engine refinement, the cars proved to be hopelessly underpowered, so it was not possible to measure any beneficial effects of the streamlined "wedge". By the end of the Grand Prix only five competitors were left running, three of the Voisins had dropped out, though Lefebvre managed fifth place in the remaining *Laboratoire*. Tragically none of these extraordinary cars has survived. Can anyone shed any light on their subsequent history?

Voisin continued building his passenger cars throughout the 'twenties, though at the same time he and Lefebvre were working on a project of great significance. This was a front wheel drive Voisin, details of which are unfortunately somewhat sparse. It seems to have been powered by a Knight sleeve valve engine, and some sources refer to a V-8. Yet again it is credited with having a Cotal gearbox, but I have read at least one reference to its being fitted with Sensaud de Lavaud automatic transmission (of which more anon). Certainly de Lavaud had close connections with Voisin, so this latter option is certainly a possibility. Apart from this involvement with Voisin, Lefebvre is also said to have been involved with another slightly obscure front wheel drive confection, Retel's AEM electric, being marketed by Chenard Walcker.

Above, the 1927 Sensaud de Lavaud car, which so impressed Andre Citroen. Left, an early Traction engine. Note the early four bladed fan and forward mounted exhaust flange. The plate on the bottom left of the engine reads "Essai" meaning "Test". Block is dated 22-1-34.

Unfortunately the economic blizzard that engulfed Europe following the Wall Street crash of 1929 (although it took a little time to reach France) severely damaged Voisin's small and tenuous market. This, coupled with Voisin's self confessed extravagant style of living resulted in the Belgium Imperia group buying him out in about 1931. Voisin therefore urged Lefebvre to leave the company and seek more productive pastures. First he joined Renault, though he is said to have remained at Billancourt for only eight days as he found the atmosphere "stultifying". He then made his way to Quai de Javel, where he found André Citroen a receptive listener to his enthusiasm for front wheel drive.

Whether Citroen had already decided on front wheel drive for his next car, or it was Lefebvre who convinced him of the wisdom of *Traction Avant* is open to conjecture. This is because there is another influence to be considered, for one of the most oft repeated facts associated with the birth of the Traction is that Citroen was impressed by a front wheel drive prototype constructed by the American Budd Manufacturing Company of Philadelphia. But what must be regarded as a "grey" area is whether he had already taken on Lefebvre at this time and the Budd car purely represented the seal of approval, or vice versa. Any observations on this chronology will be most welcome.

Fortunately there are rather more details available about this particular project than the shadowy Voisin exercise. The Budd Company, who it will be remembered, were pioneers of pressed steel bodywork for cars, had already produced a speculative front wheel drive prototype, which later emerged as the Ruxton of 1929.

Yet another front wheel drive project was soon under way, work beginning in 1929 and going on until 1931. It was largely the work of Joseph Ledwinka (a relation of Hans Ledwinka of Tatra fame?), Budd's ingenious body engineer. He was also assisted by William J. Muller, whose front wheel drive expertise went back to 1926 when he produced the aforementioned FWD prototype which emerged as the Ruxton. Muller left Budd to produce the Ruxton, though by 1931 he had returned to the fold to work on the Ledwinka car.

One of the intriguing aspects of the design was the fact that it had rubber suspension and at the front leading arms were used (á la 2CV). Muller was none too keen on this layout maintaining that it would adversely effect the braking. (It was, in fact, the reverse of the system later used on the 810 Cord of 1935.) This little car (by American standards, at least) was powered by an all aluminium V-8, the work of William Taylor who had been with Scribbs and the Kermath Marine Engine Company. In view of Budd's expertise in pressed steel bodywork, it is hardly surprising that the car displayed an ingenious body feature in that it was assembled in two halves, the join being disguised by a strip of contrasting trim.

There seems every likelihood that Citroen probably saw and may have driven this prototype. Certainly it does not bear any particular visual resemblance to the *Traction Avant* though unfortunately the only photograph I have of it is unsuitable for reproduction. Does anyone have one? What is in little doubt is Budd's influence in the construction of the bodywork of the new car, and some authorities state that the prototype TA's were built in Philadelphia.

As a result of these various influences Citroen set down his requirements. He wanted a car that would rival the Model T for lengevity and would be capable of: 100 k.p.h. (about 60 mph); seating for four passengers in comfort and safety; 30 miles per gallon; an entirely new concept in motoring.

After consulting with his technical advisers and his financial one, Charles Mauheimer (who was against it), he was informed that the earliest he could expect the car to go on sale was the October of 1936. But Citroen was adamant that this was far too long to wait and opted for the Spring of 1934. This timetable does perhaps explain that when the car finally did make its appearance in 1934, the makers claimed that it was "*Two* years ahead of its time", which must surely be one of the greatest understatements in the history of automobile advertising.

Work therefore began apace on the design of the new car, Lefebvre working with Albert Giullot, the technical director, who had once worked for Rolland-Pilain having joined Citroen in 1929. Unfortunately he died in 1933, so like his employer never lived to see the success of the car.

Another member of the talented design team established in the Rue du Theatre was Maurice Sainturat, one of France's great engineers of the inter-war years. He was responsible for engine design, having arrived at Quai de Javel in 1929 via Hotchkiss (1913). Donnets (1926) and Delage (1929). He was also Citroen's principal contact man with Budd's factory in Philadelphia. Front and rear suspension was the province of Maurice Jullien, and he also took care of the "Floating Power" engine mountings. The bodywork was the responsibility of Daininos, Cuinot and Bertoni.

Jean-Albert Gregoire's view of this team, although somewhat jaundiced, for reasons which will soon emerge is, nevertheless, of interest. "The design department ... was peculiar in that it contained too many experts ... Citroen expected the maximum results in a minimum of time, he had assembled a veritable cohort of engineers," he wrote in the English translation of his autobiography *Best Wheel Forward*.

As can be imagined there was considerable opposition within the factory against the new car, though fortunately one of the Traction's most ardent supporters was Pierre Provost, an ex-Polytechnician, who joined Citroen in 1927 as head of the experimental and test department. He was no stranger to front wheel drive having carried out some destructive testing on Gregoire's Tracta at Montlhery in 1928 when the constant velocity joints really proved their worth.

Despite this connection, it was Lefebvre who introduced Gregoire to Citroen, so heralding the Great Constant Velocity Joint Controversy. Gregoire, on the face of it was an ideal collaborator, but he did not join the company, remaining a technical adviser on a voluntary basis. As a result of this meeting it was decided to fit Tracta constant velocity joints on the new car and two prototypes were running at Montlhery in the Autumn of 1932. This gruelling test began to show up the shortcomings of the joints, for which Gregoire got the blame. The original Tracta joint of 1927 incorporated a housing that did not turn with the shaft and consequently it was known as the "stationary" type. Lucien Rosengart used them on his front wheel drive car without complaint. For reasons which I have yet to see explained, the Tracta couplings on the new Citroen were known as the rotating type, as the housing turned with the shaft. Again during testing at Montlhery in August 1933, the joints lost their lubricant and seized solid, sometimes after a 100 miles or so of running. Charles Rivolier, who as head of the French Bendix Corporation, who owned the Tracta license, redesigned the joints again, but the overheating problems with the associated loss of lubrication persisted.

As will emerge later in the story there were other problems with the transmission of the new car, all the shortcomings being exacerbated by the almost impossible time schedule. In March 1934 Charles Brull, of the experimental department, informed Gregoire that he had sent Citroen a report stating that the Tracta joints were unreliable and failed within ten minutes of being mounted on a test rig. Gregoire was given 15 days to trace the trouble and soon found that the fault was, to some extent, due to faulty machining and poor assembly. He stormed off to see Maurice Norroy, the quality director, throwing the parts on his desk and accusing him of sabotage. Probably a clash of personalities was at the root of the problem as the design team didn't have much time for Gregoire or "his" joint, having decided to replace the troublesome Tracta with the American Rzeppa constant velocity unit. Kasimir, who was assistant to Houdin in charge of machine tools first came across the Rzeppa in America in 1932. Lefebvre and Houdin gave the Rzeppa joint their blessing and special machine tools were imported from America to produce it.

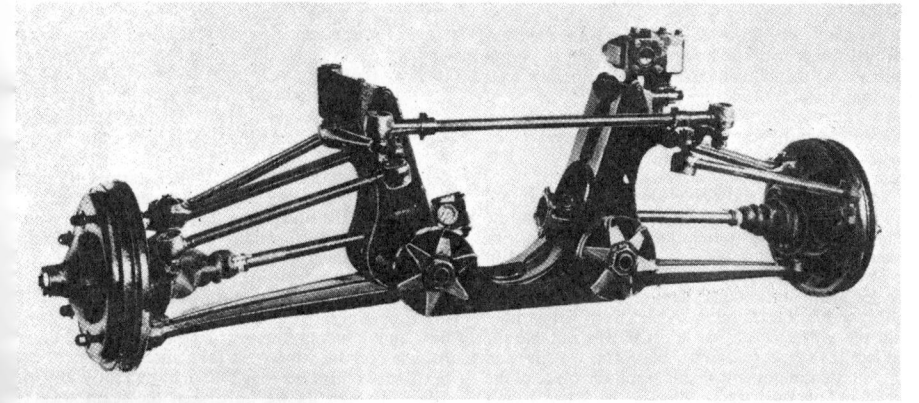

Early Traction front suspension unit fitted with friction shockabsorbers and constant velocity joints.

Right, the Sensaud de Lavaud transmission, as fitted to the 1927 car, Note the rubber rear suspension unit. Above, the Traction's Camusat designed three speed gearbox, alleged to have been produced in two weeks, but see text.

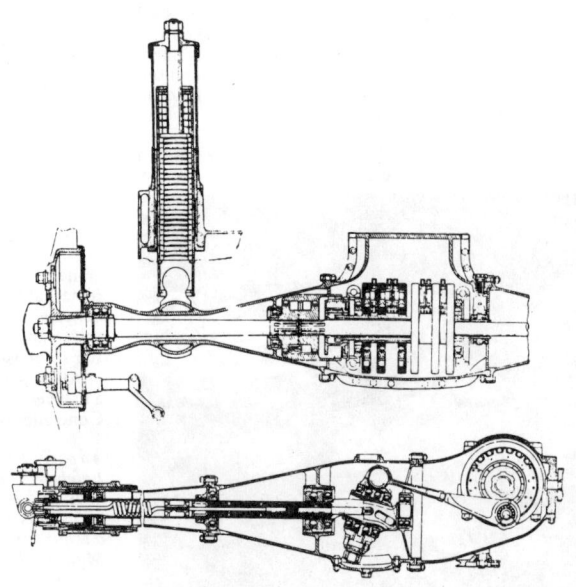

25 years in front

Unfortunately, the Rzeppa proved no better than the Tracta for when the driving wheels were turned on full lock the resulting noise was reminiscent of "someone shaking a bag of dried walnuts". Again perhaps the team at the Rue du Theatre were trying to do too much too quickly because the Rzeppa joint was successfully used on the 810 Cord of 1935. In a more modern context it is doing sterling service on millions of British Leyland front wheel drive cars.

Eventually it was decided to simplify the whole operation by opting for non constant velocity outer universals. Citroen asked his old supplier Glaenzer to come up with a design, the outcome of which was a pair of conventional Hooke joint mounted back to back. There were again lubrication problems with this bulky joint, but it was eventually standardised on the 7CV version of the Traction in 1934, though the first cars delivered to dealers had Tracta joints but these were later replaced with the Glaenzer units.

As the early Glaenzer joints were still giving a certain amount of trouble in 1934, Gregoire bounced back into the picture again with yet another variation of the Tracta cv unit. This time he had reverted to the original "stationary" type of Tracta joint on the 12CV car, which proved far more successful. But by 1935 the Glaenzer joints had been perfected, so the Tractas were superseded by those on this larger car, remaining on all the variants until the model finally ceased production in 1957.

With the business of the drive shaft joints clarified, we can now move on to the Sensaud de Lavaud Gearbox Debacle. As I've already mentioned, André Citroen relied to a great degree on *instinct* when he came to make his policy decisions. Consequently a quite unnecessary element of risk was introduced largely because Citroen himself was ignorant of the technicalities involved when he took up a new idea. There is no doubt that his decision to fit the Sensaud de Lavaud automatic gearbox to the Traction was a disastrous one which wasted much precious time and money when both were at a considerable premium.

Robert Dimitri Sensaud de Lavaud had a colourful background, to say the least. He was born in Valladolid in Spain of a French mother and Russian father in 1884, though by 1900 he was in Brazil but later emigrated to the United States. He amassed a large fortune, though how much of it derived from his coffee planting parents is difficult to say. What is not in dispute is when he arrived in Paris in 1920 this multi-millionaire embarked on a spate of inventing, no doubt having the time and money to indulge himself. A visitor to the 1923 Paris Salon, had he made his way to the Voisin stand, could have seen the outcome of one of de Lavaud's ideas. It was an automatic transmission system housed in the car's rear axle so dispensing with the conventional gearbox, although on this early example the clutch was retained; it was dispensed with on later versions.

Stripped as far as possible of its technicalities, the system consisted of six free wheels mounted on a rear axle shaft each wheel having an eccentrically mounted connecting rod attached to it. The other end of the rod was secured to a swash plate or "wabbleur" which wavered like a coin at the end of its spin so that the rods attached to it were in turn moved by the same amount. Now the inclination of the plate was determined by the resistance encountered by the road wheels, the steeper the road, the longer the stroke of the rods and the higher the ratio; the shorter the stroke, the lower the ratio. Well, that was the theory anyway! The British rights were soon snapped up by the Pickett Construction Company of 22 Queen Anne's Gate, London and it would be interesting to know if the device was ever fitted to any British cars.

De Lavaud and his transmission became regular incumbents of the Paris Salon during the 'twenties and although the design was modified over the years, the basic principle was still retained. On June 12th 1925, *The Autocar* published an article by W. F. Bradley, their Continental Correspondent, describing a comparative trial in which he participated between two 1329cc Voisins, one being fitted with a conventional transmission, the other with a de Lavaud rear axle. In all, 2,164 miles were covered embracing an excursion through the Alps at one extreme and the streets of Paris at the other. By all accounts the de Lavaud car behaved faultlessly throughout the journey, Bradley's summing up of the final test which ran from the Mount Valerien Hill in the suburbs of the capital to the gradients beyond St Germain ran thus: "the Voisin was equal to the speed of the de Lavaud on the level, but had difficulty in holding its own over a varied road when driven mercilessly." In fact, the only snag that Bradley could find was the necessity of altering the adjustment of the swash plate when changing from city traffic to the open road by "varying the compression of the Belleville washers in the rear of the housing" which was easy enough on the test car because the rear of the body was missing to permit easy access to the axle casing. One somehow feels that buyers of Voisins were hardly likely to be of the spanner wielding fraternity, though it appeared that de Lavaud had plans for a modification which would allow the necessary adjustment to be made from the driver's seat.

Sensaud de Lavaud could always be guaranteed to produce some new rabbit out of his hat and he didn't disappoint his public at the 1927 salon. For instead of displaying his transmission on a Voisin he unveiled a complete car, a mauve coupé which bristled with unorthodox features, a vehicle, as it were, for his transmission. Apart from being fitted with the aforesaid device, the car was notable for the fact that a specially cast platform of Alpax, an alloy of aluminium and silicon, took the place of the conventional channel section chassis. As if this wasn't enough to be going on with, the front axle looked rather like a banjo rear axle casing, pivoting at the centre, while the rear was, of course, full of automatic transmission. The rear suspension was rather interesting though, consisting of two vertical steel tubes either side of the axle. Each contained a rod on which slid a pile of rubber discs an arrangement which was not at all dissimilar to the Moulton-designed rubber suspension for the Duncan Dragonfly (see *T & CC*, June/July 1976) of twenty years later.

Trouble was that by all accounts André Citroen saw this device at the Salon and was won over by the automatic transmission system, no doubt due to the fact that he rarely drove and then didn't enjoy it very much! The initial outcome of this fated alliance was that a pair of C4 and C6 Citroens were equipped with the transmission as part of a publicity exercise, and a trial run to Deauville was arranged. The run, across the gentle and undemanding contours of northern France proved a great success and as a result Lefebvre was instructed to fit the first five Traction prototypes with the transmission. So in addition to his team keeping to a seemingly impossible time schedule they had to grapple with the complication of a transmission that had to be grafted on to a brand new engine.

It is possible that the systems fitted to these pre-production Tractions was a variation on the original theme as many of the recent references to the box vaguely refer to it being "hydraulic", though without offering any information to back this up. Most authorities seem to agree that Citroen was inspired by the 1927 Sensaud de Lavaud car and that certainly employed a mechanical system. Any clarification of this apparent contradiction will be most welcome.

The box certainly gave little trouble when tested at Montlhery and on flat roads though once the cars were faced with hills or mountainous slopes "the oil started to boil" and the box ceased to operate. Obviously something fairly dramatic had to be done to convince Citroen of his folly so it was decided that four prototypes should attempt to climb a steep hill at Laffrey; needless to say they all failed to reach the top. Consequently at the beginning of 1934 Charles Brull, director of engineering, submitted a report to Citroen on the subject of the gearbox condemning it in the strongest possible terms. It was then that Brull shouted at Citroen: "de Lavaud's transmission won't help you. Ditch it before you get yourself ditched because of it". As it happened, the design team were fairly certain that the transmission's days were numbered, despite Citroen's apparent intransigence and plans were secretly put in hand to design a conventional box to replace the automatic one, although they were limited in having to use the automatic's housing.

What really seems to have clinched the matter was yet another test drive to conquer the hill at Meudon, to the west of Paris. Five cars left the works, though one failed soon after starting off, two more were defeated by another steep hill, quickly followed by the fourth car. The final Traction, with its transmission oil boiling merrily, hastened back to Quai de Javel. André Citroen was contacted at his home and a meeting was arranged that afternoon. He heard the devastating news and then turning to engineer Broglie, and Lefebvre asked them how long it would to construct a conventional box. Broglie reckoned it would take about two months, but Citroen insisted that it should be ready in two weeks, though as we have seen the team wasn't completely unprepared for the news.

A simple, though ingenious gear change linkage was evolved (known as Lefebvre's Eiffel Tower), though knowing the background to the decision, it does look something of an afterthought! The gear-lever projected through the dashboard, a la Tracta, being appropriately enough referred to as "the mustard spoon" at Quai de Javel. This three speed box was destined to become the Traction's achilles heel, though strangely the company never bothered to redesign it. Thus ended the saga of the costly and disastrous flirtation with the automatic gearbox. Perhaps the last word on the subject should go to André Lefebvre who is said to have commented that de Lavaud had invented a lousy gearbox but an excellent chip fryer!

Next month I'll look at the Traction's design and launch. Meanwhile it is worth reflecting that this remarkable vehicle, in presenting Citroen's "new concept in motoring" incorporated: Front wheel drive; an overhead valve wet liner engine; torsion bar suspension; monocoque bodywork.

All of which was achieved in a little over two and a half years.

Concluded next month.

An early Traction wet liner engine in situ. Lefebvre's "Eiffel Tower" gear linkage has thicker rods and less inclination than production versions. Wing nuts on the rocker box are also unusual. Right a reminder how up to date the Citroen was in 1949. Many post war British cars designers copied the torsion bar suspension, but the Mini Minor was still 10 years away! Back inside cover: a 2.8 litre six cylinder Traction pictured in a suitably Gallic background.

CITROEN
FRONT WHEEL DRIVE

Although there has been a steady process of improvement and refinement, the front wheel drive Citroen remains basically the same as when introduced some 15 years back . . . and is still the most successful design of to-day. Unaffected by the trend towards new looks and bizarre fashioning, Citroen has steadily continued to supply discerning motorists with a car that has no equal in the ability to "get-you-there" quickly, safely and economically. Difficult route conditions make no difference to the high performance of Citroen . . . the car with the best individual and team performance in this year's arduous Alpine Rally . . . and to-day, they are proving the superiority of Citroen front drive in every part of the world.

"LIGHT FIFTEEN" Saloon
With sunshine roof £575, plus Purchase Tax £160 9 6
With fixed roof £570, plus Purchase Tax £159 1 8

"SIX CYLINDER" Saloon
With sunshine roof £857, plus Purchase Tax £238 16 1
With fixed roof £850, plus Purchase Tax £236 17 3

The cars with
INDEPENDENT FRONT SUSPENSION

TORSION BAR SPRINGING

INTEGRAL CHASSIS & BODY

DETACHABLE CYLINDER BARRELS and

FRONT WHEEL DRIVE

CITROEN CARS LTD, SLOUGH, BUCKS.   Telephone: Slough 23811   Telegrams: Citroworks

25 years in front

In the concluding section of this two part article, **Jonathan Wood** examines the Traction's launch and details its production life.

WHILE all these problems were raging on associated with the Traction's development, Citroen decided to re-build a large portion of Quai de Javel. Certainly parts of the factory were outdated having, as we have seen, been built for shell production during the First World War.

Demolition works began on 15 March, 1933 and incredibly the new assembly hall was completed by the end of July, the operation involving lowering the floor of the new complex by six feet for basements. Naturally, the idea was to involve the works in the minimum of disruption, certain sections being transferred to other parts of the Citroen empire. The main portion of the task was completed within four months and by the time the operation was completed no less than a third of the entire factory had been re-built.

Just why Citroen chose that particular moment to stretch his tottering finances yet further is difficult to say. Inevitably the introduction of the new car required new machine tools and production techniques. Perhaps there was an element of bravado in the decision, giving visual expression to his confidence in the future of the Traction and his company. Or even more likely he wanted to show his great rival Louis Renault that he could outdo the expansion going on at Billancourt at the time.

In his biography of Renault (Cassell), Anthony Rhodes writes that Renault took the unprecedented decision of inviting Citroen to his works to witness the rapid expansion that was taking place there. The year was 1931. Citroen initially rejected Renault's offer of transport though was later forced to do so to see round in time. Renault uncharacteristically then took Citroen to lunch at Maxims, which was hardly his usual stamping ground.

After Citroen had gone bankrupt, Renault is said to have told a friend, "The only dirty trick I ever did to Andre Citroen was to show him the Ile de Seguin that morning. He ruined himself afterwards by trying to do in three months what had taken me 30 years."

In February 1934 *The Automobile Engineer* published an article describing a visit to the new factory, concluding their account with the intriguing phrase: "interesting developments on the design side may be expected from the Citroen company in the near future." How right they were! For three months later Citroen called together his agents to announce the new car. According to *L'automobiliste* this took place on Wednesday, 18 April, 1934 at the Citroen exhibition hall at the Rue de Leningrad and the Rue de Berne, the meeting starting at 9.45am. However, this date is at odds with the one given in Sylvain Reiner's book *La Tragedie d'Andre Citroen* who says that the announcement took place nearly a month earlier on 21 March at the factory itself. But the later date does seem to be the correct one as it is quoted in a weighty brochure the company produced entitled "The car the world is waiting for". Also, in view of the fact that the Traction's development had not been without its headaches, it seems likely that the latest possible launch date was chosen, though as we have already seen, the constant velocity joint problems were by no means resolved.

This was coupled with the fact that Citroen's financial position was worsening day by day and the race against time meant that the drawing office staff were working ten hours a day, seven days a week. Consequently Prud'homme, chief of the test workshop, quietly pointed out to Citroen that his men were exhausted and asked for a couple of days break.

"My dear Prud'homme," replied Citroen, "it's no longer a matter of days, but of hours..."

"In that case m'sieu we shall be here on Sunday as usual," came the reply.

But to revert to details of the Traction's launch. There were hundreds of agents present brought at the company's expense from all over France, many of them ill at ease because Citroen's financial troubles were well known.

Citroen began by extolling the virtues of his rebuilt factory and followed this up with details of the sales campaign designed for the spring and summer to promote the new front-wheel-drive car. In addition, the 8, 10 and 15CV cars were to be fitted with the new torsion bar suspension. This was followed up with a film of the Citroen Central Asian Expedition made between April 1931 and March 1932.

Then at 12.30pm came the moment everyone was waiting for; the presentation of the Traction Avant. There were two examples on display, one positioned on a lift so its underside could be examined. One can only speculate on the overall reaction to the new cars which were so low you had to step *down* into them rather than up! Citroen had told the gathering: "I have to tell you that I am launching a counter attack against my enemies and you will be able to judge for yourselves the efficiency of the weapon... It is in a word a totally new and revolutionary car."

On a subsequent occasion he succeeded in countering prejudice against the new car, thus dispelling murmurs of disapproval, when he clapped his hands and demanded the presence of "Monsieur Chopski". Rather surprisingly this Polish Count worked in the export department and being over seven feet tall was able to demonstrate that it was possible to fit such a tall person into a TA, though the photographs taken on the occasion show that he entered the passenger's seat rather than the driver's.

In Britain, the first that the motoring public knew of the Citroen came in *The Autocar* of 4 May, 1934 when the magazine announced "A Citroen surprise". For when the magazine had time to catch its breath it proclaimed the new car was "exceptionally interesting and unusual throughout its construction". *The Motor* was equally impressed. "A Car We Could Not Overturn" announced the magazine, suggesting that they made a regular habit of inducing vehicles to turn turtle. "In the course of severe trials we found it to be extraordinarily stable – it proved impossible to lift the inner wheels even when sharp bends were taken fast enough to promote outward skidding on a dry surface..."

It is perhaps, an appropriate moment to reflect on just what the Citroen offered, in addition to the fact that it was driven by the front wheels. But it should be remembered that none of the features in themselves were new but that they had all been brought together in one "package" for the first time. It is, nevertheless, of great significance that practically all the "novel features" (*Motor*) have since become integral parts of many modern cars. This indicates that the Traction's advanced specifications were not novelties, but deliberately incorporated for a long production run by a team containing some of the most talented automobile engineers in Europe.

The body was, in fact, a monocoque, though not certainly new in 1934 (Lancia's Lambda had used this construction to effect). As we have seen, the Budd Manufacturing Company of Philadelphia USA had played a leading role in its creation. The ingenious "horns" at the front of the body to which the engine/transmission and wheels were attached was no doubt decided upon for the manufacturing advantages the layout offered. But as far as the public were concerned, the most impressive aspect of the car were the low body lines and attendant low centre of gravity. Cars like the Cord and

Below, animated conversation taking place around one of the first 11 cabriolets. Bottom, a prototype 11, pictured in September 1934, complete with discreetly positioned model!

Tracta in the previous decade had shown how front wheel drive, with its absence of propeller shaft and transmission tunnel, helped to create a fashionably low profile. The Traction therefore, as first encountered, must have been of particularly striking appearance with its sleek lines, raked windscreen and one of the most stylish radiator grilles to grace a mass produced car between the wars. The distinctive Citroen chevron was part of the grille, yet it was more than pure decoration; its visual contribution to the Traction's frontal prospect was immeasurable.

In addition to the general impression of lowness, the car was also significant in that it had no running boards (though later a French accessory company offered them as an optional extra!). This feature must undoubtedly caused some raised eyebrows in Britain, yet absence of running boards was a notable feature of French specialist coachbuilding in the 'twenties and early 'thirties.

But perhaps the reason for the design succeeding so well was because the designers' created a shape specifically for the monocoque process. Invariably when the angular lines of the coachbuilt saloon were transferred to pressed steel construction the results were invariably flat and uninspiring; the outcome of inflicting a shape designed for one process on to another. But the Traction body, without its reliance on a wooden framework, freed the designers from these restrictions. The result was a body that remained in production with very minor changes from 1934 until 1957. The design still retains a zestful vitality and a feeling of movement, even when the car is stationary. It is also a reminder that the results of aerodynamic influences can be pleasing to the eye in an era when this wasn't always readily apparent.

But of course there were problems. Bodies split open during tests and, worse still, after they had been delivered to agents and customers. Fortunately the main cause of the trouble was localised to the rear cross member. It was then discovered that a contributory factor was an incorrectly set up press, this causing a constructional flaw. Therefore, at the end of 1934 the cross member was replaced by a tubular one, though this was again changed in April 1935 to a cruciform. Citroen had maintained that he wanted "beauty with advanced design". The team certainly provided this and managed to save about four hundredweight in the process.

As we've already seen, Sainturat's engine was originally mated to Sensaud de Lavaud's automatic transmission, but the trials and tribulations associated with it should in no way reflect on the power unit itself. By all accounts, two engine designs were considered for the new car. Engineer Jouffroy put forward a "traditional" design, which presumably meant side valves and a conventional layout, though the power unit finally adopted was Maurice Sainturat's overhead valve wet liner unit. There were a few development problems such as lubrication troubles and overheating, but the bugs were

A prototype Traction, the 7. This was the first official photograph of the new Citroen released in the spring of 1934. Minor alterations were made to the styling. The shape of the rear wing was slightly changed as was the roof contour. Smaller bonnet handles were later fitted.

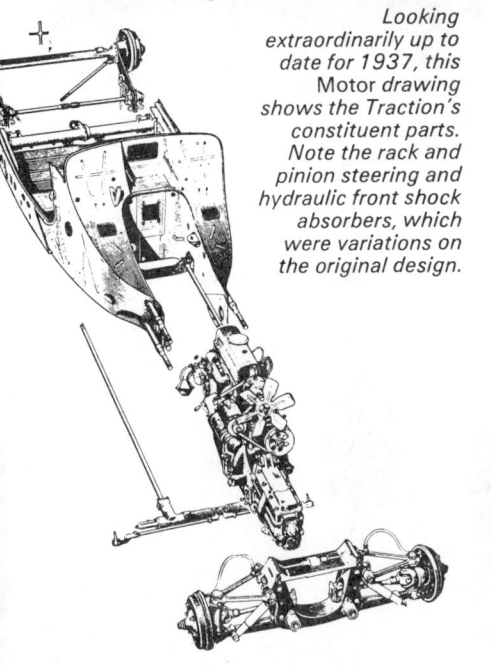

Looking extraordinarily up to date for 1937, this Motor drawing shows the Traction's constituent parts. Note the rack and pinion steering and hydraulic front shock absorbers, which were variations on the original design.

Right, Tractions being assembled at Quai de Javel's main assembly, one of the new buildings of 1933. Below, seven feet high Count Chopski shows how roomy the Traction was.

159

25 years in front

soon ironed out and by the autumn of 1933 it was running well.

But the engine was also significant in that it was the first mass produced wet liner engine in the world. Wet liners were by no means new. One thinks of the famous AC six which featured cast iron detachable liners and Geoffrey Taylor's use of them in his Alta engines, the layout permitting a quick change of capacity for competitive activities. Sainturat opted for wet liners which offered the advantage of an extremely straightforward casting for the block, though on the debit side the layout did place greater emphasis on the joint between the block and the head, accurate machining being of paramount importance. The engine was also significant for its adoption of overhead valves. At the time simple and cheap (from a manufacturing point of view) side valve engines practically completely ruled the roost in the mass produced markets. Certainly Britain in the early 'thirties regarded ohv as yet another American influence and was probably a reflection of the fact that Chevrolet, then the world's best selling car, had featured them since 1915. In France they had been the prerogative of higher priced vehicles such as Delage, Hotchkiss and Talbot. Again the Traction pointed the way to future design trends.

The torsion bar suspension was yet another example of an established design feature being incorporated into a car that Citroen hoped would rival the model T Ford for longevity of production. Parry Thomas had certainly used torsion bar assisted suspension on his Leyland Eight of 1920, while in Germany Ferdinand Porsche was a keen advocate of their use, transverse torsion bars featuring on the P-Wagen Auto Union grand prix car, which was also announced in 1934. And like the design team at Rue du Theatre, Porsche also envisaged this suspension for the mass production market, incorporating them into his people's car, the Volkswagen, the design of which so occupied his time in the immediate pre-war years. On the Citroen, longitudinal bars featured at the front of the car. The front suspension consisted of an upper wishbone which pivoted on oil less bearings. A swivel pin was mounted between this wishbone and a lower single arm which was connected to the torsion bar and damped by friction shock absorbers, a slightly archaic feature that was soon superceded by the fitting of hydraulic shock absorbers. At the rear, two transverse torsion bars were used in conjunction with the "dead" rear axle.

Although it may seem strange now, the fact that the car was fitted with hydraulic brakes was again unusual as far as the European motor industry was concerned, with the notable exception of Fiat and Morris. A Lockheed system was fitted, which may well have been prompted by the flexible demands of the independent front suspension.

This then was the car which was to provide Andre Citroen's "new concept in motoring" and on which he staked so much. As can be seen from the aforementioned details, Lefebvre and his brilliant design team had certainly produced an outstanding car, though the financial drain on the company's resources must have been enormous.

The car was officially launched in May 1934. This was designated the 7A, which was available in 1303cc form, the 7B, with a 1529cc engine which was also offered as a two-seater cabriolet and with two plus two spider bodywork in addition to the usual saloon. Then there was the 7S sporting a 1911cc engine. Again there was the 7C being a cross between the A and the S, and this soon replaced the 7B. July of 1934 saw the 11A and 11AL presented. These were bigger cars altogether, having the 1911cc engine from the 7S.

Mention should also be made of the exotic and legendary 22CV Traction Avant. Perhaps inspired by the Ford V-8, the 22 is just another example of the dilution of resources that was a recurring feature of Citroen developments at the time. If the early, and smaller Tractions suffered from teething troubles, development was even further behind with the 22. The 3·8-litre engine was made up of two 11CV blocks mounted on a common crankcase (unlike the Ford V-8 in which the block and crankcase were integral). A dramatic and beautiful range of bodies were planned for the model: cabriolet, faux cabriolet, berline, coupé de ville, coupé long and familiare. Unlike the smaller cars the headlamps were faired into the front wing line on the 22 while the larger radiator grille bearing the figure 8 distinguished it from its lower capacity brothers.

But if there were problems with the transmission of the smaller cars, one shudders to think of the troubles that must have been associated with the development of the 22. The V-8's designers were hoping for 100bhp, though it is not known whether this figure was ever achieved. There are various estimates as to how many 22s were built, though most authorities settle for a figure of 20. Certainly there are no survivors, but rumours of a Ford V-8-engined example continue to crop up.

This potentially splendid car was axed when Citroen lost control of his company at the end of 1934. Inevitably by rushing his front wheel drive car into production, the rumour soon went around that the early cars were unreliable, which they were. Therefore, by the end of the launch year Andre Citroen was completely bankrupt. He first approached Flandin, the Government's Finance Minister, but the bleak economic climate meant that there was no chance of state subsidy. Flandin suggested that Citroen should approach one of the other large automobile concerns; in other words his rivals. As it transpired he was saved that indignity, for his chief creditor, Michelin, bought the concern for very little. Andre Citroen never recovered from the blow and he died on 30 July, 1935, a broken man.

Britain got its taste of the new cars in earnest in 1935 when the first cars left the company's factory on the Slough Trading estate. These early British-assembled examples differed from their French counterparts by featuring V-shaped bumpers, 12 volt Lucas electrics and leather upholstery. A wooden dashboard was a later distinctive feature.

Top, one of the rare "22" V-8 Tractions. On this model the waist moulding was extended along the bonnet top, also being underlined by a metal "flash" along the doors. The faired in headlamps were a distinctive feature. Middle, a "light 11" of 1952. The painted radiator shell distinguished the French built cars. Left, the last Traction leaves the factory, July 1957, with DSs in the background.

25 years in front

The "12", as the 7CV car was designated in this country cost £235 which was certainly a little more expensive than the Morris of the same RAC horsepower (£196) or a Vauxhall at £205, while you would have had to have spent £275 to buy an Austin 12 during that year. But although *The Autocar* proclaimed that "never before have Citroen Cars Ltd had so striking a car to put forward", the British public approached this front wheel drive intruder with considerable caution. Perhaps the very reference to FWD was sufficient to send prospective purchasers scurrying off to the safe but unadventurous products from Longbridge and Cowley, while the sporting fraternity of the day may have been embarrassed by the fact that the new Citroens handled better than their own so-called sports cars. And, of course, insurance companies threw up their collective hands in horror at the very mention of Front-Wheel-Drive!

Many of the shortcomings that had plagued the early Tractions had been ironed out during Michellin's first year of control. The end of the year saw the hydraulic dampers take over from the earlier friction units, while 1936 saw the introduction of rack and pinion steering, another progressive feature which reached the Slough-assembled cars in May/June.

One of the more obvious external changes made to the Tractions was announced in 1937. These were the stylish and distinctive Michelin "Pilote" wheels having something of the flavour of the famous Bugatti wheels of the 'twenties and early 'thirties about them. The earliest Tractions had originally been fitted with 150 × 40 welded steel wheels. These had five or six stud fixings and open centre with the hub caps sprung in position. The tyres and rims were completely conventional, but it wasn't long before the demands of FWD caused Michelin engineers to design new rims and tyres specifically for the Traction.

The new wheels considerably reduced the unsprung weight, weighing a kilo less than the old rims, and also permitting air to circulate freely around the brake drums (that were increased from 10 to 12 inch diameter in 1936). But most important of all the Pilote's new rim was 1¾ inches wider than its predecessor, being 165 × 400 meaning that new tyres could be fitted in place of the Michelin "Superconfort" and "Stop" tyres used previously.

When a Traction fitted with the new wheels and tyres was circuit tested before the war, the tester found that he had nearly doubled his average speed, and concluded the account of his impressions thus: "Michelin's accomplishments has forced us to completely revise our criteria in all that concerns ride, handling and roadholding. As I told the designer: You have done this country [France] a great diservice, taken away our greatest superiority against foreign cars, that of roadholding – now with your Pilotes, everyone else will be able to corner as if on rails too!"

As far as the Slough assembled cars were concerned, all wheels fitted after 1936 (excluding the Pilotes) were of British manufacture with the French type solid disc or 10 and 14 hole Easi-Clean wheels also being available.

The Second World War, of course, halted Citroen production between 1940 and 1945 though if the chassis numbers are to be believed only 1000 cars (11BL) were made in 1945. The following year saw this model continued, but 1946 saw the re-introduction of the 15–6G, which had appeared in Paris and Slough in 1939/40. Although retaining the same wheelbase as the Big 15 (10ft 1¼in), this six-cylinder Traction had a wider track (4ft 10½in at the front and 4ft 9½in at the rear). The bore and stroke of this engine was 78mm × 100mm, the same as the 15hp as it meant that the same liners could be used. When *The Autocar* came to test the 15–6 they found that here was a Traction with some power (which wasn't always the case with the earlier examples). "It is obvious that there is very good power indeed, with rapid acceleration up to that commonly used range, 65–70mph with plenty of power still in reserve." Changes were made to the transmission to cope with the extra horse power. There were semi-flexible rubber joints incorporated to the inner end of the drive shaft, while a second wishbone replaced the single strut at the bottom of the front suspension unit.

The Six was designated the 15–6G, the suffix standing for Gauche meaning that the engine turned to the left when it was first announced, though by 1948 the model had been re-designated the 15–6D for Droit, showing that the engine now turned to the right. Does anyone know why?

French post-war production concentrated on the 11Bl and the 11B and the aforementioned six, with the 11C briefly creeping in for 1954. That year also saw the 15–6H announced, so designated because of its revolutionary hydropneumatic rear wheel suspension. Gone at last were the traditional torsion bars to be replaced by a system incorporating a constant levelling device which returned the rear of the car to the same static position, irrespective of the load carried, the arrangement being actuated by a pressure pump driven off the engine.

Although probably many people didn't realise it, the company was clearly experimenting with an advanced suspension system which was to flower on the DS19 which first appeared at the Paris Salon in 1955. This certainly spelled the end of Traction production as far as Slough was concerned, though the cars continued to be made in France. The 11CV had a D engine (the same as the DS) in 1956, the last Traction leaving the Paris assembly line in July 1957. In all, something like three-quarters of a million cars had been built during the 18-year production run.

As already mentioned, the Traction Avant had a profound effect on the world's attitude to front-wheel-drive, though there are plenty of instances where its influence can be found on conventionally driven British cars of the immediate pre- and post-war era.

W. O. Bentley has publicly acknowledged his debt to the Traction which he expressed in the V–12 Lagonda. His decision to opt for a wet liner engine and also the appearance of rack and pinion steering and yet again torsion bar suspension clearly reflects the debt he owed to the Citroen.

The post-war influences are even more numerous. Jaguar, like other manufacturers, bought a Citroen for test and evaluation, and the torsion bar independent front suspension which first appeared on the Jaguar Mark V saloon and soon after on the XK120 in 1948 for the 1949 season was undoubtedly inspired from the French car. It was 1951 before the rack and pinion steering appeared on a Jaguar, that being on the small production C type of 1951. This variant also featured transverse torsion bars at the rear, a la Citroen, all these features being perpetuated in the D type Jaguar of 1954. Front torsion bars lingered on, the E type being the last model so equipped to leave Jaguar's Browns Lane factory.

Another Coventry company who also looked very hard at the Traction Avant's front suspension and steering were Riley. Like Jaguar, they purchased a Citroen, so it was no surprise that the new generation of post-war Rileys were fitted with torsion bar front suspension and rack and pinion steering. Likewise Gerald Palmer's ingenious and distinctive Jowett Javelin had both front and rear torsion bar suspension.

Alec Issigonis, who was already wedded to front-wheel-drive, was a great Traction Avant enthusiast and wanted to make that most significant British post war effort, the Morris Minor, FWD. As it turned out rationalisation triumphed and the side valve engine from the pre-war Morris Series E took the place of Issigonis's flat four engine and front wheel drive. Nevertheless, Citroen's influence was present in the form of the torsion bar front suspension and rack and pinion steering. Front wheel drive had to wait for the Mini Minor of 1959.

Most British manufacturers fought shy of wet liners, though a notable exception was Standard Triumph's technical director, Ted Grintham, who decided on a wet liner layout for the company's Vanguard of 1947. This engine later soldiered on with considerable effect in the famous TR range of sports cars until the TR4A was phased out in 1967.

The sight of so many others, following in the Traction's clearly defined wheel tracks must have provided an immense source of satisfaction to the car's guiding genius Andre Lefebvre. Not that he was idle after the new car's launch, being immediately caught up in the design of the 2CV, following his chief Pierre Boulanger's instruction to create "an umbrella on four wheels". Prototypes were running in 1936, though it was 1949 before the 2CV went on sale.

By 1938 Lefebvre was involved with the designs of an adjustable pulley transmission, rather like the DAF, though using steel bands instead of belts. His post-war years were closely concerned with the design and development of the DS which clearly attempted to emulate the Traction's pioneering technology. Unfortunately Lefebvre's later years were plagued by illness and he died in 1963 at the age of 69.

But by far the greatest tragedy of the Traction story is the fact that Andre Citroen never saw the success of the car that broke his company and resulted in his untimely death. Ironically the production of three-quarters of a million cars proved that his "new concept of motoring" was right, but it was one he never lived to see. ●

Citroen four-cylinder production, 1934–1947 with British equivalents

FRENCH SYMBOL	CHARACTERISTICS	BRITISH FACTORY SYMBOL RELATING TO YEAR OF MANUFACTURE		BRITISH COMMERCIAL DESIGNATION	R.A.C. RATING
7A	72×80; track 1·32m.	7A	(1934)	"Twelve"	12·8
7B	78×80; track 1·32m.	–		–	–
7C	72×100; track 1·32m.	7A 7A1 7C 7C2 7C3 7C4	(1935) (1936) (1937) (1938) (1939) (1940)	"Twelve"	12·8
7 Economique	72×100; track 1·32m. Special gear ratio 10×31	–		–	–
11L	78×100; track 1·32m. "Horizontal" carburettor	11L 11CL 11CL2 11CL3	(1936) (1937) (1938) (1939)	"Light Fifteen"	15·1
11L "Performance"	18×100; track 1·32m. "Downdraught" carburettor	11CL4 11CL6	(1940) (1946/47)	"Light Fifteen"	15·1
11	78×100; track 1·45m. "Horizontal" carburettor	11A 11A1 11C 11C2 11C3	(1935) (1936) (1937) (1938) (1939)	"Big Fifteen"	15·1
11 "Performance"	18×100; track 1·45m. "Downdraught" carburettor	11C4	(1940)	"Big Fifteen"	15·1

Heavy metal

Philip Llewellin discovers why André Citroen's Big Fifteen is such a classic

THE MAN'S BRIGHT eyes, brisk manner and razor-sharp memory belied his years. But he was old enough to have spent more than half a century in the garage trade and could remember when there were only two cars — "An early Wolesley and a De Dion Bouton" — in the whole of the sprawling parish on the Welsh border. The delight radiating from Hughes the Bus was a joy to behold that sunny morning in the sleepy village street; "Haven't seen one of these for years. You've really made my day. Fantastic!" he enthused.

The car was a 1955 Citroen, one of the last Big Fifteens assembled in Slough and an enduring memorial to the organisational genius whose engineers started work on the *traction avant* concept 50 years ago. A family friend used to run one of the beefy six-cylinder versions when I was a child. He was a very talented engineer who had worked at Citroen before the war and learned to drive with true Gallic panache. Meticulous tuning had extracted a lot of extra power from the 2.8-litre engine; "I may not be able to match the Jaguars for sheer speed," he once told me, "but in terms of handling and roadholding I can certainly drive the bastards to distraction!"

Driving a traction was one of the many things I had never quite got round to organising until Dame Fortune intervened. One day in the early summer of 1980 I went to the local station to meet the lady who had come to collect a road-test Citroen GSA. The train had been delayed, enabling her to spend a lot of time chatting with a bearded, 22-year-old computer expert by the name of Alan Moore. I offered him a lift home and discovered he had just bought PYT 779 for £1600 from a dealer near Peterborough; "I think I must be a genuine enthusiast," he grinned, "because it's the first car I've ever owned, new or old.

"I just fancied a nice old saloon and was browsing through one of the motoring magazines when I spotted an advert for this Big Fifteen. I went to see it and immediately fell in love. It was very definitely one of those spur-of-the-moment decisions, because I don't think I'd ever even seen a traction before. I liked the comfort, the space and the styling. My parents recoil in horror, of course, and an uncle was convinced I had gone mad. Most of my friends thought it was great, but wouldn't consider buying one themselves. At times I've felt inclined to agree with them, when things go wrong, but on the other hand I've covered up to about 200 miles in a day without any problems whatsoever."

Various factors such as a defunct starter motor, binding brakes and the need to have the gearbox rebuilt last winter — the car was off the road for several months — resulted in more than a year passing before I eventually found myself behind the wheel. Alan's is definitely not a concours car — he uses it for everyday transport and has to watch the pennies — but its overall condition is now remarkably good. The most obvious flaw, unless you are a nit-picking traction expert, is the absence of Citroen's chrome-plated double chevron emblem behind the radiator grille. That famous symbol, incidentally, was inspired by the V-shaped gears which Andre Citroen's small factory started producing in 1913.

Transporting a family's holiday luggage must have been a major problem without resorting to a roof rack, but the Big Fifteen, several inches longer and wider than its Light Fifteen stablemate, lives right up to its name in every other respect. You could almost hold a barn dance in the limousine-like rear compartment. Most modern saloons make me feel cramped if I 'sit behind myself' — I am six feet tall with a long body and relatively short legs — but my head was well clear of PYT 779's roof lining and there was a gap of ten inches twixt knees and driver's seat.

Indeed, an immense Austrian friend who is 6ft 8ins tall and built like the Empire State Building has travelled comfortably in Alan's car. The body is, in fact, the same as that which clad the six-cylinder tractions whose ultimate versions featured an independent, self-levelling rear suspension system that was destined to be used on the futuristic DS19, announced in 1955.

I slid into the bench seat, divided by a folding arm rest, and looked out past the twin-spoke steering wheel, a huge affair all of 18 inches in diameter. The white bonnet seemed to go on for ever, flanked by elegant wings and Lucas headlights. Instruments set in a wooden facia included a 95mph speedometer, but Alan reckons around 50mph to be a reasonable cruising speed that can produce 33-34mpg figures on a three-star diet. The overall average is more like 26-27mpg. Another feature, unknown today, is an advance and retard switch for the ignition which can be operated to generate a little more steam on really steep hills.

I checked the mirror — rearward vision is comically poor — turned the key, pressed the starter button and mentally crossed by fingers. It is always just a little nerve-wracking to set off in a completely strange car — particularly an old and cherished one — when the proud owner is sitting alongside you, ears tuned to every clunk and crunch.

One of the traction's most unusual characteristics is the lever for the three-speed gearbox. It emerges from a slot like a large, vertical letterbox, in the middle of the facia, and has a shift pattern far removed from the conventional 'H' layout. First is towards you and down, with the unguarded reverse slot immediately above. So you ease the lever up, push it across through neutral — "A bit of double-declutching helps at that stage," Alan advised — then go up for second and straight down to engage top. It sounds a mite tricky, rather like the back-to-front shifts on some heavy commercial vehicles, but comes naturally after a few minutes. Or should do. I recall a friend, one of the world's most nervous and inept drivers, whose after-dark routine involved striking a cigarette lighter to help him sort the Citroen cogs out into the correct sequence. I kid you not.

Gently down on the accelerator pedal. Ease the clutch home. No problem! I just nicked reverse on the way to second, but the Big Fifteen's 90lb/ft of torque — developed by a four-cylinder engine of 1,911cc — soon had us bowling sedately along in top with the sunroof open and additional ventilation coming from the top-hinged windscreen that was cracked open just a couple of turns. What could be better, on a fine sunny morning, than motoring through the countryside in one of the world's most notable cars, a genuine trendsetter that had been demonstrating the virtues of front-wheel drive for 25 years before the Mini made its debut?

It was towards the end of 1931 that 53-year-old Andre Citroen sowed the seeds of the traction avant. Although not lacking in technical novelties, his early cars, built since 1919, had been designed along more or less traditional lines with a girder-type chassis and a front-mounted engine driving the rear wheels. Now, in sharp contrast, his engineers were directed to cast convention aside and conceive a genuinely new light car that would carry four adults in comfort — the Big Fifteen will take five or six — exceed 60mph and average 28-30mpg.

It was a fine example of a multi-talented team, backed by a visionary with a gambler's streak, being handed the proverbial blank sheet of paper and told to get on with the job. There were many problems in the early stages — several apparently bright ideas had to be abandoned — but the end product was a masterpiece that set new standards.

Known as the 7A, the first traction boasted a specification whose outline still sounds remarkably modern. It featured an all-steel monocoque hull, independent front suspension, hydraulic brakes and a 1.3-litre engine whose 32bhp, developed at 3200rpm, powered the Michelin-shod front wheels. The low build made aerodynamic sense, helped provide excellent roadholding qualities and also enabled the long-serving step-up-and-enter running boards to be done away with. Inside, a flat floor emphasised another advantage of front-wheel drive. Launched in Paris in 1934, when the adverts rightly hailed it as an 'entirely new concept', the traction was a car whose advanced specification inevitably attracted criticism as well as praise. But its inherent brilliance swiftly forced the sceptics to go on a diet of humble pie. They were eventually proved wrong by more than 750,000 customers.

Unfortunately, developing and tooling-up for the newcomer had stretched Citroen's financial resources to the limit. The company would almost certainly have survived intact, however, had the supplier of steering wheels — an American based in Paris — not demanded payment in full. Citroen's inability to find sufficient money at the drop of a hat triggered off a panic. The traction would have died at birth had Michelin not hastened to the rescue after being given a sharp prod by the French government. The tyre manufacturers cleared the debts and took the company over, holding the reins until Peugeot moved in some 40 years later. Andre Citroen's great enterprise had been pulled back from the brink, but the man himself was doomed. He died of cancer on July 23, 1935, at the tragically early age of 57.

But the traction went on and on. Bigger and better engines were developed — including the rare 3-litre V8 — and variations on the theme ranged from rakish convertibles to a long-wheelbase 'familiale' with enough space for nine people. Production at the French and Belgian plants continued until 1957 — 23 years after the car's debut — but the need to adapt the British plant to cater for the physically bigger DS19 had ended the traction's career in Slough two years earlier.

The original 7A had worm-and-roller steering, but 'rat-and-pigeon' soon became an option and was later standardised. Alan's car certainly responded nicely as we left Shropshire behind and headed up the River Tanat's lovely valley into the rolling Welsh hills. Speed never exceeded 50, and was closer to 45 for most of the time, but PYT 779 swept round bends in fine style with its torsion bar suspension providing a good ride; "I think the steering's good — accurate, responsive and a lot lighter than you might expect — but the lock's terrible," Alan commented. "The turning circle's something like 40 feet and gives you an idea of what it must be like juggling a big truck round in a confined space."

I turned right to tackle a fairly long and steep hill, wondering if bottom gear would be needed. You have to stop to engage it, I had been warned, although techniques dating from my previous incarnation as an Austin Seven exponent might have done the trick. But there was no need. Four long-stroking, hard-slogging pistons pulled us over the crest without a hint of protest. It was the same story a little later, on a much steeper gradient with a couple of tight corners whose high hedges ruled out a fast, wide approach; "C'est un bon sloggeur," I muttered in an accent that must have made even Winston Churchill rotate in his grave.

A long downhill straight sent us swooping towards the end of the route and gave the brakes an opportunity to demonstrate their efficiency; "They're good for their age," Alan agreed, "but you do need to keep alert in traffic."

I thoroughly enjoyed my jaunt, but wondered how Francois Lecot must have felt when he reached Monaco in his traction on July 25, 1936. The 56-year-old restaurateur had spent the previous year shuttling endlssly between his home in central France, Paris and Monte Carlo, driving for 19 hours a day and averaging a steady 40mph. During that time he covered roughly 250,000 miles and established a solo endurance record that has still to be beaten. □

Avant Garde

Citroen celebrates 50 years of front wheel drive

By Sam Brown

Top: An F. Gordon Crosby cutaway drawing of the 1934 12 h.p. Citroen. The gearbox was placed ahead of the engine with clutch and differential in between. Right: A Citroen graphic of the time, showing the advantages of front-wheel drive. Above: La mode Parisienne, by the Seine with the 7CV in 1934

WHILE there may be some arguments about the French claim to have invented the motor car 100 years ago this year, there is no doubt at all over the notable and undeniably French half-centenary that also falls in 1984 – the introduction of Citroen's famous Traction Avant.

While other manufacturers had dabbled with front-wheel drive before, the Citroen 7A, officially unveiled in May 1934, was certainly the first car with this drive configuration to go into real volume production, although the products of the burgeoning German industry of the time included a small front-wheel-drive DKW and the bigger Adler and Brennabor, although these were made in very small numbers.

The 7A caused a sensation. André Citroen and his team of brilliant engineers and designers at the Quai de Javel in Paris conceived, designed, built and launched the car inside 13 months. It electrified a French motor industry that had been battling grimly through a recession and drew praise from a usually cynical press for its brave new approach.

This bravery was backed by an extensive re-organization of the factory at Javel, designed to allow more than 1,000 of the new cars to be produced each week. Production of rear-wheel-drive models tapered off.

While *"le patron"* himself provided the driving force for the project, the leader of the engineering team, Andre Lefebvre, could claim to be regarded as the other "father" to the 7A. Lefebvre, formerly an aero engineer with Voisin, was recruited by Citroen after two years at Renault.

The A7 was revolutionary in many respects. Its aerodynamic bodywork was the first example of mass-production using the monocoque system. Suspension was by torsion bars, and at the front the suspension was built up on a subframe to which engine and transmission was attached, before the whole pack was mounted between forward extensions of the monocoque hull.

The engine design also broke new ground for this class of car with its overhead valve arrangement and removable wet liners. Power output of that first 1,303 c.c. engine was, however, unremarkable. The 32 bhp it produced was only enough to propel *The Autocar* road test car to a maximum speed of 61.1 mph and accelerate from rest to 60 mph in 24 seconds. Later models had engines of up to 1,911 c.c. producing 46 bhp. A V8-engined model was shown in 1934 but never produced.

Citroen's original plans for the Traction transmission included the use of the automatic gearbox and torque converter de signed by Sensaud de Lavaud. But it was insufficiently developed and at the last minute was switched in favour of a three-speed manual gearbox.

This box had its share of troubles too, early cars suffering from split casings and drive joints that broke up. Stronger casings and a change to Hardy Spicer outboard joints, cured the problems.

Sadly, Citroen's innovations were not appreciated by everyone – especially his dealers. The early difficulties with the gearbox helped to cool the ardour of the public and the banks failed to provide the backing needed to carry on. The result was that the company were brought to their knees and Citroen himself was broken in health before Michelin stepped in to put them back on the road. The original Traction design, begun with the 7A, continued until 1957, but the total number of Citroen front-wheel-drive cars produced since 1934 has topped the 16 million mark.

Left: The revolutionary monocoque bodyshell. Leaving off the running board was daring for the time. A subframe, carrying front suspension, engine and transmission was attached to the pair of catwalks formed as extensions to the scuttle

Above: Cabriolet 11CV of 1935. Left: Another cabriolet, the 11B of 1938, posed before the most famous landmark in France

Top: A half-century in reflection. The 7CV, first of the line, in a double image with the latest fwd Citroen, the BX

Autocar

The "7C", 1935. (Citroën 30.170-7)

Date	Type	Colours		
		Body	Wings	Wheels
1934 1935 1936	7	pinkish beige pearl grey AC 126 navy blue wine red black AC 201	black	black
	11	black and pinkish beige black and pearl grey black and navy blue black and wine red	black AC 201	hubcap in colour matching body
1937	7	navy blue wine red black AC 201	navy blue wine red black	black
	11	navy blue wine red black AC 201 iridescent beige iridescent grey AC 109	navy blue wine red black iridescent beige iridescent grey	navy blue wine red black iridescent beige iridescent grey
1938 1939 1940 1941	11 B saloon and commercial	black AC 201	black	red
	11 BL saloon	black AC 201	black	yellow
	11 B 6-windows	black AC 201 iridescent grey AC 109	black iridescent grey	red iridescent grey
	11 B family model	black AC 201 iridescent beige iridescent green	black iridescent beige iridescent green	red iridescent beige iridescent green
	7 and 11 convertible	navy blue olive green wine red black AC 201 iridescent grey AC 109	matching body, or black	matching body, or red with black body
	15 six	black AC 201	black	ivory AC 113
1945 1946 1947	11 B and 11 BL	reseda green AC 500 dark iridescent grey AC 106 light iridescent grey AC 105 black AC 201	reseda green dark iridescent grey light iridescent grey black	pearl grey AC 126
	15	iridescent grey AC 109 black AC 201	iridescent grey black	ivory AC 113
1948 1949 1950 1951 1952	11 B and 11 BL	black AC 201	black	ivory AC 113
	15 six	metallic grey AC 109 black AC 201	metallic grey black	ivory AC 113
1953 to April 1954	11 B, 11 BL and 15 six	RAF blue AC 130 pearl grey AC 126 black AC 201	RAF blue pearl grey black	ivory AC 113
April 1954 to 1957	11 B and 11 BL	heather grey AC 131 pearl grey AC 126 night blue AC 601 Iceland blue AC 122 black AC 201	heather grey pearl grey night blue Iceland blue black	pearl grey pearl grey pearl grey Iceland blue ivory AC 113
	15 six	heather grey AC 131 pearl grey AC 126 night blue AC 601 Iceland blue AC 122 black AC 201 smoky grey AC 124	heather grey pearl grey night blue Iceland blue black smoky grey	pearl grey pearl grey pearl grey Iceland blue ivory AC 113 smoky grey

PROFILE
Citroën Traction Avant

The 'classic' Traction Avant, shown here in French-built, Onze normale *guise. The famous Double Chevron grille gives the car immeasurable charisma*

Traction attraction

Introduced in 1934, Citroën's advanced front wheel drive saloon was new from stem to stern. Peter Nunn looks back at this enduring French legend

It's a dangerous, perhaps contentious, thing to describe a car as a masterpiece, especially in a world in which so many dull machines are habitually referred to as 'classics'. Yet there can be few who can refute the claim that Citroën's remarkable *Traction Avant* (or front wheel drive) cars, built between 1934 and 1957, should be afforded that accolade.

Beloved of politicians, poets, painters, businessmen, farmers, rally drivers, record breakers, Nazi officers, The Resistance, not to mention Gallic bank robbers – and of course Inspector Maigret – the 'Traction' (as it became known) has had many roles and admirers during its 49-year life.

Purists continue to rate the car for its technical advances, historians note approvingly that it was the first mass-produced car to wed monocoque construction with front wheel drive and torsion bar springing, while enthusiasts, together with those with only a slight interest in cars, tend to remember nostalgically the days when Big Black Citroëns were a common sight on the roads. Many still consider the Traction to have unmatched style and character. In short, then, there's no way André Citroën's technological breakthough (incredibly, the Traction was launched as being just 'two years ahead of its time') can be dismissed as just another old car.

New throughout from front to back, the Traction represented a watershed for Citroën and possibly, for the world's motor industry as a whole. Singlehandedly, it took the company's famous Double Chevron motif and overnight gave it a whole new meaning. Up until the Traction's introduction to Citroën's dealers on March 24, 1934, the cars built at Quai de Javel had been renowned for their bargain prices, sturdiness and simplicity but not for their innovation. The Traction changed all that. From then on, Citroëns would be avant-garde, unconventional and in some cases, brilliant.

Born in 1878, André Citroën was an inveterate gambler but had an undoubted flair for production techniques and for making things sell. A former Mors employee, he began making cars at the end of the Great War, his Paris-based factory having successfully turned out armaments by the million for the French forces during the conflict. Using the latest American mass-production methods the Quai de Javel works had been set up by Citroën to produce shells quickly and efficiently, and thanks to Citroën's own ability for streamlining manufacture, output soared impressively.

Inspired by Ford's achievements

With the war over, Citroën, inspired no doubt by Henry Ford's Model T achievements in America, turned his attention to cars. The first Citroën, the Type A, set the scene. Launched in May 1919, it was a straightforward, easy to build machine that some might say was virtually a French-speaking Ford. Nevertheless, it sold well. A bigger-engined version, the B2, followed in June 1921, and proved even more popular. At the Paris Salon that year, the immortal 5CV *petite citron* was introduced to great acclaim. Another winner for Javel.

In October 1924 Citroën utilised Budd's *tout acier* (all steel) body pressings for his transitional, monocoque B10 model; its B12 successor, which appeared a year later, also benefitted from this US-patented principle. Four wheel brakes and a new front suspension were two other B12 improvements. Due to Citroën's slick production methods, which gave customers the choice of no fewer than 28 different body styles, and the fact that the car looked, and indeed was, good, the B14, the next model, was yet another big seller.

The C series Citroëns, phased in after the 1928 Salon, were available in two basic forms, as a 'four' or a 'six'. It was during 1931 that the larger C6 became the first recipient of Citroën's newly-acquired 'Floating Power' engine mountings, Citroën himself having bought the patents for these vibration damping mounts from Chrysler during one of his frequent visits to the United States. But despite this bonus to the cars' refinement, the undistinguished C series gave way, in October 1932, to the more ambitious Rosalie range, the Cs having run out of serious development potential by that stage.

Named after the never-say-die record breaking *Petite Rosalie*, a car that raced around Monthléry for 134 days, setting up 106 world records and 19 international records in the process, the newcomer comprised of three models, the 8, 10 and 15 (1452cc, 1767cc and 2650cc respectively), available in a host of open and closed styles. Bodies, though similar in style to the Cs, were newly-designed and allegedly tougher to boot. They were also easier to manufacture in quantity yet in spite of these assets, and the eventual adoption of torsion bar front suspension on all three versions, the Rosalies were still rather mundane machines. Competent yes, exciting no. The next step was to be the Traction Avant, a car that would ultimately devour André Citroën's wealth, break his company and tragically push poor old Citroën himself onto his death bed. That said, it' ironic that the Traction went on to become one of the company's greatest triumphs, bettered only, som

y, but the 2CV and DS. But Citroën never lived to
e his dream come true.

During the past decade, interest in Tractions has
veloped to the point where the background story
the car's hectic development has been thoroughly
d expertly chronicled on several occasions. We
w know that André Citroën, having seized the
tion of a front wheel drive successor to the
osalies, insisted on an impossibly tight schedule to
t his new car into production. It's said that during
e of his trips to the US and to Budd's Philadelphia
emises, he saw and maybe even drove, a front
heel drive, unitary prototype, designed by Joseph
edwinka (later of Tatra fame). A 3.2-litre, fwd
achine created by Gabriel Voisin and André
efebvre a short while before, also comes into the
cture at this point, as it was Lefebvre who was later
red by Citroën to overlord the Traction project in
s entirety. And with the aid of Maurice Sainturat,
efebvre (known by *Tractionistes* as *le père de la
raction Avant*) managed to do just that – but only
st.

loating Power vibration dampers

Time was against the Traction right from the
eginning and there's little doubt the car was releas-
d before the gifted, yet hard pressed, Lefebvre was
ady. However, from the Rosalie came monocoque
onstruction, torsion bar suspension (front and
ack) and the celebrated 'Floating Power' vibration
ampers in the engine bay. But the Traction's power
nit and transmission were new, the Sainturat-
esigned, short-stroke four-cylinder engine boasting
three-bearing crankshaft, pushrod overhead valves
d wet cylinder liners, the three-speed gearbox
mounted in front of the engine) synchromesh on
econd and third and a pleasing dashboard 'change.
dd hydraulic four-wheel brakes, a low, wide,
eautifully streamlined body on a flat floor with no
unning boards and, of course, front wheel drive,
d you arrive at a car literally years in advance of any
ompetitor. True, these characteristics were not *that*
volutionary taken one by one, but combined in one
ar they formed a unique recipe, and one totally
nlike Citroën had offered beforehand.

Going into this pioneer car's technical specifica-
on more closely, we can see the tough monocoque
hell located the detachable front suspension cradle
ecurely via two extended arms, the cradle also
olding the engine, gearbox and radiator all in place.
he Traction's front end consisted of upper wish-
ones, torsion bar-attached lower radius arms and,
n early cars, friction dampers and worm and roller
teering. Later versions adopted hydraulic shockers
nd a rack and pinion system. At the back, a trailing
xle was fitted along with twin, transverse torsion
ars, longitudinal radius arms and hydraulic
ampers.

David Boyd's beautifully-restored, pre-war Traction roadster shown, top left, on the move and, above, in its element

On the transmission front, the gears inside the Traction's fragile 'box were placed ahead of the integral crown wheel and pinion assembly. The drive thus passed from the clutch, to the gear cluster, back along towards the differential and out to the road wheels by means of Cardan drive-shafts, incorporating Hooke joints at the differential ends and CV joints at the axles. During the Traction's prototype days, the choice of drive shafts had caused Lefebvre and his team considerable cause for concern as the original Tracta CV joints overheated and broke down with predictable regularity.

The driveshaft problems were excabeated by the need to complete development work in time for André Citroën's self-appointed spring 1934 launch date. To meet this deadline, which Citroën insiders maintained was a good two years premature, Lefebvre's team frequently had to work long hours, right around the clock.

Nor did Citroën's insistence that the Traction be fitted with an elaborate Sensaud de Levaud automatic transmission ease the situation – in fact it made it worse – as the system, like the early drive shafts, soon proved disastrously weak under stress. Given time, it's probable the faults could have been erradicated but an eleventh hour switch to a less complex manual set-up (constructed in just 15 days, according to Traction legend) saved the day.

Citroën discretely ousted

Incredibly, Lefebvre had given Citroën his *voiture revolutionnaire* in just 18 months. But by then it was too late; Citroën's money had run out and the company was in trouble. Following a number of desperate attempts to raise capital, one of which included a demonstration of the Traction to interested backers, the company went into liquidation.

Although Michelin, Citroën's chief creditor, took over control and André himself was discretely ousted (only to die the following July, a broken man), the Traction quickly took off in style, creating a wave of interest wherever it went. The first one, the 1300cc 7A was available only as a Berline (saloon) yet at the 1934 Paris Salon, where the Traction was the star of attention, cabriolet and faux cabriolet (fixed head coupé) cars were shown, along with the still-born V8 Traction, the ephemeral 22. The Michelin men killed off this exciting *grand routier* and all the cars were broken up (or were they?) but the extremely pretty drophead and coupé Tractions, both of which featured quaint rear dickey seats, survived in superb style up until the outbreak of war.

As far as the Traction family tree is concerned, the 1911cc 7 Sport version, made only during 1934, supplemented the earlier engines and gave way to the 'classic' Onze family in October that year. Available in France in two distinct styles, as an Onze *legère*, or as the longer and wider Onze (later known as the *normale*), the cars were also made up, soon after, at Citroën's Slough headquarters here in Britain and sold as Sports Twelves and Fifteens respectively, the numbers referring to the cars' RAC horsepower rating. From late '37, the familiar Light Fifteen and Big Fifteen names came into use.

Simplifying the Traction's complicated production run still further, one can say that, like the Rosalies, both four- and six-cylinder versions were offered – but as the 1938-1955 six-cylinder Tractions are now both rare and comparatively expensive to buy (especially the late-model 15H with hydropneumatic rear suspension) this survey is to concentrate on the more common-place 'fours'.

Contemporary road testers, for the most part, greeted the 'Super Modern' Citroën with unstinting praise, extolling, in particular, its roadholding, cornering and hillclimbing abilities. Ride comfort, too, received high marks. Yet it was the Traction's ability to travel over rutted roads or less than smooth surfaces in a relaxed, sure-footed manner that excited one *Autocar* test team, the report adding that the car's prowess at taking bends at what appeared to be foolhardy speeds had no peers. Performance was respectable rather than spectacular.

Drive a Traction today (I had the privilege of driving three before writing this) and one begins to appreciate the dramatic effect the Citroën must have had when it was first announced. Quite simply, properly set up and on the move, the car hardly feels its age at all until, that is, the time comes to manoeuvre, say, a nine-seat *Familiale* into a confined or badly-lit parking space! At that point in time, the sheer size of the Traction – and some of them are *very* large – combines with minimal rearward visibility to create all sorts of headaches. And the much-criticised steering doesn't help much either. When moving, however, the steering lightens instantly and isn't nearly as heavy or unwieldly as popular opinion would have you believe. It is high geared, though, but there's plenty of feel.

Remarkably, the refinement that *Autocar* commented on back in the thirties is still there to be experienced. The ride, for example, is drama-free, road noise is virtually non-existent and the car's famed roadholding and handling qualities really do command respect, even in this day and age. With a little acclimatisation it's possible to hustle a Traction through a bend more quickly than you might have thought possible although don't expect too much from the standard four-cylinder engine which has its work cut out to propel a Traction Berline at anything over 70mph.

Depending on adjustments, the Traction's idiosyncratic gearchange can be delightful yet changes are usually slow, deliberate, and not rushed.

Inside, as well as outside, the Traction has real charm and character, the British cars having the edge on the French-built versions should wood, leather and other such refinements take your fancy. *Normale* rear seat passengers are afforded an overwhelming degree of legroom and ride comfort on an Traction is first class.

Citroën's Traction Avant stayed in production fo 23 years, which in itself is a telling tribute to th foresight, or should that be bravery, of Andr Citroën. Over three-quarter of a million were buil (708,339 'fours' and 50,518 'sixes') and in its day, th Traction sold well on both sides of *La Manche*. Tha the car bristled with technical innovation was signifi cant enough, but the fact that its technicalitie worked, and worked well enough to have a profoun influence on Europe's motor manufacturers, does even greater credit. *Les Chevrons de la Glorie*, indeed

Production history

On the face of it, the Traction's life story is accept ably straightforward. Introduced in May 1934, th car was built in France up until July 1957 in four cylinder form, and until October 1955 in Britain Bodystyles on both sides of the Channel took i saloons, fixed head coupés and open roadsters bu the latter two styles didn't survive the war.

So far so good. But the fact that British and Frenc versions of the same model were known by differen names, that there were five basic engine option (four, four-cylinder types and one 'six') and tha saloons came in three wheelbase lengths and fou distinct shapes – the French, for example, kne those shapes as *legères*, *normales*, *familiales* an *commerciales* (the world's first hatchback?) – doe make model identification very confusing.

A late-model, French-built Onze legère, with big boot

The nine-seater, long wheelbase familiale Traction is huge

Fixed head coupés were built only between 1934 and 1938

Traction roadsters are very glamorous and much coveted

French designation	Engine capacity	French production	Slough designation	Slough production
7A	1303cc	May 1934-Sep 1934	Super Modern Twelve	Aug 1934-Jan 1935
7B	1529cc	May 1934-Dec 1934	–	–
7S	1911cc	May 1934-Sept 1934		
7C	1628cc	Oct 1934-Sep 1938	Super Modern Twelve	Feb 1935-Apr 1940
11(A)	1911cc	Oct 1934-Dec 1936	Super Modern Fifteen	Feb 1935-Aug 1937
11(A)L	1911cc	Oct 1934-Dec 1936	Sports Twelve	Feb 1935-Aug 1937
11AM	1911cc	Oct 1936-Sep 1937	–	–
11BL	1911cc	Oct 1936-Sep 1939	Light Fifteen	Sep 1937-Apr 1940
11B	1911cc	Oct 1936-Sep 1939	Big Fifteen	Sep 1937-Apr 1940
7C Economy	1628cc	Oct 1938-June 1940	Popular Twelve	Jan 1939-Apr 1940
11BL Performance	1911cc	Oct 1938-Oct 1940	–	–
11B Performance	1911cc	Oct 1938-Oct 1940	–	–
11 Legère	1911cc	Jun 1945-Jul 1957	Light Fifteen	Oct 1945-Oct 1955
11 Normale	1911cc	Jun 1945-Jul 1957	Big Fifteen	Oct 1952-Oct 1955

Slough's model availability and specifications varied from Quai de Javel cars in several respects. The *commerciale* Traction, for example, was never offered in England and the Large seven/nine-seater was not available from Slough post-war. And if you wanted a new Big Fifteen in Britain between 1945 and 1952 (the model wasn't officially listed), a converted lhd example came over from France.

On the specification front, Slough cars were given vee-shaped bumpers (up to 1936), a wooden dash, leather trim, (optional) sliding roof, 12 volt Lucas electrics and, of course, right-hand-drive. French Tractions were more austere, offering cloth trim, minimal instrumentation and a dearth of chrome.

Although the Traction's basic configuration and shape never changed, there were numerous improvements and minor updates introduced along the way. By and large, changes in Paris followed on at Slough a few months later. Significant improvements included the adoption, in France, of hydraulic front dampers and a cruciform rear axle in May 1935 and an externally opening boot later that year.

In May 1936 rack and pinion steering arrived and in October (for the '37 model year) the special edition 11AM featuring metallic paint was offered. January 1938 marked the introduction of Michelin's stylish Pilote wheels but September that year saw the dropping of the coupés from the range together with the rear-wheel drive UA models, kept in production to appease customers frightened by the Traction's unconventionality.

Just before the war, the 56bhp 'Performance' model was phased in and, although not shown in the table, it was available from Slough during 1939, by all accounts. After the war production resumed quickly in France (in June 1945, according to one authority) using the 'Performance' engine as standard but only saloons with plain wheels were on offer.

In July 1952 (or October, the date varies!), Citroën specified the Big Boot for its ageing beauty, the earlier 'bobtail' rear end being dispensed with. And from October 1953, French buyers could at last buy a different coloured Traction; the 1945-53 cars were all black, but now grey and blue were on offer. The year also saw the re-introduction of the *familiale* in France.

In May 1955, the Traction received the 60bhp 11D engine from the newly-announced DS but it isn't thought this update was adopted by Slough. UK production ceased during October that year in any case but in France manufacture continued rather longer, the final Traction of all – a *familiale* – leaving the Quai de Javel on July 25, 1957.

Accurate production figures are difficult to supply as Slough's archives were destroyed some years back. Citroën say, however, that 708, 339 four-cylinder Tractions were built between 1934 and 1957. Breaking that down, it's estimated that Slough's production of Traction 'fours' reached approximately 35,000 units (some 10,000 pre-war, 25,000 post-war).

Buyers spot check

First, Bryn Hughes of Classic Restorations (see 'Clubs, specialists and books') sets the scene.

'10 years ago we were breaking Tractions because we literally had nowhere to keep them. But now it's different; then there were no spares and the cars were worth nothing, but today Traction restorations have become much more viable.

His partner, John Gillard, continued: 'In those days, Tractions were largely unappreciated – rebuilt, rechromed and run into the ground, especially by the French who generally don't appreciate old things. In 1983, the odds are stacked more in the Traction's favour. However, one still has to be careful – and it doesn't pay at all to underestimate the amount of work involved in a 'basket case' restoration.

'Structural soundness is very important, as is originality. If the car has a good monocoque, you're half way there. But detailing is equally significant.

The ravages of countless British winters take their toll on Slough-built Tractions principally around the monocoque's centre and rear sill areas. South of France cars don't suffer so much in this respect but *in extremis* their bodies tend to crack. The condition of the sills is important, though, as to a large extent,

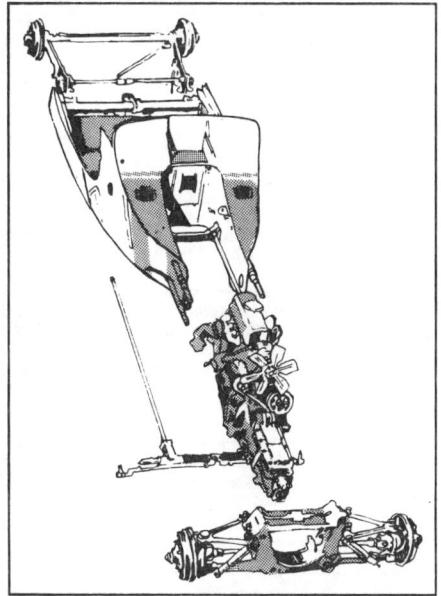

Note the mono shell, transposed engine/gearbox, rack and pinion steering and detachable front suspension

The Citroën's monocoque body has great built-in strength

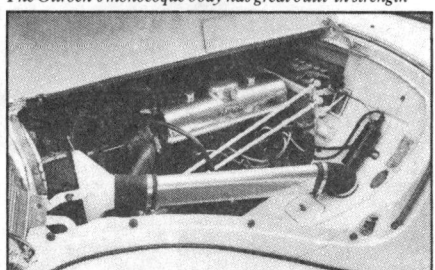

Traction four-cylinder engine, in 'classic' 1911cc guise

they determine the structural strength of the car.

Checking the sills are sound isn't always easy as the rust works from the inside out. Examine the door frames carefully, then; if the doors refuse to shut properly or their edges are scuffing the body, the monocoque is beginning to sag. The cure is to drop the rear suspension and fabricate new sill sections on either side – not an easy job, considering each sill comes in three sections (outer covering, inner box section and inner sill).

At the front of the body, the triple-skinned front arms rust around the steering rack mountings but this isn't too serious if caught in time. Because the monocoque has considerable inherrent strength in this and other regions (André Citroën once had a Traction pushed over a cliff to demonstrate the fact), a securely-welded patch is an acceptable repair, we're told.

Other common rust points include the door bottoms, guttering, boot floor, front wings – glass-fibre replacements are available but to date have had few, if any, takers – and the floor pan. Water drips through the front scuttle and, on Slough cars, the semaphore slots, while the floorpan edges tend to suffer on all versions. However help is at hand, as the British Traction Club can supply ready-made repair sections for the floorpan as well as outer sills.

Engines are long-lived and generally reliable. Long-in-the-tooth examples succumb to big end wear, at which point most owners today elect to drop in the later 11D engine which looks identical, develops greater power and, most importantly, has shell-type big ends. Remetalling one of the earlier engines fitted with white metal bearings has now become a specialist (read expensive) art so the 11D conversion has, understandably, won many friends.

On the subject of engine rebuilds, John and Bryn recommend the fitment of a piston and liner set every time and not a rebore.

The drive-line is the Traction's weak point. Gearboxes, differentials and driveshafts need careful maintenance to avoid trouble, which usually manifests itself by assorted clunks, howls and whining noises. A badly worn crown wheel and pinion set is a serious and time-consuming problem; with age it will howl consistently and alarmingly and, if the car is bump or tow started too often, the teeth will soon strip. Gearbox maladies usually occur due to the effects of worn second and third gear phosphor bronze bushes so check that the gearbox doesn't jump out of gear or crash its synchromesh. A driveshaft in need of attention will knock rhythmically while on the move and clunk on a tight lock. Test for wear by twisting the shaft against itself, and don't forget to keep the gaiters in trim and the grease nipples well lubricated – that's vitally important.

A violent judder on take-up could indicate worn Silentbloc front suspension bushes but a check 'off the car' is the only conclusive test for this. Check the front suspension ball joints in the usual manner but don't worry unduly about the brakes; they're conventional in operation and all parts are available. Spares also exist to counteract steering rack wear.

Regarding spares (always a big worry for an old car owner) the position is better now than has been for some time, thanks to a comprehensive and well-organised remanufacturing programme initiated by Traction clubs in Britain, Holland, Sweden and Germany, and the wealth of secondhand parts.

Clubs, specialists and books

Membership details of the Traction Owners Club, the group formed in 1976 to cater for all enthusiasts of breed, can be obtained from Steve Hedinger of 3 St Catherine's Court, 190 Clarence Road, Windsor, Berkshire.

The club, which issues a glossy, high quality magazine *Floating Power* every three months or so, organises a variety of social events around the country, ranging from pub meets and picnics to the annual TOC rally and concours contest.

The origins of the Traction OC can be traced back to that eminent organisation, the Citroën Car Club which began life as long ago as 1949. Most CCC activities seem to revolve around the club's legendary monthly magazine *The Citroënian*, that's always full of interesting information and tit-bits. If you would like to know more about this all-embracing club (all Citroën enthusiasts are made welcome), write to David Saville at 49 Mungo Park Way, Orpington, Kent BR5 4EE.

Without a doubt, most day-to-day Traction affairs in Britain tend to end up outside the doors of a railway arch in Waterloo, London. The Classic Restorations firm inside, run by long-time Traction fans John Gillard and Bryn Hughes, can advise on all Traction enquiries and problems. John and Bryn operate a parts mail order service, hire out manuals and tools, undertake car sales and evaluations and oversee restorations from beginning to end.

Classic Restorations can be found at Arch 124, Cornwall Road, London SE1 (tel: 01-928 6613) but note that visitors without an appointment are not encouraged during normal working hours.

Northern Traction drivers might like to make a note of John Howard's Traction Avant Engineering company, based in Paradise Street, Bradford BD1 2RR (tel: Bradford 309093). Moving further south, BA Autos, the Guildford concern run by Barry Annels, son of Fred Annels (*the* doyen of Traction experts) is situated at Unit 3, 47/53 Station Road, Shalford, Guildford (tel: Guildford 576672).

Traction bibliomanics will at least need to have *La Traction, Un Roman d'Amour* by Jacques Borge and Nicolas Viasnoff and *L'album de la Traction* by the same authors on their bookshelves.

The first, available in both hard and softback editions, is a fascinating review of this charismatic French car. *L'album* goes one step further by including several hundred beautifully-produced photos, showing the Traction in every conceivable situation. Both books have French text.

Serious students will also have to buy a copy of Michael Sedgwick's superb history of the TA published back in the sixties by Profile Publications. And David Owen's Citroën Story, in *Automobile Quarterly* Vol 13, no 2, is strongly recommended, too.

Pierre Dumont has been responsible for a number of EPA Citroën books. His first, the acclaimed *Citroën the Great Marque of France* has been reprinted several times and is now available with an English translation. Other Dumont titles include *Quai de Javel, Quai André Citroën* (in two parts) and *Citroën* in the Autohistory pocket book series. *André Citroën, Les Chevrons de la Glorie* (by Fabian Sabates and Slyvie Schweitzer) sticks in the memory because it originally included a gramophone recording of André Citroën's voice! But cheap it is not.

Rene Bellu's fantastic *Toutes Les Citroëns* work must be the Citroën bible to end all bibles. It's big, very comprehensive and ferociously expensive.

Citroën, written by Raymond Broad in 1975, is a pithy analysis of the subject, strong on comment but short on pictures. As it is, however, the sole English history of the marque to date – discounting the Dumont translation which is expensive and oddly worded in places – it obviously has its merits. Brooklands Books' *Citroën Traction Avant 1934-57* exists as a useful round-up of original road tests.

Rivals when new

Following on after the front-drive DKWs and Audis in Europe, the Cords in America, the Traction achieved prominence on a number of levels; it was widely-available, mass-produced and remarkable value for money. It also had a long production run. No other car, introduced during 1934-1955 could quite equal these attributes, nor the Traction's pioneer technology.

Before the war, Renault's Celaquatre, Primaquatre and Vivaquatre faced up to the *legère* and *normale* quite strongly, as did Peugeot's 301 and 402. The Peugeots and the Renaults were often faster than the Citroën which had superior roadholding as compensation.

In post-war 1945 Britain, the Traction changed its appeal, its side-opening bonnet, exposed headlights and separate wings giving it an almost 'vintage' charm. Nevertheless, only the flat-four Jowett Javelin came close as a true rival, the Bradford car sharing rack and pinion steering, torsion bar suspension and aerodynamic styling. Pricing was comparable, too, but the Javelin survived only between 1947 and 1952 and lacked the Traction's cult appeal.

Prices

Fortunately, the old car boom hasn't elevated Traction prices into the 'silly' category. Although the days of really cheap Tractions have all but gone, the good news is that prices have remained near enough constant over the past few years.

That said, Traction saloons are still not especially valuable and, in general terms, only *very* early 'fours', or 'sixes' of any age, are deemed collectable by Traction buffs. Pre-war cars are rarer, and therefore slightly dearer than post-1945 variants but they're not necessarily more sought-after. Coupés and roadsters, on the other hand, are much coveted, in any condition.

According to John Gillard of Classic Restorations, ... reflects the amount of work ... back to scratch. Or to put it ... according to what has, or has ... e car's overall presentation is ... stressed two or three times. ... at £150/£200 in the UK. But ... you'll get only maybe a box of

SPECIFICATION	Light 15 (1937-1940)
Engine	In-line 'four'
Construction	Iron block and head
Main bearings	Three
Capacity	1911cc
Bore × stroke	78×100mm
Valves	Pushrod ohv
Compression	6.25:1
Power	46bhp at 3200rpm
Torque	88lb.ft at 2000rpm
Transmission	Three-speed manual
Final drive	Spiral bevel, 3.44:1 ratio
Brakes	Hydraulic drums, front and rear
Suspension F.	Ind. by torsion bars, radius arms, upper wishbones, hydraulic dampers
Suspension R.	Trailing axle, radius arms, torsion bars, Panhard rod, hydraulic dampers
Steering	Rack and pinion
Body	Monocoque, all steel
Tyres	165 × 400
DIMENSIONS	
Length	14ft 0in
Width	5ft 5¼in
Height	4ft 11¼in
Wheelbase	9ft 6½in
Turning circle	40ft
Kerb weight	20½cwt
PERFORMANCE	
Max speed	70mph
0-50mph	20½sec
Fuel con.	21/25mpg

This Rosengart 'Supertraction' is one of many special-bodied Tractions

Tractions weren't noted for their speed – but this one's going quite fast!

Austere French interior with minimal switchgear

A Slough Traction, complete with polished wooden dash

bits, a rusty shell and some usable panels. For £500, expect a sound, original car that probably doesn't run but could soon be made to move.

An average Traction with an MoT, that one can buy and instantly drive away, falls in the £500/£1000 bracket. It will more than likely be scruffy around the edges and have a tatty interior, though reliability should be in its favour. A car that's been on the receiving end of a partial restoration could command anything between £1000 and £2000. From this point onwards you're talking of anything between £2000 and £4000, the latter figure representing the cost of a proper, front-to-back professional restoration.

As coupé and roadster Tractions rarely change hands or come onto the market, quoting retail values might seem rather meaningless. Even in lamentable condition, these cars can fetch £3000 and fully restored, a coupé could top £8000, with the strikingly beautiful roadster commanding maybe £2000 more when in prime condition. We did hear, however, of a mint open-top Traction realising £12,000 – is this the record?

On average, Traction prices in France are a shade lower than in Britain but the cost of Import Duty and of making the car roadworthy (there's no MoT in France) can erode the price differential.

Just a few of the 450 Tractions that converged on Breda, Holland for the fifth International Car Club Rally, 1981. Next year sees even more elaborate events taking place around Europe, to celebrate the 50th anniversary of the Traction's introduction

Slough's model availability and specifications varied from Quai de Javel cars in several respects. The *commerciale* Traction, for example, was never offered in England and the Large seven/nine-seater was not available from Slough post-war. And if you wanted a new Big Fifteen in Britain between 1945 and 1952 (the model wasn't officially listed), a converted lhd example came over from France.

On the specification front, Slough cars were given vee-shaped bumpers (up to 1936), a wooden dash, leather trim, (optional) sliding roof, 12 volt Lucas electrics and, of course, right-hand-drive. French Tractions were more austere, offering cloth trim, minimal instrumentation and a dearth of chrome.

Although the Traction's basic configuration and shape never changed, there were numerous improvements and minor updates introduced along the way. By and large, changes in Paris followed on at Slough a few months later. Significant improvements included the adoption, in France, of hydraulic front dampers and a cruciform rear axle in May 1935 and an externally opening boot later that year.

In May 1936 rack and pinion steering arrived and in October (for the '37 model year) the special edition 11AM featuring metallic paint was offered. January 1938 marked the introduction of Michelin's stylish Pilote wheels but September that year saw the dropping of the coupés from the range together with the rear-wheel drive UA models, kept in production to appease customers frightened by the Traction's unconventionality.

Just before the war, the 56bhp 'Performance' model was phased in and, although not shown in the table, it was available from Slough during 1939, by all accounts. After the war production resumed quickly in France (in June 1945, according to one authority) using the 'Performance' engine as standard but only saloons with plain wheels were on offer.

In July 1952 (or October, the date varies!), Citroën specified the Big Boot for its ageing beauty, the earlier 'bobtail' rear end being dispensed with. And from October 1953, French buyers could at last buy a different coloured Traction; the 1945-53 cars were all black, but now grey and blue were on offer. The year also saw the re-introduction of the *familiale* in France.

In May 1955, the Traction received the 60bhp 11D engine from the newly-announced DS but it isn't thought this update was adopted by Slough. UK production ceased during October that year in any case but in France manufacture continued rather longer, the final Traction of all – a *familiale* – leaving the Quai de Javel on July 25, 1957.

Accurate production figures are difficult to supply as Slough's archives were destroyed some years back. Citroën say, however, that 708, 339 four-cylinder Tractions were built between 1934 and 1957. Breaking that down, it's estimated that Slough's production of Traction 'fours' reached approximately 35,000 units (some 10,000 pre-war, 25,000 post-war).

Buyers spot check

First, Bryn Hughes of Classic Restorations (see 'Clubs, specialists and books') sets the scene.

'10 years ago we were breaking Tractions because we literally had nowhere to keep them. But now it's different; then there were no spares and the cars were worth nothing, but today Traction restorations have become much more viable.

His partner, John Gillard, continued: 'In those days, Tractions were largely unappreciated – rebuilt, rechromed and run into the ground, especially by the French who generally don't appreciate old things. In 1983, the odds are stacked more in the Traction's favour. However, one still has to be careful – and it doesn't pay at all to underestimate the amount of work involved in a 'basket case' restoration.

'Structural soundness is very important, as is originality. If the car has a good monocoque, you're half way there. But detailing is equally significant.

The ravages of countless British winters take their toll on Slough-built Tractions principally around the monocoque's centre and rear sill areas. South of France cars don't suffer so much in this respect but *in extremis* their bodies tend to crack. The condition of the sills is important, though, as to a large extent,

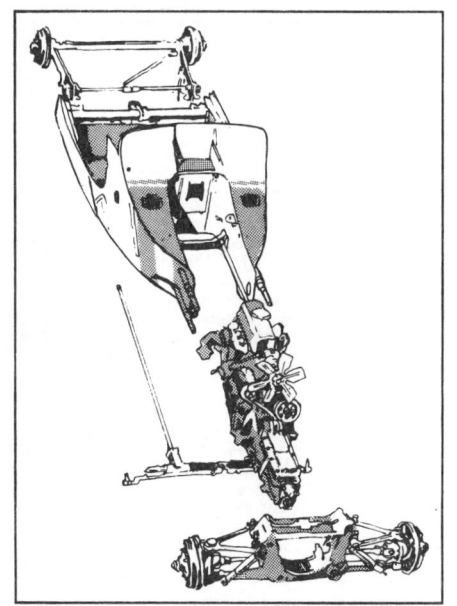

Note the mono shell, transposed engine/gearbox, rack and pinion steering and detachable front suspension

The Citroën's monocoque body has great built-in strength

Traction four-cylinder engine, in 'classic' 1911cc guise

they determine the structural strength of the car.

Checking the sills are sound isn't always easy as the rust works from the inside out. Examine the door frames carefully, then; if the doors refuse to shut properly or their edges are scuffing the body, the monocoque is beginning to sag. The cure is to drop the rear suspension and fabricate new sill sections on either side – not an easy job, considering each sill comes in three sections (outer covering, inner box section and inner sill).

At the front of the body, the triple-skinned front arms rust around the steering rack mountings but this isn't too serious if caught in time. Because the monocoque has considerable inherent strength in this and other regions (André Citroën once had a Traction pushed over a cliff to demonstrate the fact), a securely-welded patch is an acceptable repair, we're told.

Other common rust points include the door bottoms, guttering, boot floor, front wings – glass-fibre replacements are available but to date have had few, if any, takers – and the floor pan. Water drips through the front scuttle and, on Slough cars, the semaphore slots, while the floorpan edges tend to suffer on all versions. However help is at hand, as the British Traction Club can supply ready-made repair sections for the floorpan as well as outer sills.

Engines are long-lived and generally reliable. Long-in-the-tooth examples succumb to big end wear, at which point most owners today elect to drop in the later 11D engine which looks identical, develops greater power and, most importantly, has shell-type big ends. Remetalling one of the earlier engines fitted with white metal bearings has now become a specialist (read expensive) art so the 11D conversion has, understandably, won many friends.

On the subject of engine rebuilds, John and Bryn recommend the fitment of a piston and liner set every time and not a rebore.

The drive-line is the Traction's weak point. Gearboxes, differentials and driveshafts need careful maintenance to avoid trouble, which usually manifests itself by assorted clunks, howls and whining noises. A badly worn crown wheel and pinion set is a serious and time-consuming problem; with age it will howl consistently and alarmingly and, if the car is bump or tow started too often, the teeth will soon strip. Gearbox maladies usually occur due to the effects of worn second and third gear phosphor bronze bushes so check that the gearbox doesn't jump out of gear or crash its synchromesh. A driveshaft in need of attention will knock rhythmically while on the move and clunk on a tight lock. Test for wear by twisting the shaft against itself, and don't forget to keep the gaiters in trim and the grease nipples well lubricated – that's vitally important.

A violent judder on take-up could indicate worn Silentbloc front suspension bushes but a check 'off the car' is the only conclusive test for this. Check the front suspension ball joints in the usual manner but don't worry unduly about the brakes; they're conventional in operation and all parts are available. Spares also exist to counteract steering rack wear.

Regarding spares (always a big worry for an old car owner) the position is better now than has been for some time, thanks to a comprehensive and well-organised remanufacturing programme initiated by Traction clubs in Britain, Holland, Sweden and Germany, and the wealth of secondhand parts.

Clubs, specialists and books

Membership details of the Traction Owners Club, the group formed in 1976 to cater for all enthusiasts of breed, can be obtained from Steve Hedinger of 3 St Catherine's Court, 190 Clarence Road, Windsor, Berkshire.

The club, which issues a glossy, high quality magazine *Floating Power* every three months or so, organises a variety of social events around the country, ranging from pub meets and picnics to the annual TOC rally and concours contest.

The origins of the Traction OC can be traced back to that eminent organisation, the Citroën Car Club which began life as long ago as 1949. Most CCC activities seem to revolve around the club's legendary monthly magazine *The Citroënian*, that's always full of interesting information and tit-bits. If you would like to know more about this all-embracing club (all Citroën enthusiasts are made welcome), write to David Saville at 49 Mungo Park Way, Orpington, Kent BR5 4EE.

Without a doubt, most day-to-day Traction affairs in Britain tend to end up outside the doors of a railway arch in Waterloo, London. The Classic Restorations firm inside, run by long-time Traction fans John Gillard and Bryn Hughes, can advise on all Traction enquiries and problems. John and Bryn operate a parts mail order service, hire out manuals and tools, undertake car sales and evaluations and oversee restorations from beginning to end.

Classic Restorations can be found at Arch 124, Cornwall Road, London SE1 (tel: 01-928 6613) but note that visitors without an appointment are not encouraged during normal working hours.

Northern Traction drivers might like to make a note of John Howard's Traction Avant Engineering company, based in Paradise Street, Bradford BD1 2RR (tel: Bradford 309093). Moving further south, BA Autos, the Guildford concern run by Barry Annels, son of Fred Annels (*the* doyen of Traction experts) is situated at Unit 3, 47/53 Station Road, Shalford, Guildford (tel: Guildford 576672).

Traction bibliomanics will at least need to have *La Traction, Un Roman d'Amour* by Jacques Borge and Nicolas Viasnoff and *L'album de la Traction* by the same authors on their bookshelves.

The first, available in both hard and softback editions, is a fascinating review of this charismatic French car. *L'album* goes one step further by including several hundred beautifully-produced photos, showing the Traction in every conceivable situation. Both books have French text.

Serious students will also have to buy a copy of Michael Sedgwick's superb history of the TA published back in the sixties by Profile Publications. And David Owen's Citroën Story, in *Automobile Quarterly* Vol 13, no 2, is strongly recommended, too.

Pierre Dumont has been responsible for a number of EPA Citroën books. His first, the acclaimed *Citroën the Great Marque of France* has been reprinted several times and is now available with an English translation. Other Dumont titles include *Quai de Javel*, *Quai André Citroën* (in two parts) and *Citroën* in the Autohistory pocket book series. *André Citroën, Les Chevrons de la Glorie* (by Fabian Sabates and Slyvie Schweitzer) sticks in the memory because it originally included a gramophone recording of André Citroën's voice! But cheap it is not.

Rene Bellu's fantastic *Toutes Les Citroëns* work must be the Citroën bible to end all bibles. It's big, very comprehensive and ferociously expensive.

Citroën, written by Raymond Broad in 1975, is a pithy analysis of the subject, strong on comment but short on pictures. As it is, however, the sole English history of the marque to date – discounting the Dumont translation which is expensive and oddly worded in places – it obviously has its merits. Brooklands Books' *Citroën Traction Avant 1934-57* exists as a useful round-up of original road tests.

Rivals when new

Following on after the front-drive DKWs and Audis in Europe, the Cords in America, the Traction achieved prominence on a number of levels; it was widely-available, mass-produced and remarkable value for money. It also had a long production run. No other car, introduced during 1934-1955 could quite equal these attributes, nor the Traction's pioneer technology.

Before the war, Renault's Celaquatre, Primaquatre and Vivaquatre faced up to the *legère* and *normale* quite strongly, as did Peugeot's 301 and 402. The Peugeots and the Renaults were often faster than the Citroën which had superior roadholding as compensation.

In post-war 1945 Britain, the Traction changed its appeal, its side-opening bonnet, exposed headlights and separate wings giving it an almost 'vintage' charm. Nevertheless, only the flat-four Jowett Javelin came close as a true rival, the Bradford car sharing rack and pinion steering, torsion bar suspension and aerodynamic styling. Pricing was comparable, too, but the Javelin survived only between 1947 and 1952 and lacked the Traction's cult appeal.

Prices

Fortunately, the old car boom hasn't elevated Traction prices into the 'silly' category. Although the days of really cheap Tractions have all but gone, the good news is that prices have remained near enough constant over the past few years.

That said, Traction saloons are still not especially valuable and, in general terms, only *very* early 'fours', or 'sixes' of any age, are deemed collectable by Traction buffs. Pre-war cars are rarer, and therefore slightly dearer than post-1945 variants but they're not necessarily more sought-after. Coupés and roadsters, on the other hand, are much coveted, in any condition.

According to John Gillard of Classic Restorations, the price of a Traction reflects the amount of work needed to bring the car back to scratch. Or to put it another way, you pay according to what has, or has not, been done. And the car's overall presentation is important, a point John stressed two or three times.

Traction prices start at £150/£200 in the UK. But for that kind of money, you'll get only maybe a box of bits, a rusty shell and some usable panels. For £500, expect a sound, original car that probably doesn't run but could soon be made to move.

An average Traction with an MoT, that one can buy and instantly drive away, falls in the £500/£1000 bracket. It will more than likely be scruffy around the edges and have a tatty interior, though reliability should be in its favour. A car that's been on the receiving end of a partial restoration could command anything between £1000 and £2000. From this point onwards you're talking of anything between £2000 and £4000, the latter figure representing the cost of a proper, front-to-back professional restoration.

As coupé and roadster Tractions rarely change hands or come onto the market, quoting retail values might seem rather meaningless. Even in lamentable condition, these cars can fetch £3000 and fully restored, a coupé could top £8000, with the strikingly beautiful roadster commanding maybe £2000 more when in prime condition. We did hear, however, of a mint open-top Traction realising £12,000 – is this the record?

On average, Traction prices in France are a shade lower than in Britain but the cost of Import Duty and of making the car roadworthy (there's no MoT in France) can erode the price differential.

SPECIFICATION	Light 15 (1937-1940)
Engine	In-line 'four'
Construction	Iron block and head
Main bearings	Three
Capacity	1911cc
Bore × stroke	78×100mm
Valves	Pushrod ohv
Compression	6.25:1
Power	46bhp at 3200rpm
Torque	88lb.ft at 2000rpm
Transmission	Three-speed manual
Final drive	Spiral bevel, 3.44:1 ratio
Brakes	Hydraulic drums, front and rear
Suspension F.	Ind. by torsion bars, radius arms, upper wishbones, hydraulic dampers
Suspension R.	Trailing axle, radius arms, torsion bars, Panhard rod, hydraulic dampers
Steering	Rack and pinion
Body	Monocoque, all steel
Tyres	165 × 400
DIMENSIONS	
Length	14ft 0in
Width	5ft 5¼in
Height	4ft 11¼in
Wheelbase	9ft 6½in
Turning circle	40ft
Kerb weight	20½cwt
PERFORMANCE	
Max speed	70mph
0-50mph	20½sec
Fuel con.	21/25mpg

This Rosengart 'Supertraction' is one of many special-bodied Tractions

Tractions weren't noted for their speed – but this one's going quite fast!

Austere French interior with minimal switchgear

A Slough Traction, complete with polished wooden dash

Just a few of the 450 Tractions that converged on Breda, Holland for the fifth International Car Club Rally, 1981. Next year sees even more elaborate events taking place around Europe, to celebrate the 50th anniversary of the Traction's introduction